1975

UNDERSTANDING CHEMISTRY:

FROM ATOMS TO ATTITUDES

UNDERSTANDING CHEMISTRY:
FROM ATOMS TO ATTITUDES

T. R. Dickson

JOHN WILEY & SONS
New York   London   Sydney   Toronto

Library of Congress Cataloging in Publication Data:

Dickson, Thomas R.
  Understanding chemistry: from atoms to attitudes.

  Bibliography: p.
  1. Chemistry. I. Title.

QD31.2.D532   540   73-12695
ISBN 0-471-21285-7

Printed in the United States of America

10 9 8 7 6 5 4 3 2 1

TO CHILDREN OF THE EARTH

PREFACE

This book is designed for a beginning or survey course in chemistry for non-science or parascience students. The major objective of the book is to provide the student with a chemical view of the environment. The basic concepts and models of chemistry are fundamental to the development of such a chemical description of the environment. That is, the ideas of atoms, molecules, and ions as structural components of matter and the rearrangement of these particles in chemical reactions is the basis of a description of the static and dynamic aspects of physical reality. Once such a chemical view is established, it is possible to explore how natural chemical processes and the use of chemistry and chemical technology by humans can affect the environment.

To facilitate this approach, basic chemical concepts are discussed followed by expositions of some of the chemical factors which influence the world. The reason for this is that the principles of chemistry provide an elegant description of reality and are intellectually interesting apart from environmental concerns. However, the practical significance of these principles certainly revolves around the chemical phenomena occurring in the environment which influence our lives. Consequently, these chemical principles are used to discuss such topics as air pollution, water pollution, energy sources, agricultural endeavors, population control, biochemistry of life processes, and medicines. The discussions are presented to serve as a foundation for the development of values and attitudes concerning environmental dilemmas. Knowledgeable attitudes are a necessity for citizens of today who will undoubtedly be involved in many political and social decisions revolving around environmental issues.

Some mathematical concepts are fundamental in chemistry. The metric system, atomic weights, and the mole concept involve numbers. Furthermore, a quantitative description of chemistry and chemical processes involves numbers. We are all constantly being fed numerical information about the things that go on around us. In this book the use of such numbers is explained where applicable. Some of the mathematical concepts are also explained, but these descriptions carefully separated from the other material.

Each chapter has a selective bibliography of books, pamphlets, and articles. These references are provided as sources of information on environmental topics from popular and semitechnical books and journals. The bibliography at the end of Chapter 1 includes books of general interest.

(Note to Instructors: Some of the questions at the end of certain chapters are designed to have the students express their feelings concerning various topics. Since these questions involve personal feelings, the decision to reveal any responses to the questions should be left to the individual. The intent of these questions is to help the students recognize their values, and the privacy of individual feelings should be guarded.)

I would like to thank all of those individuals involved in the preparation of this book. They have helped me express my hope for the world through this message to students who comprise the world's most precious resource.

Aptos, California   T. R. Dickson

CONTENTS

4   CHEMICAL ELEMENTS IN THE ENVIRONMENT

5   ENERGY AND THE ENVIRONMENT

6 NUCLEAR ENERGY

7 GASES

12 ORGANIC CHEMISTRY

13 THE CHEMISTRY OF LIFE: BIOCHEMISTRY

14 CHEMISTRY: HEALTH, MEDICINE, AND DRUGS

15 PEOPLE, AGRICULTURE, AND FOOD

UNDERSTANDING CHEMISTRY:
FROM ATOMS TO ATTITUDES

THE PHYSICAL ENVIRONMENT

1

1-1   ENVIRONMENTAL ATTITUDES

Of the many creatures on earth, humans alone have been able to rise above the forces of nature to exercise a certain amount of control over the environment. In a sense, humans have found it possible to manipulate the natural environment in a fashion which appears to be, and often is, beneficial. It cannot be denied that many of the triumphs of humans over nature are necessary and provide needed comfort. On the other hand, it cannot be denied that humans in their dominion over nature have assaulted and upset the environment. As we are beginning to realize, many of the intricate and interrelated processes which are continuously occurring in nature are being affected by human activity. Such realizations come from a study of the structure of the environment and an understanding of relevant human activities which affect the environment. Citizens of today will be involved in numerous decisions involving the environment. You may be involved in deciding whether or not a nuclear power plant should be built near your community or if a particular insecticide should be banned. Are these decisions to be left to the experts? Realistically, we will all participate in solving various environmental dilemmas.

When dealing with environmental problems you will be confronted with a variety of physical facts, statistics, opinions or value judgments, and predictions of the future. For example, it is a fact that carbon monoxide is a poisonous gas which can accumulate in the atmosphere and be a dangerous air pollutant. It is a statistic that 59 percent of man-made carbon monoxide produced in the United States comes from automobiles. What should be done about this air pollution problem is a matter of opinion, values, and priorities. Most of us have attitudes or values concerning such problems, or we can develop opinions when we learn about a problem. This book is intended to provide you with facts and information, but it does not presume to give solutions to our problems.

The statement that carbon monoxide emissions from automobiles are expected to rise from the current 65 million tons per year to 100 million tons per year by 1980 is a prediction of the future based on current statistics and trends of the past. Keep in mind that such predictions may or may not prove valid and, thus, should be viewed as educated opinions. Usually, too many factors and variables are involved in a problem to allow accurate predictions of the future. However, we often use these predictions to form opinions, priorities, and modes of action. In fact, it is likely that with the installation of emission control devices on automobiles, stimulated by federal emission standards established by the 1970 amendments to the Clean Air Act, carbon monoxide emissions should be drastically reduced. It is important to be aware that predictions made by scientists, nonscientists, government agencies, industrial representatives, and politicians are opinions. By collecting facts and sorting out various statements and predictions concerning an environmental issue we can develop our own attitudes and values upon which decisions can be based.

1-2   SCIENCE AND TECHNOLOGY

Science is the attempt to describe physical reality in a systematic and logical manner. Physical reality refers to the physical objects which exist and phenomena which occur within our space–time environment. The practice of science seems to spring from the desire of humans to know about and describe physical reality and to accumulate knowledge. Scientific thought has evolved within the realm of human thought to the highly organized, extensively communicated, and somewhat domineering discipline which includes all fields of science today.

Pure science and scientific research are in a sense concerned only with the development of new knowledge. However, the evolution of scientific knowledge provides a better understanding of physical reality and reveals ways in which the environment can be manipulated to serve humans. **Technology** is the application of science to the manipulation of the physical environment especially for industrial or commercial purposes. It can be argued that technological advances have improved the lives of humans, but it also can be argued that technology has been used to destructive ends.

It is important to distinguish between pure science and technology which is applied science. These two aspects of science are closely allied and practically inseparable. That is, many scientific advances have stimulated technological applications of the advances. Similarly, many technological advances have stimulated scientific thought. For instance, scientific knowledge of the chemical nature of materials and the chemical function of the body has stimulated the technology of the production of man-made chemicals to fight disease. An example of the influence of technology on science is the development of the steam engine, which stimulated the creation of a field of scientific thought called thermodynamics.

Throughout the ages, the application of science has been a two-edged sword. The discovery of the technology of the refining of iron ore to make iron established the iron age, and iron serves as a backbone of the industrial societies which exist today. However, iron brought new dimensions to the production of weapons of war. The technological production of ammonia is necessary to the production of fertilizers needed by a hungry world, but the technique was developed for purposes of making explosives. The development of the knowledge of nuclear energy has led to many promising uses of this energy, but the development was stimulated by the quest for nuclear bombs as the ultimate weapons of war. The list of useful and reprehensible applications of technology could go on and on. It is true that technology can be used for constructive and destructive purposes, but certainly the decisions concerning such uses are made by humans.

It is somewhat nonproductive to argue whether or not science and technology have been responsible for the many environmental dilemmas which exist today. It is important to realize that scientific advances and the proper use of technology will play an important part in the resolution of environmental problems.

1-3 CHEMISTRY

Chemistry as a science is based on observations of events that occur and objects (matter) that exist within our space–time environment. Chemistry is a dynamic science which deals with the structure and behavior of matter. Chemistry involves finding out what things are made of and how they undergo changes. We are surrounded by material objects, and human beings as living organisms have a material existence. Chemistry is concerned with investigating the nature of all matter ranging from substances such as water to complex biological material such as deoxyribonucleic acid (DNA). The science of chemistry has developed from the desire of man to describe what objects are made of and how the structure of these objects causes them to have certain properties. From a practical point of view, desirable properties of matter are sought. Substances that are useful (to cure disease, explode, intoxicate, and smell or taste good) and capable of being fabricated into things (clothing, utensils, and tools) are isolated from nature or manufactured from other substances. Chemists explore nature and experiment with substances to develop and refine theories concerning the structure and behavior of matter.

We all benefit from the practical applications of chemistry. Synthetic and natural chemicals are used as drugs and medicines. Agricultural productivity is greatly enhanced by the use of chemical fertilizers and pesticides. Many important consumer products such as foods, gasoline, and plastics are derived from chemical processing. Actually, modern living would not be as convenient as it is without chemical technology. Moreover, we are learning that when we utilize substances in our environment for convenience, the environment is altered and can become polluted. An understanding of environmental problems requires knowledge of the chemical processes involved. Furthermore, solutions to problems will require the development of appropriate chemical technology. It is important to learn some chemistry to understand the nature of our environment and the forces that threaten it.

1-4 MATTER AND MASS

We use our senses—sight, hearing, taste, smell, and touch—to observe those objects which surround us and the changes which continuously occur in and around us. We see the sizes and colors of objects and see changes in the positions of objects (i.e., a moving automobile). We hear the sounds of certain phenomena involving the interaction of objects (i.e., the sound of a musical instrument or a jet airplane). We taste and smell pleasant and obnoxious things (i.e., good food and foul air). We detect heavy or light objects using the sense of touch. We also use our sense of touch to discern hot and cold objects.

A simple description of what we see in our environment is that it consists of objects having substance. We call such objects matter. **Matter,** of course, makes up the material things around us which we conceive of as occupying space and having mass. **Mass** is a property of matter that determines its resistance to being set in motion or resistance to any change in motion. A baseball is much easier to throw than a more massive shotput ball and, obviously, more desirable to catch. It is not too difficult to develop feeling for the idea of mass, since we are familiar with objects in our environment. The useful aspect of mass is that it can be used to compare different objects. For example, when you pick up two objects, you can often say that one feels heavier and therefore has a mass which is greater than the other.

As we observe, the environment is not static but rather involves dynamic changes. Day and night come and go, sunlight replaces rain and snow, plants and animals grow. Another aspect of our environment which is related to the dynamic phenomena is energy. To understand energy we need the concept of force. A **force** is generally an action that may cause some effect. When you push on a boulder you are applying a force in an attempt to set it in motion. There are various types of forces. We are familiar with the force of gravity—the force that arises when one mass is near another mass; for example, the attraction between you and the earth. In fact, the weight of an object is the force that results from the gravitational attraction of the earth for the mass of the object. When we jump we have to exert a force great enough to overcome the force of gravity. Water flowing from a higher level to a lower level can apply a force which results in the motion of a water wheel or a turbine. Expanding hot gases such as steam can exert forces which make engine pistons move.

Magnetic forces are familiar to most of us. We know that in the immediate vicinity of a typical magnet there is a magnetic force field. This force field results in certain types of objects being attracted toward the magnet when they are introduced into the field. Electrical forces are also important. Electrical forces involve the existence of electrical charges. There are two types of electrical charge: positive and negative. Objects with electrical charges have electrical force fields about them. Objects of like charge tend to repel one another, and objects of opposite charge tend to attract one another. These are called electrostatic forces of repulsion and attraction. The flow of electricity in wires involves repulsive electrostatic forces.

1-5   ENERGY

Energy is involved in dynamic changes used to perform work. However, the basis for dynamic changes is unbalanced forces. When a force is exerted and overcomes any opposing force, motion or some related phenomena occurs and it is said that work has been done or an energy flow has taken place. For instance, when we pick up an object we must exert a force to overcome the force of gravity. When we pick it up we have done work or expended energy. The object we have picked up possesses stored energy or the potential for doing work. If we release the object, the force of gravity sets it in motion and expends its stored energy as it falls. If the object were dropped so that it struck a paddle wheel, the stored energy could do work by setting the wheel in motion.

Energy appears in many interrelated forms. Observations of energy changes have shown that energy can be stored and transformed from one form to another but cannot be created or destroyed. Let us consider an example of how energy is involved in our environment. Certain processes occurring on the sun produce energy. Some of this energy takes the form of solar or **radiant energy** (sunlight) which travels through space and strikes the earth. Some of this radiant energy is absorbed by water in the ocean and is converted to **heat energy.** This heat energy causes some water to evaporate and ultimately to form clouds which drift inland. Some of the water is deposited in mountain areas as rain. This water now possesses **potential energy** (energy of position) since it is attracted back to sea level by gravity. As the water flows through the rivers toward the sea, some of it is trapped behind dams. The **energy of motion (kinetic energy)** of the water as it is released from the dams is used to turn a mechanical device called a turbine. This results in the conversion of kinetic energy to mechanical energy. The turbine is designed to convert the **mechanical energy** to electricity **(electrical energy).** The electricity is transported to populated areas and is used to perform various types of work. For example, electricity is converted to mechanical energy in an electric motor.

Other portions of the original solar energy are absorbed by plants. Through a chemical process called photosynthesis, chemicals are produced by the plants in which some of the energy is captured in the form of chemical energy. As we shall learn, **chemical energy** is energy which is stored as certain kinds of chemical substances. Men and animals eat parts of the plants and, through chemical processes, gain some of the energy that is used in life processes (metabolism) and for performing useful work.

The point of the above example is to demonstrate the significance of energy and energy transfers. In chemistry we are concerned with energy transfers involved in chemical processes. Energy use is fundamental to an industrial society. Thus, the concept of energy is of great importance. Energy use is discussed in Chapter 5.

1-6 SCIENTIFIC LAWS AND THEORIES

To obtain a better view of the environment, scientists have developed instruments and techniques to extend the realm of our normal senses. Distant objects can be viewed with telescopes, while small objects can be viewed with microscopes. Masses of objects can be determined with balances. The degree of hotness or coldness of an object can be measured with a thermometer.

The increased sophistication of these instruments allows for more precise observations and, thus, more precise descriptions of environmental objects and events.

Through the use of such methods scientists are able to make consistent and reliable observations. In this manner any patterns and consistencies within the environment can be established. By repeatedly observing phenomena, it is sometimes possible to make a general statement about the reliability of certain phenomena. For example, throughout centuries scientists have been observing energy transformations and have noted that no energy is ever created or destroyed but is rather transformed from one form to another. Such a generalization based upon years of observation is called a physical principle or law, or simply a **Law.** The observation mentioned above is called the **Law of Conservation of Energy:** Energy cannot be created or destroyed but only transformed from one form to another. Physical laws are expressions of an apparently consistent pattern in nature which never seems to have an exception. Another such physical law is the **Law of Conservation of Matter:** Matter cannot be created or destroyed by normal processes occurring on earth.

Physical laws provide a way to describe phenomena which occur around us. Furthermore, such physical laws have significance in our interaction with our environment. For example, the Law of Conservation of Energy tells us that if we try to convert energy in one form to electrical energy and if the conversion is not completely efficient, then the leftover energy must take on some other form. Many electrical generating plants utilize the burning of coal or oil to obtain heat energy which is ultimately converted to electrical energy. Another physical law (the Second Law of Thermodynamics) reveals that it is not possible to have 100 percent conversion of heat to work to provide electricity. Much of the heat energy produced in an electrical generating plant is lost as waste heat. This is unavoidable and is expected. However, such waste heat changes the temperature of the immediate environment of the plant. When this waste heat disrupts the animal and plant life around the generating plant, it is called thermal pollution.

As another example, consider that the Law of Conservation of Matter tells us that we cannot destroy matter. When we use some material and then throw it away it does not disappear but rather can accumulate in inconvenient ways. In other words, matter will always exist on earth and our using it merely moves it from one place to another. It has been said that "it is impossible to throw anything away, because there is no away."

In addition to the physical laws, we have sought a satisfactory explanation of

Figure 1-1
An Example of a
model or theory. A
theory or model is a
mental picture or
explanation of an
aspect of reality which
is unobservable with
normal senses. The
example here shows
that we can observe
the external result of
the watch functioning.
To explain how the
watch functions
without looking
inside, we can
develop a model or
theory. Many theories
are possible, but often
most can be refuted
since they are not
substantiated by the
external observations.

(a)

(b)

(c)

Observations of the external operations of a watch: Hour hand moves more slowly than minute hand and watch needs periodic winding.

Theory one: Small animal is running a wheel which turns gears which control the hands.

Theory two: Falling sand turns paddle wheel which turns gears which control the hands.

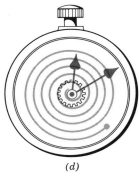

(d)

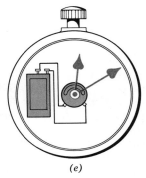

(e)

Theory three: Force of unwinding spring turns gears which control the hands.

Theory four: Small battery runs a small motor which controls the hands.

why objects behave as they do and why certain processes occur or can be made to occur. Such explanations involve the development of a description of reality which provides a view of our environment and will allow the prediction of what may happen under specific circumstances. Such a view of reality involves the establishment of concepts or mental pictures of aspects of reality that are unobservable with normal senses. In fact, modern science is based mainly upon such conceptualizations of reality.

A concept which serves as a description of why material objects behave as they do and/or why certain processes occur is called a **theory** or a **model.** As we shall see, there are many important theories and models in chemistry. As an analogy, consider a case in which you are able to observe a watch from the outside only. By observing you see what the watch does. To explain how it works you have to develop a mental picture of what might be inside the watch which would produce the external effects. (See Figure 1-1.) A mental picture serves as an explanation of something which is not directly observable.

1-7 THE INTERNATIONAL SYSTEM OF UNITS (METRIC SYSTEM)

The qualities of our environment which we can perceive with our senses can be used to describe objects and phenomena. We can recall that matter is defined as any object which has mass and occupies space. Thus, one of the apparent qualities is that objects have mass. The concept of space results from our ability to observe distances between points. This provides a quality of our spatial environment which we call **length.** Our sense of touch conveys to us that some objects are hotter or colder than we are. This degree of hotness or coldness provides another quality called **temperature.** Still another quality arises from our ability to observe durations between events and the irreversibility of a sequence of events. This quality is called **time.** To illustrate, consider that if you drop a glass, a certain duration of time passes before it shatters on the floor, and there is no way to reverse the time sequence so that the pieces restructure the glass. With the added dimension of time, we refer to our space–time environment.

Using these four qualities of mass, length, time, and temperature, we can describe what exists around us and the events that are occurring. To give precise descriptions it is necessary to establish a reference unit for each of these qualities. For instance, if we want to measure the length of an object we need a reference to compare the length to. In other words, we need to give some definite length (distance between two points) a name (i.e., the inch) so that other distances can be measured in terms of this defined unit of length. Throughout the centuries many different kinds of reference units have been defined for the four qualities of length, mass, time, and temperature. The cubit, a biblical measurement, corresponded to the length from the elbow to the end of the middle finger. The foot originated from the length of the human foot and the inch from the width of a human thumb. A mile was first defined by the Romans as 1,000 double paces and later defined as 5,280 feet by the English. The grain originated from the weight of a grain of wheat. Scientists have developed more logical units of reference for convenience.

The most common system of reference units used to day is called the **International System of Units (SI)** or the **metric system.** This set of units is used internationally and, thus, provides universal agreement in measurements. The defined units of the International System are listed below, and the official definitions are given in Table 1-1.

Quality	Unit	Abbreviation
Mass	**Kilogram** or **gram** (the gram is not the basic defined unit of mass in the SI, but it is the unit of mass most often used; 1 gram is 1/1,000 of a kilogram)	**kg, g**
Length	**Meter**	**m**
Time	Second	**sec** or **s**
Temperature	**Degree Kelvin** or **degree Celsius** (the Celsius scale is not the basic temperature scale in the SI, but it is often used to express temperature measurements)	**K, °C**

1-8 MEASUREMENTS

We can describe a measurement of a quality of an object by stating that it is a certain multiple or fraction of the basic unit. Of course, the unit used will indicate the quality described by the measurement. Remember that a measurement must always consist of a number (fraction or multiple) and a unit. For example, if we measure the length of an object and state that it has a length of 2.25, we have failed to convey any meaningful information. We must include the

Table 1-1
Definitions of the
Metric Units

Quality	Unit name and abbreviation	Definition
Mass	Kilogram (kg) (the gram, g, which is 1/1,000 of a kilogram is often used as the basic reference unit, even though the kilogram is the basic defined unit of mass)	*Original:* mass of 1,000 cm³ of water at 4°C *Current:* mass of the prototype kilogram number one kept at the International Bureau of Weights and Measures at Sevres, France
Length	Meter (m)	*Original:* 1/10,000,000 of the quadrant of the earth's meridian passing through Barcelona and Dunkirk *Current:* 1,650,763.73 times the wavelength corresponding to a specified transition occurring in excited atoms of krypton-86 under specific conditions
Time	Second (sec or s)	*Original:* 1/86,400 of the mean solar day *Current:* the duration of 9,192,631,770 cycles of the radiation corresponding to a specified transition in the excited atoms of cesium-133
Temperature	Degree Kelvin (K) or Degree Celsius (°C)	*Original:* 1/100 of the interval between the freezing point of water and the boiling point of water *Current:* 1/273.16 of the thermodynamic temperature of the thermal state corresponding to the triple point of water; this is the thermal state at which solid, liquid, and gaseous water exist in equilibrium

unit involved in the measurement. So, we could say that an object has a length of 2.25 m. (Read m as meter.) This would not be ambiguous, but would indicate that the length of the object is 2.25 times the standard unit of length called the meter.

When expressing measurements, it is permissible to use abbreviations for the units. You must become familiar with the ones that are most often used.

Using the proper measuring instrument we can measure the properties of an object and, thus, describe it. An example of such a description is given in Figure 1-2. Sometimes we use a combination of measurements to express a property.

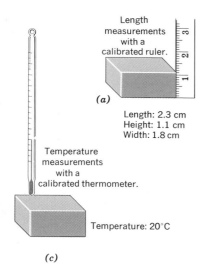

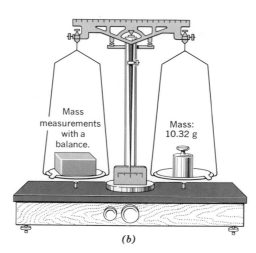

Figure 1-2
The measurement of the basic qualities of an object. From *Introduction to Chemistry* by T. R. Dickson, John Wiley & Sons, New York, 1971.

For instance, the volume of a rectangular solid is the product of its length, height, and width, as in Figure 1-2. When an object is moving with a certain speed, we can express this by stating the distance (length) traveled in a certain period of time. That is, we can express a speed in terms of miles per hour. As we shall see, many other properties and phenomena can be expressed in terms of some combination of basic units.

Occasionally, it is necessary to express measurements of very small or large entities. For instance, the distance between two cities may be 200,000 meters, or the mass of an object may be 0.0023 gram. To conveniently express such measurements, the metric system uses a set of prefixes which denote large or small multiples of the basic units. Three of the most commonly used prefixes are:

Prefix	Abbreviation	Meaning
kilo	k	1,000 times unit
centi	c	1/100 times unit
milli	m	1/1,000 times unit

These prefixes are used in front of the unit and represent the indicated multiple or fraction. The distance 200,000 meters can be expressed as 200 kilometers (km) and the mass 0.0023 grams can be expressed as 2.3 mg (read mg as milligrams). Note that the prefixes centi and milli have the same meaning as when they are used in our monetary system. One cent is 1/100 of a dollar and one mil is 1/1,000 of a dollar. To illustrate the relationship between these prefixes and the base unit, consider that if we divide a distance of 1 meter into 100 equal parts, each part corresponds to 1 centimeter (cm). One hundred centimeters equals 1 meter. If the meter distance is divided into 1,000 equal parts, each part is a millimeter (mm). One thousand millimeters equals a meter.

The prefixes can be used with any metric unit whenever desired. A few examples of measurements expressed in prefixed units are given in Table 1-2.

One of the basic units of the metric system is that of length. However, to express the amount of space occupied by an object, called the **volume,** we need to indicate the dimensions of the object. For instance, the volume of a rectangular solid is found by multiplying the length by the height and by the width. Thus,

Table 1-2
Some Examples of
Prefixed Units

	SI	English
Dimensions of postage stamp	22.4 mm × 25.4 mm	7/8 inch × 1 inch
Diameter of United States quarter	2.4 cm	15/16 inch
Distance between New York City and San Francisco	4,868 km	3,025 miles
Mass of 200 pound man	90.7 kg	200 pounds
Mass of 5 pounds of apples	2.27 kg	5 pounds
Mass of a 5 grain aspirin tablet	32.4 cg	0.000714 pound
Mass of a fly's wing	2 mg	0.0000044 pound

the units of volume are cubic length such as cubic feet, cubic inches, cubic meters, or cubic centimeters. A special unit for volume, called the liter, is used in the metric system. The **liter** (ℓ) is defined as the volume occupied by 1,000 cubic centimeters:

$$1 \ \ell = 1,000 \ cm^3$$

The cm³ is a symbolic way of stating cubic centimeters. Likewise, we use in.³ for cubic inch, ft³ for cubic feet, and m³ for cubic meter. The volumes of substances are often expressed in terms of liters. One liter is slightly larger than 1 quart.

The metric prefixes can be used with the liter. If 1 liter of volume is divided into 1,000 equal parts, each part is 1 milliliter (ml). There are 1,000 ml in 1 liter. Note that since 1 liter is also equivalent to 1,000 cubic centimeters, then the milliliter and cubic centimeter correspond to the same amount of volume:

$$1 \ ml = 1 \ cm^3$$

Since the milliliter and cubic centimeter have the same meaning, they are often used interchangeably.

The English system of units (inches, feet, pounds, etc.) is used in the United States. Legislation is pending in Congress which will make the SI the legal system of units in this country. Some comparisons of SI units and English units of measurement are shown in Figure 1-3.

1-9 METRIC AND ENGLISH CONVERSIONS

When measurements are made in the metric system it is sometimes necessary to change from the base unit to a prefixed unit or from a prefixed unit to the base unit. This is accomplished by a conversion process in which the meaning of the prefix is used. To illustrate, consider the relation between the meter and the common prefixes used with the meter. (Similar relations can be expressed for any unit.)

$$1,000 \ mm = 1 \ m \qquad 100 \ cm = 1 \ m \qquad 1,000 \ m = 1 \ km$$

These relations can be expressed as ratios in the form:

$$\left(\frac{1,000 \ mm}{1 \ m}\right) \qquad \left(\frac{1 \ m}{1,000 \ mm}\right)$$

$$\left(\frac{100 \ cm}{1 \ m}\right) \qquad \left(\frac{1 \ m}{100 \ cm}\right)$$

$$\left(\frac{1,000 \ m}{1 \ km}\right) \qquad \left(\frac{1 \ km}{1,000 \ m}\right)$$

Such ratios can be used as conversion factors to change a measurement from a unit to a prefixed unit or vice versa.

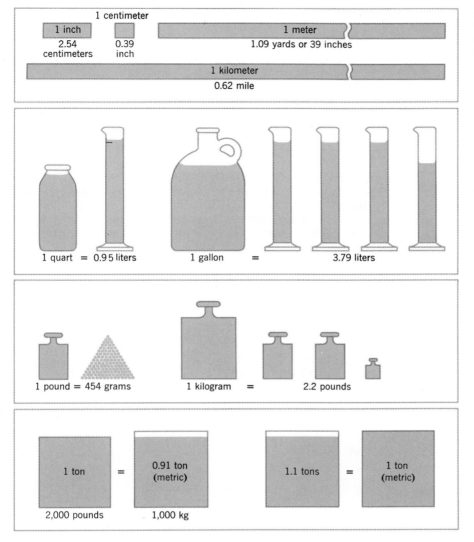

Figure 1-3
Comparison of SI
(metric) and English
units.

To convert a measurement in a prefixed unit to the base unit, we just multiply by the factor corresponding to the meaning of the prefix. The distance 15 kilometers expressed in terms of meters is found by multiplying by the factor 1,000 meters per kilometer:

$$15 \text{ km} \left(\frac{1,000 \text{ m}}{1 \text{ km}}\right) = 15,000 \text{ m}$$

Note that the kilometer units cancel, leaving the m unit.

The distance 2.5 millimeters expressed in terms of meters is found by multiplying by the factor 1 meter per 1,000 millimeters:

$$2.5 \text{ mm} \left(\frac{1 \text{ m}}{1,000 \text{ mm}}\right) = 0.0025 \text{ m}$$

To express 200,000 meters in terms of kilometers, we note that for each kilometer there are 1,000 meters. Thus, multiplying the distance in meters by this factor should convert the measurement to kilometers:

$$200,000 \text{ m} \left(\frac{1 \text{ km}}{1,000 \text{ m}}\right) = 200 \text{ km}$$

To express the measurement 0.056 meters in units of centimeters, we note that there are 100 centimeters in 1 meter. Thus, multiplying the distance in meters by this factor will convert the measurement to centimeters:

$$0.056 \text{ m} \left(\frac{100 \text{ cm}}{1 \text{ m}}\right) = 5.6 \text{ cm}$$

In the United States we often use the English system of units to make measurements. However, since the SI system is the commonly used technical system of units, it is necessary to convert measurements made in English units to metric units, or vice versa. To accomplish such conversions it is necessary to use conversion factors which relate the units in each system. Table 1-3 lists some of the most common conversion factors. To convert a measurement we just multiply by the proper factor. The factor used eliminates one unit and replaces it with a desired unit. To express the length 10 inches in terms of centimeters, we note that there are 2.54 centimeters per 1 inch. Thus, we multiply the 10 inches by this factor:

$$10 \text{ in.} \left(\frac{2.54 \text{ cm}}{1 \text{ in.}}\right) = 25.4 \text{ cm}$$

Table 1-3
Comparison of English
and SI (Metric) Units[a]

	English to SI	SI to English
Length		
1 inch (in.) = 2.54 centimeters (cm)	$\left(\dfrac{2.54 \text{ cm}}{1 \text{ in.}}\right)$	$\left(\dfrac{0.394 \text{ in.}}{1 \text{ cm}}\right)$
1 foot (ft) = 0.305 meter (m)	$\left(\dfrac{0.305 \text{ m}}{1 \text{ ft}}\right)$	$\left(\dfrac{3.27 \text{ ft}}{1 \text{ m}}\right)$
1 yard (yd) = 0.914 meter (m)	$\left(\dfrac{0.914 \text{ m}}{1 \text{ yd}}\right)$	$\left(\dfrac{1.09 \text{ yd}}{1 \text{ m}}\right)$
1 mile (mi) = 1.61 kilometers (km)	$\left(\dfrac{1.61 \text{ km}}{1 \text{ mi}}\right)$	$\left(\dfrac{0.62 \text{ mi}}{1 \text{ km}}\right)$
Volume		
1 pint (pt) = 0.47 liter (ℓ)	$\left(\dfrac{0.47 \text{ ℓ}}{1 \text{ pt}}\right)$	$\left(\dfrac{2.11 \text{ pt}}{1 \text{ ℓ}}\right)$
1 quart (qt) = 0.95 liter (ℓ)	$\left(\dfrac{0.95 \text{ ℓ}}{1 \text{ qt}}\right)$	$\left(\dfrac{1.06 \text{ qt}}{1 \text{ ℓ}}\right)$
1 gallon (gal) = 3.8 liters (ℓ)	$\left(\dfrac{3.8 \text{ ℓ}}{1 \text{ gal}}\right)$	$\left(\dfrac{0.26 \text{ gal}}{1 \text{ ℓ}}\right)$
Mass (weight)		
1 pound (lb) = 454 grams (g)	$\left(\dfrac{454 \text{ g}}{1 \text{ lb}}\right)$	$\left(\dfrac{0.0022 \text{ lb}}{1 \text{ g}}\right)$
1 ounce (oz) = 28.3 grams (g)	$\left(\dfrac{28.3 \text{ g}}{1 \text{ oz}}\right)$	$\left(\dfrac{0.0353 \text{ oz}}{1 \text{ g}}\right)$
1 pound (lb) = 0.454 kilogram (kg)	$\left(\dfrac{0.454 \text{ kg}}{1 \text{ lb}}\right)$	$\left(\dfrac{2.20 \text{ lb}}{1 \text{ kg}}\right)$

[a] To remember how the SI and English systems compare, note the following: There are about 2.5 centimeters per inch. One meter is nearly the same as a yard. One kilometer is about 0.6 of a mile. One liter is slightly larger than a quart. One kilogram is 2.2 pounds.

To express 80 kilograms in units of pounds, we multiply by the factor 2.2 pounds per 1 kilogram.

$$80 \text{ kg} \left(\frac{2.2 \text{ lb}}{1 \text{ kg}}\right) = 176 \text{ lb}$$

To express 4.0 quarts in units of liters, we multiply by the factor 0.95 liters per 1 quart.

$$4.0 \text{ qt} \left(\frac{0.95 \; \ell}{1 \; qt}\right) = 3.8 \; \ell$$

To express the speed 100 kilometers per hour in miles per hour, we multiply the speed by the fact that there are 0.62 miles per 1 kilometer.

$$\left(\frac{100 \text{ km}}{1 \text{ hr}}\right)\left(\frac{0.62 \text{ mi}}{1 \text{ km}}\right) = \frac{62 \text{ mi}}{1 \text{ hr}}$$

1-10   TEMPERATURE AND HEAT

As noted earlier, the various forms of energy can be interconverted. One form of energy which we experience often in our environment is heat. We associate heat with hot bodies and we shall find out later that the heat energy of a body is mainly the result of the motion of the very tiny constituent particles of the body. Heat is a convenient form of energy because it can be transferred from one body to another as a result of contact between the bodies. We say that when a body becomes hotter it gains heat energy. We say that a body is at a certain temperature and when it gains heat energy it changes to a higher temperature.

In order to state the temperatures of bodies, it is necessary to establish a relative scale of comparison called a **temperature scale.** To establish a temperature scale, it is necessary to define certain states as corresponding to certain temperatures. Then the difference between one state and another can be measured with a thermometer and the difference can be divided into a certain number of parts called degrees. For example, the Celsius scale (formerly called the centigrade scale) was originally defined by establishing the freezing point of water as zero degrees (0°) and the boiling point of water as 100 degrees (100°). The **Celsius degree,** °C, was defined as 1/100 of the temperature difference between the freezing point and boiling point of water.

There is an important difference between the concepts of heat and temperature. The amount of heat associated with a body depends on the amount of material involved, while the temperature of the body may be the same for different amounts. To illustrate the difference, consider that much more heat is associated with a bonfire than a lighted match, while the temperatures may be the same.

The Celsius scale is very often used in science to express temperatures. In the United States we use the Fahrenheit scale as a nontechnical temperature scale. The Fahrenheit scale was established centuries ago using reference temperatures that were not reliable. Today it is a secondary scale and is defined with reference to the scientific temperature scales. The symbol °F is used to represent a degree on the Fahrenheit scale.

It is sometimes necessary to convert a temperature measurement made in one scale to the other scale. To accomplish this, the two scales must be compared. (See Figure 1-4.) As can be seen in the figure, the Fahrenheit degree is smaller than the Celsius degree. Between the freezing point and the boiling point of water there are 100°C and 180°F. Thus, there are 5°C for every 9°F:

$$\left(\frac{100°C}{180°F}\right) = \left(\frac{5°C}{9°F}\right) \quad \text{or} \quad \left(\frac{180°F}{100°C}\right) = \left(\frac{9°F}{5°C}\right),$$

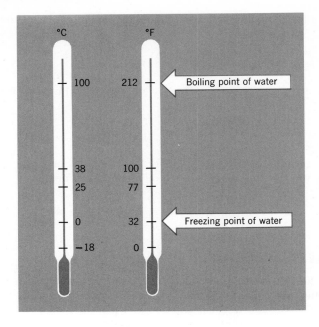

To convert a temperature in one scale to the other scale, the degree ratios can be used as conversion factors. However, as can be seen in Figure 1-4, the zero points on the two scales do not coincide and, in fact, differ by 32°F. Consequently, the zero points must be adjusted in the conversion process.

To convert a temperature measured in degrees Fahrenheit to degrees Celsius, we first subtract 32°F to adjust the zero points and then multiply by the factor 5°C per 9°F.

For example, 68°F expressed in degrees Celsius is

$$68°F - 32°F = 36°F$$

$$\left(\frac{5°C}{9°F}\right) 36°F = 20°C$$

To convert a temperature measured in degrees Celsius to degrees Fahrenheit, we multiply by the factor 9°F per 5°C and then add 32°F to adjust to the Fahrenheit zero point. For example, 20°C expressed in degrees Fahrenheit is

$$\left(\frac{9°F}{5°C}\right) 20°C = 36°F$$

$$36°F + 32°F = 68°F$$

1-11   Exponential Notation

In chemistry we often deal with large and small numbers. These numbers are usually written in a special manner. Before discussing this, however, we should consider how multiples of ten are represented. Multiples of ten, the base of our number system, can be represented by expressing ten raised to a power or exponent. For instance,

$$100 = (10)(10) = 10^2$$
$$1,000 = (10)(10)(10) = 10^3$$
$$10,000,000,000 = (10)(10)(10)(10)(10)(10)(10)(10)(10)(10) = 10^{10}$$

Fractions of ten can be represented in a similar manner:

$$\frac{1}{100} = 0.01 = 10^{-2}$$

$$\frac{1}{100,000} = 0.000001 = 10^{-6}$$

The power to be used is determined by counting the number of positions moved when the decimal point is moved to the right of the one:

$$10,000 = 10^4 \quad \text{(positive exponent moving left)}$$
$$0.001 = 10^{-3} \quad \text{(negative exponent moving right)}$$

When we write a large or small number, it is necessary to repeat many zeros:

$$602,000,000,000,000,000,000,000 \qquad 0.00052$$

It is possible to represent such numbers in a form called **exponential notation** by moving the decimal point so that only one digit is on the left (i.e., the number is between 1.00 and 9.99):

$$6.02000000000000000000000 \qquad 0005.2$$

Then, all digits are written with the decimal point in this position and these digits are multiplied by ten raised to an exponent corresponding to the number of positions the decimal has been moved:

$$6.02 \times 10^{23} \qquad 5.2 \times 10^{-4}$$

If the decimal is moved to the left, the power of ten is positive, and if it is moved to the right, the power is negative:

$$523,000 = 5.23 \times 10^5$$
$$0.00721 = 7.21 \times 10^{-3}$$

To convert a number from exponential notation to the normal form, we just move the decimal point the number of positions indicated by the power of ten.

$$7.23 \times 10^6 = 7,230,000. \quad \text{(move decimal 6 places to the right)}$$
$$4.7 \times 10^{-2} = 0.047 \quad \text{(move decimal 2 places to the left)}$$

1-12 PERCENTAGE

Percentage is a common way of expressing what fraction one quantity is of another. **Percentage** means parts per hundred and is represented by %. Thus, 20% means 20 parts per 100 parts or 20/100. A percentage can be expressed as a decimal fraction by dividing by 100. So, 20% as a decimal fraction is 0.20. To calculate what percentage one quantity, a, is of another quantity, b, a is divided by b and multiplied by 100:

$$\text{percentage} = \left(\frac{a}{b}\right) 100$$

Percentages can use various bases of comparisons. For example, if it is found that in a large group of people 3 out of every 100 are left-handed, we say that 3% are left-handed. The basis of comparison is numbers of people. Very often percentages are expressed on a basis of mass or weight. A deposit of iron ore may contain 20 grams of iron in each 100 grams of ore. Thus, we say that the ore contains 20% iron by mass. It is essential to state the basis of comparison when giving a percentage. When this is not done, the context in which the percentage is used must indicate the basis of comparison.

BIBLIOGRAPHY

BOOKS

American Chemical Society. *Cleaning Our Environment: The Chemical Basis For Action.* Washington, D.C.: American Chemical Society, 1969, 249 pp.

Commoner, B. *The Closing Circle.* New York: Knopf, 1971.

Day, J. A., F. F. Fost, and P. Rose. *Dimensions of the Environmental Crisis.* New York: Wiley, 1971. 212 pp.

Goldman, M. I. *Ecology and Economics: Controlling Pollution in the 70's.* Englewood Cliffs, N.J.: Prentice-Hall, 1972. 234 pp.

Holum, J. R., and R. B. Boolootian. *Evnironmental Science: An Introduction to Their Topics and Terms.* Boston: Little, Brown, 1972.

Ehrlich, P. (ed.) *Man and Ecosphere.* San Francisco: W. H. Freeman, 1971. 307 pp.

Odum, E. P. *Fundamentals of Ecology.* 3rd ed. Philadelphia: W. B. Saunders, 1971. 574 pp.

Sax, J. L. *Defending the Environment: A Strategy for Citizen Action,* New York: Knopf, 1971.

ARTICLES AND PAMPHLETS

Astin, A. V. "Standards of Measurement." *Scientific American* **218,** 50 (June 1968).

Brief History and Use of the English and Metric Systems of Measurements. Washington, D.C.: National Bureau of Standards Special Publication 304A, 1969.

Brower, L. P. "Ecological Chemistry." *Scientific American* **220** (2), 22–29 (1969).

Paul, M. A., "International System of Units (SI)" *Chemistry* (Oct. 1972).

Ritchie-Calder, Lord. "Conversion to the Metric System." *Scientific American* **223,** 17 (July 1970).

Scientific American (Sept. 1970). Entire issue devoted to "The Biosphere."

QUESTIONS AND PROBLEMS

1. Give a definition of chemistry.

2. Reread Section 1-5 and list the forms of energy mentioned.

3. Give a statement of the Law of Conservation of Energy.

4. Give a statement of the Law of Conservation of Matter.

5. What is a theory or model?

6. List the four qualities of our environment for which units are defined in the SI system. Give the names of the basic SI units of these qualities.

7. What is a liter?

8. A person has the following body measurements: 38 inches, 22 inches, 36 inches. What would these body measurements be in centimeters?

9. Calculate your height in meters. Convert your weight from pounds to kilograms.

10. In the Olympics, sprinters run the 100 meter dash. How many yards is 100 meters?

11. In Europe a speed limit sign reads 80 kilometers per hour. What is this speed limit in the United States?

12. Suppose the fuel use of a car is 20 miles per gallon. Express this in terms of kilometers per liter. [*Hint:* Convert the miles to kilometers and one gallon to liters. Then divide.]

13. Normal body temperature is 98.6°F. What is normal body temperature in degrees Celsius? If a person has a temperature of 100°F, what is his temperature in degrees Celsius?

14. If the air temperature is 70°F, what is the temperature in degrees Celsius?

15. What is the temperature 20°F in degrees Celsius?

16. Adoption of the SI or metric system by the United States will have certain advantages and disadvantages. One disadvantage will be the expense of changing the machinery and tools of industry and the various trades. Another is public education and phasing-in of new units of measure. Advantages include the monetary gain as the result of increased export of machinery and tools and compatibility of United States measurements with the measurements of other countries of the world. Considering these factors and any other information or opinions you might have, rank the following according to your feelings (1 — strongly agree, and so on):

    _____ The SI system should be phased-in over a period of 10 years.
    _____ The SI system should not be adopted.
    _____ The SI system is of no concern to me.

CHEMICAL ELEMENTS

2

2-1 THE STATES OF MATTER

As you look around at the water, air, and earth, you see that matter can take on three different physical forms. These forms are called **physical states** or **phases.** The three states of matter are the **solid state,** the **liquid state,** and the **gas** or **vapor state.** (See Figure 2-1.) Matter in the solid state, such as a rock or a piece of metal, has a definite volume and a definite shape or firmness associated with it. A liquid, such as water or oil, also occupies a specific volume, but it requires a container. A given volume of a liquid will take the shape of the container which holds it. A gas has neither a definite shape nor volume and, thus, it must be contained in a sealed vessel for storage. By considering an air-filled balloon, you can see that a gas occupies the entire volume into which it is placed.

2-2 ELEMENTS, COMPOUNDS, AND MIXTURES

Chemists have always been interested in finding out what things are made of. To do this they observe the properties of the various forms of matter found in our environment. These characteristics or properties serve to distinguish one type of matter from another. Most matter that we see is made up of mixtures of materials. For example, as we look closely at soil we see the different parts. Concrete obviously has different parts. Food is made up of the varied forms of plant and animal cells. The grain in a piece of wood reveals that wood is a mixture. Such mixtures are called **heterogeneous mixtures.** Heterogeneous means made up of different parts.

It is possible to separate mixtures into simpler parts. The components of mixtures are pure substances. **Pure substances** are forms of matter which have the same properties throughout and have definite chemical compositions. Actually, there are two kinds of pure substances: elements and compounds. The chemical **elements** are the fundamental forms of matter and are components of all matter. Elements are pure substances that cannot be separated into simpler substances. **Compounds** have definite properties and composition, but, under certain circumstances, they can be separated into two or more chemical elements. In other words, compounds are pure substances made up of combinations of chemical elements. For instance, the compound water is composed of the chemical elements hydrogen and oxygen. Chemistry is concerned with the study

Figure 2-1
The three common
states of matter. From
*Introduction to
Chemistry* by T. R.
Dickson, John Wiley &
Sons, New York, 1971.

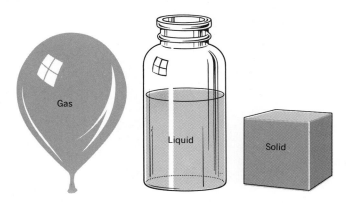

of the properties of the elements and with the ways in which the elements form compounds.

Chemical compounds are composed of elements, but a given compound will always contain the same elements in definite proportions by mass. For instance, the compound water always contains hydrogen and oxygen and any sample will contain 8 grams of oxygen to every 1 gram of hydrogen. Compounds are characterized by this kind of definite composition. You might wonder if mixing sugar in water produces a compound, since the resulting mixture appears to be pure and homogeneous. Mixtures of this kind do appear to be homogeneous, but do not have definite composition. That is, sugar and water mixtures can contain variable amounts of sugar in water. Since such variable compositions occur, these mixtures are not compounds. Homogeneous mixtures of variable composition are called **solutions.** Solutions are mixtures of substances which dissolve in one another to become single-phase (i.e., solid, liquid, or gas) mixtures. Hold on, you may say, we can see the parts of a heterogeneous mixture, but how can a solution which looks pure be a mixture. First of all you have to realize that matter can be broken up into tiny pieces and mixed together. Some mixtures, like milk, appear to the eye to be pure substances but are actually heterogeneous mixtures. The difference between heterogeneous mixtures and solutions is that solutions involve very small particles of matter (far below even microscopic observation) that are intimately mixed, and heterogeneous mixtures involve larger aggregates of such particles. Solutions are quite common and include such liquids as wine and ocean water. Solutions are discussed in Chapter 9.

Since the elements are the fundamental building units of matter, it is important to understand their natures and differences. Much time and effort have been devoted to the isolation, purification, and description of the chemical elements. To date, 105 different elements have been identified. Most of these elements occur in nature in the uncombined form or as constituents of compounds. Some of the elements are not found in nature, but have been synthesized by nuclear scientists. These man-made elements are discussed in Chapter 6. Many of the elements which occur in compounds are separated from these compounds and used in numerous ways. For example, many metals such as iron and aluminum are obtained (refined) from compounds (contained in ores) that are found in nature. Eleven elements are gases and two occur as liquids at normal conditions. The rest of the elements occur as solids.

Each of the elements has been given a unique name and **symbol.** Since the elements have been isolated and identified over a period of centuries, these names and symbols have historical origin. An alphabetical list of the element names and corresponding symbols is given in Table 2-1. As you look over the list, note that most of the symbols are derived from the names of the elements. However, in some cases the symbols are derived from Latin names. For instance, iron has the symbol Fe and mercury has the symbol Hg. Knowledge of the names of the elements and the symbols that represent them is fundamental to any discussion of chemistry. It is necessary to memorize some of the element names and symbols. A list of the most important ones is given in Table 2-2.

Many elements are found on earth in very small amounts, while some occur in much greater amounts. The distribution of the elements on earth is discussed in Section 4-1.

2-3 THE NATURE OF ELEMENTS

Since the beginnings of reflective thought man has wondered about the nature of matter at a submicroscopic, unseeable level. If you imagine dividing a piece of a pure substance into smaller and smaller bits, will you soon reach a characteristic particle that cannot be further broken down, or could this subdivision continue indefinitely? The Greek philosophers pondered this question. Some philosophers, realizing that they could feel the wind blow and that bricks in the street seemed to wear away in unseeable pieces, considered matter to be made up of tiny units or particles. The Greek philosopher Democritus (about 400 BC) proposed that matter was composed of tiny, indestructible particles which he called **atoms,** from the Greek word for indestructible. However, the dominant philosophers such as Aristotle and Plato rejected this idea and considered matter to be continuous and composed of only one substance. The Aristotelian view prevailed for centuries. In the middle ages, the alchemists investigated and experimented with matter with the intention of trying to convert other elements into gold. This feat was never accomplished, but many compounds and elements were discovered in the process which later contributed to the understanding of matter.

In the 1600s alchemistry was well on the decline, and the pioneers of modern chemistry, the first chemists, began to develop the science of chemistry. These pioneers included such men as Robert Boyle (1627–1691), Joseph Priestly (1733–1804), Antoine Lavoisier (1743–1794), and John Dalton (1766–1844). In the next few hundred years significant development of chemistry occurred. Chemists actually experimented with substances and were able to identify many elements and compounds. They were able to decompose compounds into the constituent elements and to form compounds from the elements. Furthermore, with the development of weighing methods, they were able to weigh samples of elements and compounds. This allowed the precise observation of the relative masses of elements combined in compounds and provided evidence for the development of a theory concerning the submicroscopic nature of matter.

Let us consider some of this evidence. It was realized that when elements were combined to form compounds or compounds were decomposed to form elements, no mass was lost or gained. For example, when 56 grams of iron were compounded with 32 grams of sulfur, the resulting compound had a mass of 88 grams. This realization indicated that in a chemical process mass is conserved, and this was stated as the **Law of Conservation of Matter:** Matter is not created or destroyed in a chemical process. This idea was surprising to many, since they thought that, for example, when a candle burned, the matter was lost. It can be shown that when a candle is burned in a closed container of air the mass of the candle and air in the container is not changed. Actually, the products of the burning candle are given off as gases.

Another observation which developed was the fact that a specific compound always contained the same elements. This idea was mentioned previously, but let us consider it in more detail. The common compound salt (or sodium chloride) will be used as an example. When salt was produced in the laboratory or

| **Table 2-1** | | |
The Elements	Name	Symbol
	Actinium (Gr. *aktis, aktinos,* beam or ray)	Ac
	Aluminum (L. *alumen,* alum)	Al
	Americium (the Americas)	Am
	Antimony (L. *antimonium, stibium,* mark)	Sb
	Argon (Gr. *argon,* inactive)	Ar
	Arsenic (L. *arsenicum;* Gr. *arsenikon,* yellow orpiment—identified with *arsenikos,* male, from the belief that metals were different sexes—Arab. *az-zernikh,* the orpiment from Persian *zerni-zar,* gold)	As
	Astatine (Gr. *astatos,* unstable)	At
	Barium (Gr. *barys,* heavy)	Ba
	Berkelium (Berkeley, home of University of California)	Bk
	Beryllium (Gr. *berryllos,* beryl; also called Glucinium or Glucinum, Gr. *glykys,* sweet)	Be
	Bismuth (Ger. *Weisse Masse,* white mass; later Wismuth and Bisemutum)	Bi
	Boron (Arab. *Buraq;* Persian *Burah*)	B
	Bromine (Gr. *bromos,* stench)	Br
	Cadmium (L. *cadmia;* Gr. *kadmeia*—ancient name for calamine, zinc carbonate)	Cd
	Carbon (L. *carbon,* charcoal)	C
	Cerium (named for the asteroid Ceres)	Ce
	Cesium (L. *caesius,* sky blue)	Cs
	Chlorine (Gr. *chloros,* greenish-yellow)	Cl
	Chromium (Gr. *chroma,* color)	Cr
	Cobalt (Ger. *Kobold,* goblin or evil spirit)	Co
	Copper (L. *cuprum,* from the island of Cyprus)	Cu
	Curium (Pierre and Marie Curie)	Cm
	Dysprosium (Gr. *dysprositos,* hard to get at)	Dy
	Einsteinium (Albert Einstein)	Es
	Erbium (Ytterby, a town in Sweden)	Er
	Europium (Europe)	Eu
	Fermium (Enrico Fermi)	Fm
	Fluorine (L. and F. *fluere,* flow or flux)	F
	Francium (France)	Fr
	Gadolinium (gadolinite—a mineral named for Gadolin, a Finnish chemist)	Gd
	Gallium (L. *Gallia,* France)	Ga
	Germanium (L. *Germania,* Germany)	Ge
	Gold (Sanskrit *Jval;* Anglo-Saxon *gold;* L. *aurum,* shining dawn)	Au
	Hafnium (L. *Hafnia,* Copenhagen)	Hf
	Helium (Gr. *helios,* the sun)	He
	Holmium (L. *Holmia,* Stockholm)	Ho
	Hydrogen (Gr. *hydro,* water, and *genes,* forming)	H
	Indium (from the brilliant indigo line in its spectrum)	In
	Iodine (Gr. *iodes,* violet)	I
	Iridium (L. *iris,* rainbow)	Ir
	Iron (Anglo-Saxon, *iron;* L. *ferrum*)	Fe
	Krypton (Gr. *kryptos,* hidden)	Kr
	Lanthanum (gr. *lanthanein,* to lie hidden)	La
	Lawrencium (Ernest O. Lawrence, inventor of the Cyclotron)	Lr
	Lead (Anglo-Saxon *lead;* L. *plumbum*)	Pb
	Lithium (Gr. *lithos,* stone)	Li
	Lutetium (Lutetia, ancient name for Paris)	Lu
	Magnesium (Magnesia, district in Thessaly)	Mg
	Manganese (L. *magnes,* magnet)	Mn
	Mendelevium (Dmitri Mendeleev)	Md
	Mercury (planet Mercury; hydrargyrum, liquid silver)	Hg

Name	Symbol
Molybdenum (Gr. *molybdos,* lead)	Mo
Neodymium (Gr. *neos,* new, and *didymos,* twin)	Nd
Neon (Gr. *neos,* new)	Ne
Neptunium (planet Neptune)	Np
Nickel (Ger. *Nickel,* Satan or "Old Nick," and from *kupfernickel,* Old Nick's copper)	Ni
Niobium (Niobe, daughter of Tantalus)	Nb
Nitrogen (L. *nitrum;* Gr. *nitron,* native soda, and *genes,* forming)	N
Nobelium (Alfred Nobel)	No
Osmium (Gr. *osme,* a smell)	Os
Oxygen (Gr. *oxys,* sharp, acid, and *genes,* forming, acid former)	O
Palladium (named after the asteroid Pallas; Gr. *Pallas,* goddess of wisdom)	Pd
Phosphorus (Gr. *phosphoros,* light-bearing; ancient name for the planet Venus when appearing before sunrise)	P
Platinum (Sp. *platina,* silver)	Pt
Plutonium (planet Pluto)	Pu
Polonium (Poland, native country of Marie Curie)	Po
Potassium (English, potash—pot ashes; L. *kalium;* Arab. *quali,* alkali)	K
Praseodymium (Gr. *prasios,* green, and *didymos,* twin)	Pr
Promethium (Prometheus, who, according to mythology, stole fire from heaven)	Pm
Protactinium (Gr. *protos,* first)	Pa
Radium (L. *radius,* ray)	Ra
Radon (from radium)	Rn
Rhenium (L. *Rhenus,* Rhine)	Re
Rhodium (Gr. *rhodon,* rose)	Rh
Rubidium (L. *rubidius,* deepest red)	Rb
Ruthenium (L. *Ruthenia,* Russia)	Ru
Samarium (Samarskite, a mineral)	Sm
Scandium (L. *Scandia,* Scandinavia)	Sc
Selenium (Gr. *Selene,* moon)	Se
Silicon (L. *silex, silicis,* flint)	Si
Silver (Anglo-Saxon, *Seolfor, siolfur;* L. *argentum*)	Ag
Sodium (English, *soda;* Medieval L. *sodanum,* headache remedy; L. *natrium*)	Na
Strontium (Strontian, town in Scotland)	Sr
Sulfur (Sanskrit *sulvere;* L. *sulphurium*)	S
Tantalum (Tantalus, mythological character—father of Niobe)	Ta
Technetium (Gr. *technetos,* artificial)	Tc
Tellurium (L. *tellus,* earth)	Te
Terbium (Ytterby, village in Sweden)	Tb
Thallium (Gr. *thallos,* a green shoot or twig)	Tl
Thorium (Thor, Scandinavian god of war)	Th
Thulium (Thule, the earliest name for Scandinavia)	Tm
Tin (Anglo-Saxon *tin;* L. *stannum*)	Sn
Titanium (L. *Titans,* the first sons of the Earth, mythology)	Ti
Tungsten (Swedish *tung sten,* heavy stone)	W
Uranium (planet Uranus)	U
Vandium (Scandinavian goddess, Vanadis)	V
Xenon (Gr. *xenon,* stranger)	Xe
Ytterbium (Ytterby, village in Sweden)	Yb
Yttrium (Ytterby, village in Sweden)	Y
Zinc (Ger. *Zink,* of obscure origin)	Zn
Zirconium (Arab. *zargum,* gold color)	Zr

From *Introduction to Chemistry* by T. R. Dickson, John Wiley & Sons, New York, 1971; Adapted from *Handbook of Chemistry and Physics.* Chemical Rubber Publishing Co.

Element	Symbol	Element	Symbol
Aluminum	Al	Iodine	I
Arsenic	As	Iron	Fe
Bromine	Br	Lead	Pb
Cadmium	Cd	Mercury	Hg
Calcium	Ca	Nitrogen	N
Carbon	C	Oxygen	O
Chlorine	Cl	Phosphorus	P
Copper	Cu	Potassium	K
Fluorine	F	Sodium	Na
Helium	He	Sulfur	S
Hydrogen	H	Uranium	U

obtained from a natural source, it was always found to be a compound of the elements sodium and chlorine. Furthermore, pure compounds were always found to have the same amounts of each element. For instance, 100 grams of salt would always contain 39.4 grams of sodium and 60.6 grams of chlorine (39.4 percent sodium and 60.6 percent chlorine). The observation became known as the **Law of Constant Composition:** A compound always contains the same elements, and they are present in definite proportions by mass.

An additional observation was that a given element would tend to form specific kinds of compounds with other elements. Sometimes, two or more elements would form similar compounds with another element. Sodium and potassium were found to form similar compounds with chlorine. Also, on occasion, it was found that two elements could form more than one kind of compound. Hydrogen and oxygen commonly form the compound water but can form the compound hydrogen peroxide. Water contains 88.8 percent oxygen and 11.2 percent hydrogen by mass, and hydrogen peroxide contains 5.9 percent hydrogen and 94.1 percent oxygen.

All these observations indicated that elements must consist of some kind of building units that could join to form compounds.

2-4 THE ATOMIC THEORY

In 1803, John Dalton, drawing from the work of many early scientists, proposed a theory or model of the submicroscopic nature of matter. This is the **Atomic Theory,** which can be stated as follows (see Figure 2-2):

1. Elements are composed of tiny, fundamental particles of matter called atoms. (Dalton used the term atom as recognition of the brilliant suggestion of Democritus 2,000 years before.)
2. Atoms of the same element are the same but differ from atoms of other elements. (Each kind of element has a unique kind of atom which has mass.)
3. Atoms enter into combinations with other atoms to form compounds.

The atomic theory was and is very important, since it provides a mental picture of matter. An atom can be pictured as a tiny spherical particle. An **atom** is the smallest representative particle of an element. Atoms are indeed very tiny particles. A penny contains about 30,000,000,000,000,000,000,000 (3×10^{22}) copper atoms. The masses of the atoms of the various elements differ from one another, but all atoms have masses in the range of 10^{-24}–10^{-23} gram. As you look at a sample of a compound, such as some salt, you can imagine that it consists

Element 1   Element 2

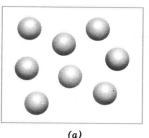

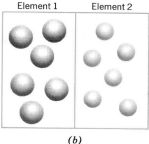

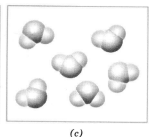

(a)   (b)   (c)

Elements are made up of tiny particles called atoms.

The atoms of a given element are alike but have different properties than the atoms of other elements.

Compounds involve atoms combined in definite arrangements so that there is a specific amount of each element present in the compound.

Figure 2-2
Illustration of the atomic theory. From *Introduction to Chemistry* by T. R. Dickson, John Wiley & Sons, New York, 1971.

of numerous combined atoms of sodium and chlorine. Even as you view a rock or a flower, you can imagine that it consists of a complex combination of atoms. The natures of such atomic combinations are discussed in Chapter 3.

Let us consider how the atomic theory explains some of the observations mentioned previously. The Law of Constant Composition becomes clear when it is realized that compounds involve definite combinations of atoms. That is, a

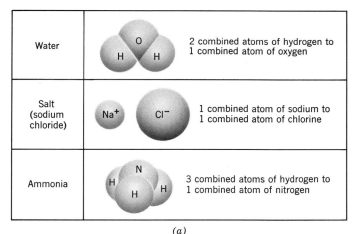

(a)

A compound always contains the same elements in the same relative amounts.

Figure 2-3
Compounds viewed using the atomic theory.

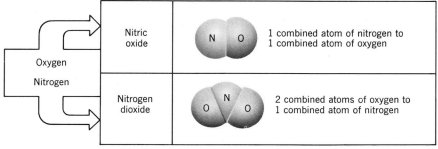

(b)

Two elements may form more than one kind of compound.

71421

given compound will always include the same atoms in the same amounts. (See Figure 2-3.) For instance, water always contains two combined atoms of hydrogen for every combined atom of oxygen. Since the atoms of elements differ from one another, they form different kinds of compounds with other elements. In some cases it is possible to have more than one combination of the atoms of two elements. (See Figure 2-3.)

The atomic theory served as the foundation for the development of chemistry and is still, at this time, one of the most fundamental models in chemistry. Once the idea of the atom and the idea that elements were unique had become established, chemists searched nature for the various elements. Over the years, numerous elements were discovered, and now 105 elements are known. Once the wide variety of the elements had been discovered in nature, another question arose. "If the elements are composed of atoms, what must the conceivable structure of atoms be which will allow them to form compounds?" In other words, the question of the structure of the atom became important.

2-5   ATOMIC STRUCTURE

The atom as a basic particle of matter is composed of still smaller particles called **subatomic particles.** Ordinary atoms are neutral, i.e., they do not carry an electrical charge. However, under certain circumstances atoms can become electrically charged. You have probably noticed that a comb can sometimes become electrically charged when you pass it through your hair. To observe the effect of this charge, place several grains of salt (sodium chloride) on a piece of paper, pass a comb through your (dry) hair a few times, and then place the comb just above the salt. Notice the attractive and repulsive forces that arise between the salt and the comb. Atoms of some elements can take on electrical charges. Such charged atoms are called **ions.** Some elements form positive ions, called **cations,** and some elements form negative ions, called **anions.** Since atoms are able to form ions, it was proposed that atoms contain basic units of negative charge which could be lost to form positive ions or gained to form negative ions. These units of charge were found to be the same units of charge which characterized electricity. Such units of negative charge became known as **electrons.** Since the electron is apparently the smallest unit of electrical charge, we shall refer to its charge as −1 unit charge or quantum charge. A +1 charge refers to a charge which is equal in magnitude to an electron charge, but positive rather than negative. When plus or minus charges are used in further discussion, unit charges will be intended.

The electron is one of the subatomic particles. The presence of such a negatively charged particle as part of a neutral atom suggests the presence of positively charged particles in atoms. That is, positive charge is necessary to neutralize the negative electrons in a neutral atom. It was discovered that such positive subatomic particles were present in the atom, and these subatomic particles were called **protons.** A proton has a +1 unit charge. Surprisingly, a third subatomic particle was found to be a constituent of atoms. This particle carries no electrical charge (is neutral) and is known as the **neutron.** An explanation of the distribution of such an aggregation of subatomic particles in atoms was established in the early 1900s by Lord Rutherford and Niels Bohr, among others.

According to this model, the atom consists of a small, massive, positively charged **nucleus** (containing protons and neutrons) surrounded by a swarm of negatively charged electrons. The term nucleus means center. This view of the atom became known as the **nuclear model of the atom.** (See Figure 2-4.) The nuclear model of the atom provides a picture of the atom in which a small positively charged nucleus is surrounded by electrons in motion. In a neutral atom the charge on the nucleus (i.e., number of protons in the nucleus) must be the same as the number of electrons in motion about the nucleus. Furthermore,

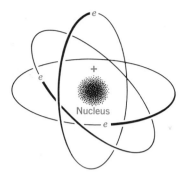

Figure 2-4
The nuclear atom.
Electrons are in
motion about a
massive positively
charged nucleus.
From *Introduction to
Chemistry* by T. R.
Dickson, John Wiley &
Sons, New York, 1971.

most of the mass of the atom is found in the nucleus. A question which arises is: "What is the nature of the nucleus?" The structure of the nucleus is a topic currently being investigated by nuclear physicists.

The atoms of the various elements differ as a result of having different numbers of electrons, protons, and neutrons. The nuclei of atoms of a given element have the same number of protons and, therefore, the atoms of the element must also have the same number of electrons. The number of protons in

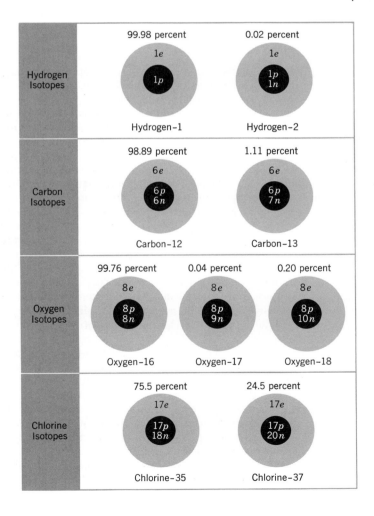

Figure 2-5
The compositions of some common isotopes found in nature. The percentages indicate the contribution of the isotopes of each element to the make-up of the naturally occurring element;
e = electron,
p = proton,
n = neutron.

the nucleus of an atom is called the **atomic number.** Each element has a unique atomic number. It is possible to determine the atomic number of elements, and the elements can be classified according to these atomic numbers.

An atom of a given element always contains the same number of protons and electrons (this is the atomic number). For instance, hydrogen atoms have 1 proton and 1 electron, and oxygen atoms have 8 protons and 8 electrons. Upon intensive study of the atoms of elements, it was found that for most elements there are two or more kinds of atoms. The difference between these kinds of atoms of the same element is that they contain different numbers of neutrons. Such atoms are called **isotopes**—atoms of the same element with the same number of protons but different numbers of neutrons in the nuclei. (See Figure 2-5.) Of the elements that exist in nature, 22 have no naturally occurring isotopes and, thus, exist in the form of one type of atom. The other elements have two or more isotopes. For instance, hydrogen has two natural isotopes, oxygen has three, carbon has two and tin has ten. Even though an element has isotopes, all of the atoms of that element behave as atoms of that element. This is so because the chemical behavior of an atom is related to the number of electrons in the atom. The elements with higher atomic numbers have more subatomic particles than elements with lower atomic numbers. That is, an atomic number of 1 (hydrogen) indicates atoms containing one proton and one electron (and possibly some neutrons in isotopes), while an atomic number of 105 (hahnium) indicates atoms containing 105 protons and 105 electrons and many neutrons.

2-6   THE MODERN VIEW OF THE ATOM

The study of the abilities of atoms to combine to form compounds revealed that compound formation involves changes in the distributions of electrons of the atoms. A theory describing the distribution of the electrons in atoms was developed in the early 1900s. This theory of atomic structure is called the **quantum mechanical model** and was developed in the late 1920s by Erwin Schrodinger, P. A. M. Dirac, and Werner Heisenburg, along with the contributions of many other scientists.

This model describes how the electrons in an atom are distributed around the nucleus. The quantum mechanical model does provide a satisfactory view of the atom even though it is somewhat complex. The electrons in an atom are said to possess energy as a result of motion about the nucleus (kinetic energy) and position with respect to the nucleus (potential energy). According to the quantum model there are several possible electron locations surrounding the nucleus. These locations are called **energy levels.** The electrons occupy these levels according to their energies. However, as shown in Figure 2-6, there are limitations to the number of electrons which may occupy the various levels. Each level consists of one or more energy states called **energy sublevels.** Furthermore, each sublevel consists of one or more specific electron energy states called **electron orbitals.** (See Figure 2-7.) The electrons can occupy the orbitals which comprise an energy level.

Each orbital can only accommodate a maximum of two electrons (an orbital may contain no electrons, one electron, or two electrons). The electrons in an atom are distributed among the possible energy states. The lower energy states or orbitals are filled first and then the next higher states. In this way the electrons are distributed in a specific pattern. Simple atoms, such as those of oxygen, involve only a few occupied orbitals while atoms of elements of higher atomic weight involve numerous occupied orbitals. The occupied orbitals can be viewed as volumes of space which represent regions in which the electrons are likely to be found in residence about the nucleus of an atom. Such orbitals have specific three-dimensional shapes and are sometimes called **electron clouds.** (See Figure 2-8.)

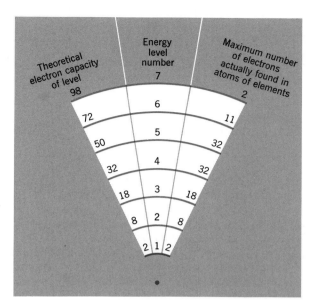

Figure 2-6
Each energy level has a theoretical electron capacity. However, these levels fill with electrons according to a certain pattern, which gives rise to the distribution of electrons in the levels as shown.

Theoretical electron capacity of level	Energy level number	Maximum number of electrons actually found in atoms of elements
98	7	2
72	6	11
50	5	32
32	4	32
18	3	18
8	2	8
2	1	2

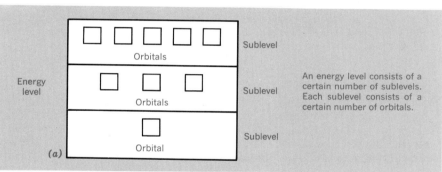

Figure 2-7
A representation of the quantum model of the atom. From *Introduction to Chemistry* by T. R. Dickson, John Wiley & Sons, New York, 1971.

(a) An energy level consists of a certain number of sublevels. Each sublevel consists of a certain number of orbitals.

Energy level — Orbitals — Sublevel
Orbitals — Sublevel
Orbital — Sublevel

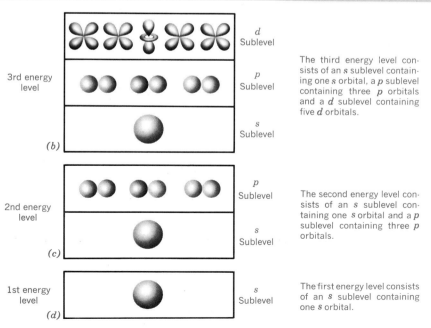

(b) 3rd energy level

d Sublevel
p Sublevel
s Sublevel

The third energy level consists of an *s* sublevel containing one *s* orbital, a *p* sublevel containing three *p* orbitals and a *d* sublevel containing five *d* orbitals.

(c) 2nd energy level

p Sublevel
s Sublevel

The second energy level consists of an *s* sublevel containing one *s* orbital and a *p* sublevel containing three *p* orbitals.

(d) 1st energy level

s Sublevel

The first energy level consists of an *s* sublevel containing one *s* orbital.

Figure 2-8
Shapes of electron orbitals. The electron orbitals or electron clouds are volumes of space about the nucleus in which the electrons are found. These orbitals have shapes much like the representative orbitals shown here.

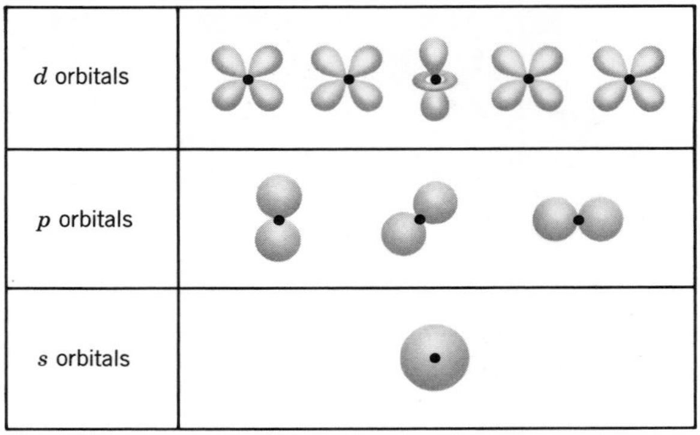

The quantum mechanical model provides a pattern for electron distribution in atoms and a pictorial representation of the electron orbitals. This pattern is illustrated in Figure 2-9. As can be seen in the figure, elements of higher atomic number and, thus, greater numbers of electrons, have more complex distribution of electrons. These electron distributions are called the **electronic structures** or **configuration** of the elements.

Electronic distributions allow the visualization of simple atoms as shown in Figure 2-10. The atoms of each element contain a certain number of electrons which are furthest from the nucleus. These electrons are called outer energy level electrons.

2-7 PERIODIC TABLE

Certain groups of elements have been found to have very similar properties. That is, they form similar kinds of compounds with other elements. For example, the elements lithium, sodium, and potassium each form a compound with chlorine which involves one combined atom of the element for every one combined atom of chlorine. The elements oxygen, sulfur, selenium, and tellurium each form a compound with hydrogen involving two combined atoms of hydrogen to every combined atom of the element. For many years prior to an understanding of atomic structure, such similarities in elements were observed. Chemists tried to arrange the elements in a tabular fashion according to this repetition in properties. Around 1860, L. Meyer and D. I. Mendeleev developed such a table based upon a repeating pattern in the observed properties of the elements. This table, called the **Periodic Table of the Elements** (shown in Figure 2-11), is one of the foundations of chemistry. The periodic table is based upon what has become known as the periodic law, which can be stated as follows:

> The properties of the elements are a periodic function of the atomic numbers.

The law indicates that as we compare the elements we find a pattern in which the elements of higher atomic number have similar properties to elements of lower atomic number. For instance, the eleventh element, sodium, is found to be similar to the third element, lithium; the twelfth element, magnesium, is similar to the fourth element, beryllium; the thirteenth element, aluminum, is similar to the fifth element, boron; the fourteenth element, silicon, is similar to the sixth element, carbon; and so on. When the elements are arranged in columns according to this periodic repetition in properties, the periodic table results.

It was not until electronic structure was related to the properties of the elements that an explanation of the table became apparent. It was found that the elements with the same number of electrons in the outer energy level of the

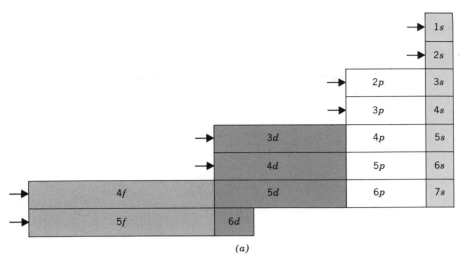

(a)

The boxes correspond to the various sublevels of the electron energy levels. For instance, $1s$ is the s sublevel of the first energy level and $2p$ is the p sublevel of the second energy level. The general order of distribution of electrons in the various sublevels is that shown by the arrows.

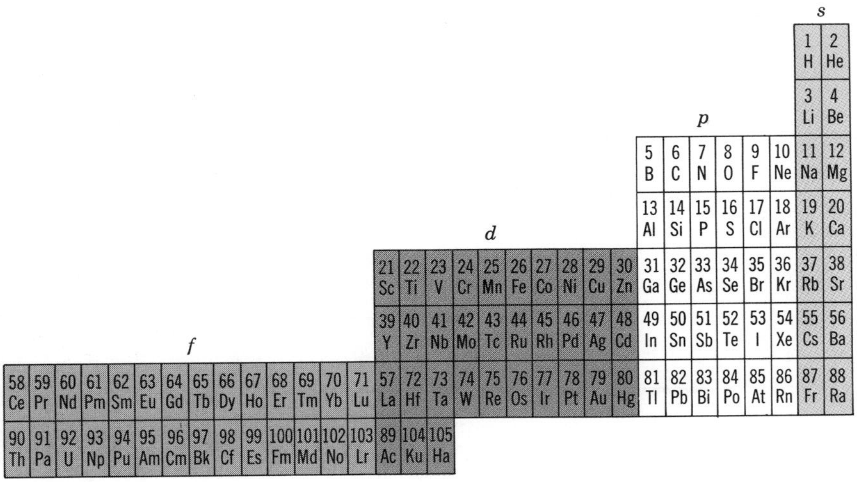

																		s	
																		1 H	2 He
													p					3 Li	4 Be
											5 B	6 C	7 N	8 O	9 F	10 Ne		11 Na	12 Mg
											13 Al	14 Si	15 P	16 S	17 Cl	18 Ar		19 K	20 Ca
21 Sc	22 Ti	23 V	24 Cr	25 Mn	26 Fe	27 Co	28 Ni	29 Cu	30 Zn		31 Ga	32 Ge	33 As	34 Se	35 Br	36 Kr		37 Rb	38 Sr
39 Y	40 Zr	41 Nb	42 Mo	43 Tc	44 Ru	45 Rh	46 Pd	47 Ag	48 Cd		49 In	50 Sn	51 Sb	52 Te	53 I	54 Xe		55 Cs	56 Ba

d (label above the 21–48 block), f (label above the 58–105 block)

58 Ce	59 Pr	60 Nd	61 Pm	62 Sm	63 Eu	64 Gd	65 Tb	66 Dy	67 Ho	68 Er	69 Tm	70 Yb	71 Lu	57 La	72 Hf	73 Ta	74 W	75 Re	76 Os	77 Ir	78 Pt	79 Au	80 Hg	81 Tl	82 Pb	83 Bi	84 Po	85 At	86 Rn	87 Fr	88 Ra
90 Th	91 Pa	92 U	93 Np	94 Pu	95 Am	96 Cm	97 Bk	98 Cf	99 Es	100 Fm	101 Md	102 No	103 Lr	89 Ac	104 Ku	105 Ha															

(b)

The numbers given above are the atomic numbers and correspond to the number of electrons in the atoms of the element given by the symbol. According to the pattern, each subsequent element has the same electron distribution as the previous element except that one more electron has been added. (There are known exceptions to this simple pattern.) As an example, consider that oxygen atoms have 8 electrons distributed as

sublevel	$1s$	$2s$	$2p$
electrons	2	2	4

fluorine atoms have 9 electrons distributed as

sublevel	$1s$	$2s$	$2p$
electrons	2	2	5

and neon atoms have 10 electrons distributed as

sublevel	$1s$	$2s$	$2p$
electrons	2	2	6

Figure 2-10
Pictorial visualizations of some atoms. The electrons are distributed about the nucleus of an atom in the various sublevels as illustrated. However, this distribution is such that the atom can be visualized as having essentially a spherical or ball-like shape. Consequently, a simple picture of atoms of the elements is that they are spheres of various sizes.

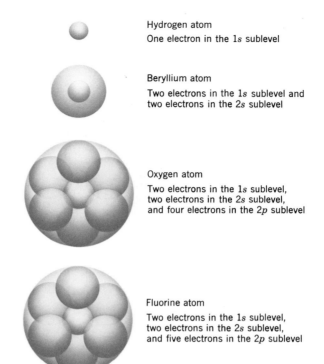

Hydrogen atom
One electron in the $1s$ sublevel

Beryllium atom
Two electrons in the $1s$ sublevel and two electrons in the $2s$ sublevel

Oxygen atom
Two electrons in the $1s$ sublevel, two electrons in the $2s$ sublevel, and four electrons in the $2p$ sublevel

Fluorine atom
Two electrons in the $1s$ sublevel, two electrons in the $2s$ sublevel, and five electrons in the $2p$ sublevel

Figure 2-11
A periodic table of the elements. (Atomic weights are given to only three digits in this table. However, atomic weights are generally known to more digits.)

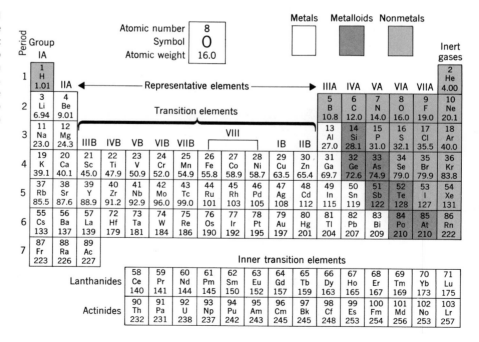

Group number

IA Alkali metals	IIA Alkaline earth metals	IIIA Boron– aluminum group	IVA Carbon group	VA Nitrogen group	VIA Oxygen group	VIIA Halogens
Lithium 3 Li	Beryllium 4 Be	Boron 5 B	Carbon 6 C	Nitrogen 7 N	Oxygen 8 O	Fluorine 9 F
Sodium 11 Na	Magnesium 12 Mg	Aluminum 13 Al	Silicon 14 Si	Phosphorus 15 P	Sulfur 16 S	Chlorine 17 Cl
Potassium 19 K	Calcium 20 Ca	Gallium 31 Ga	Germanium 32 Ge	Arsenic 33 As	Selenium 34 Se	Bromine 35 Br
Rubidium 37 Rb	Strontium 38 Sr	Indium 49 In	Tin 50 Sn	Antimony 51 Sb	Tellurium 52 Te	Iodine 53 I
Cesium 55 Cs	Barium 56 Ba	Thallium 81 Tl	Lead 82 Pb	Bismuth 83 Bi	Polonium 84 Po	Astatine 85 At
Francium 87 Fr	Radium 88 Ra					

Figure 2-12
Some common groups or families of elements.

atoms have similar properties. That is, those elements with one outer energy level electron are similar, those with two are similar, those with three are similar, and so on. If we arrange those elements with the same number of outer energy level electrons into columns in a table, we will end up with the periodic table as shown in Figure 2-11. The elements which occur in a vertical column of the table are those with similar properties. Such a column is called a **family** or **group** of elements. The important groups of elements are described in Figure 2-12. A horizontal sequence of elements in the table is called a **row** or **period.** There are two distinct types of elements in the table. (See Figure 2-11.) A large portion of the elements have metallic properties—good electrical conductors, flexible enough to be deformed, possess metallic luster—and are called **metals.** Some elements do not have these properties and are called **nonmetals.** A few elements display both metallic and nonmetallic properties. These are called the **metalloids.** (See Figure 2-11.)

The elements which comprise the A groups (Roman numeral designations) are called the **representative elements** and include many important elements. The B group elements between the two A groupings are called the **transition metals.** The elements displaced to the bottom of the table are called the **inner transition elements,** the **rare earth elements,** or the **actinides** and **lanthanides.** (See Figure 2-11.)

2-8   ELECTRON DOT SYMBOLS

The chemical behavior of the elements is directly related to the electronic configurations of the atoms and especially the configuration of the outer energy level (the outermost) electrons. It is these outer energy level electrons, called **valence electrons,** that are involved in the formation of compounds by the elements. Compound formation is discussed in Chapter 3.

A simple method has been devised to show the number of valence electrons in an atom of a representative element (A groups in the periodic table). This is done by writing what is called the **electron dot symbol** of the element. The electron dot symbol gives the symbol of the element surrounded by a number of dots corresponding to the number of valence electrons. The number of valence electrons for a representative element is given by the Roman numeral group number in the periodic table. When writing a dot symbol, imagine a square around the symbol of the element and put a dot on each side until all the valence electrons are used. Double up electrons only when necessary. For example, the electron dot symbol for sodium (Group IA) is

$$Na \cdot$$

The electron dot symbol for carbon (Group IVA) is

$$\cdot \overset{\displaystyle \cdot}{C} \cdot$$

The electron dot symbol for oxygen (Group VIA) is

$$: \overset{\displaystyle \cdot \cdot}{\underset{\displaystyle \cdot}{O}} \cdot \quad \text{or} \quad \cdot \overset{\displaystyle \cdot \cdot}{\underset{\displaystyle \cdot \cdot}{O}} \cdot \quad \text{or} \quad \cdot \overset{\displaystyle \cdot \cdot}{\underset{\displaystyle \cdot \cdot}{O}} :$$

The electron dot symbol for fluorine (Group VIIA) is

$$: \overset{\displaystyle \cdot \cdot}{\underset{\displaystyle \cdot \cdot}{F}} \cdot$$

(The side used for pairing dots is not important.)

The dot symbols for most of the representative elements are given in Table 2-3. Note carefully that the dot symbols for all the elements in a group are the same since the number of valence electrons is the same. Although these dot symbols are simple representations of the distributions of valence electrons, they are extremely useful in the discussion of the chemical bonding between atoms when elements form compounds.

Table 2-3
Electron Dot Symbols of Some Representative Elements

IA	IIA	IIIA	IVA	VA	VIA	VIIA	
Li·	·Be·	·B·	·C·	·N·	:O·	:F·	H·
Na·	·Mg·	·Al·	·Si·	·P·	:S·	:Cl·	
K·	·Ca·		·Ge·	·As·	:Se·	:Br·	
Rb·	·Sr·				:Te·	:I·	

2-9 ATOMIC WEIGHTS

One of the important differences between the atoms of the various elements is that these atoms have different masses. The mass of an atom depends upon how many protons, neutrons, and electrons it contains. Thus, atoms of elements of higher atomic numbers have greater masses than atoms of elements with lower atomic numbers. Oxygen atoms have more mass than hydrogen atoms but less mass than sulfur atoms. Atoms are such tiny particles that it is not possible to weigh individual atoms. Nevertheless, it is possible to determine the relative masses of the atoms of the elements. One method of determination of relative masses can be illustrated by an example. Oxygen and sulfur (both Group VIA elements) form a similar compound with carbon. Oxygen can combine with carbon to form carbon dioxide, which contains the maximum amount of oxygen that can combine with carbon. Sulfur forms carbon disulfide with carbon, which contains the maximum amount of sulfur that can combine with carbon. Decomposition of these compounds allows for analysis of the amount of each element present. This analysis gives the following results:

carbon–oxygen compound   carbon–sulfur compound

$$\left(\frac{2.67 \text{ grams O}}{1 \text{ gram C}} \right) \qquad \left(\frac{5.34 \text{ grams S}}{1 \text{ gram C}} \right)$$

Since the two compounds are similar in the sense that they contain the maximum amount of each element that can combine with carbon and must be made up of large collections of atoms, what is true for a large collection will be true for individual atoms. Considering this, the analysis shows that an atom of sulfur must have a mass which is twice as great as an atom of oxygen:

$$\frac{5.34 \text{ grams S}}{2.67 \text{ grams O}} = \frac{2.00 \text{ grams S}}{1 \text{ gram O}}$$ Sulfur atoms must have twice the mass of oxygen atoms to give the 2:1 ratio

Much effort has been expended by chemists to determine the relative masses of the atoms of the elements.

As an example of the significance of relative masses, consider the comparison of the weights of two individuals. The first weighs 110 pounds and the second weighs 132 pounds. Their relative weights are

$$\left(\frac{132 \text{ pounds}}{110 \text{ pounds}}\right) = \left(\frac{1.2 \text{ pounds}}{1 \text{ pound}}\right)$$

The relative weights are 1.2:1, which means that the second individual weighs 1.2 times as much as the first. Now suppose we express the weight of the first in kilograms. He weighs 50 kilograms. The second individual must weigh 60 kilograms (1.2×50 kilograms $= 60$ kilograms). Once we know the relative weights we can easily express the weights in any units desired.

The relative masses of atoms tell us how the masses of the atoms of one element compare with the masses of the atoms of other elements. However, to conveniently refer to the masses of atoms, we need to use a unit of mass. The unit used is called the **atomic mass unit (amu).** It is agreed upon by chemists to define the mass of an atom of the common isotope of carbon (carbon-12) as exactly 12 amu. The masses of the atoms of all other elements can be expressed relative to carbon-12. For instance, the relative mass of fluorine atoms to carbon-12 atoms is 1.6:1. Consequently, the mass of a fluorine atom is 19 amu (1.6×12 amu $= 19$ amu). In a like manner, the masses of all the atoms can be expressed in atomic mass units. The choice of carbon-12 is arbitrary but convenient, since none of the atoms of the elements will have a mass of less than 1 amu.

The relative masses of the atoms of the elements expressed in atomic mass units are called **atomic weights.** For an element with isotopes, the atomic weight expresses the average mass of an atom of the element considering the contribution of each isotope. A list of the atomic weights of the elements is given inside the front cover of this book. Note that oxygen has an atomic weight of about 16 amu and sulfur has an atomic weight of about 32 amu. This corresponds to the 2:1 relative mass mentioned above. Turn to the periodic table inside the back cover of this book and note that the atomic weights of the elements are given in this table under the symbol of the element. The periodic table is a handy source of atomic weights. Atomic weights are useful when we want to refer to the masses of atoms of the elements. This is done by chemists quite often, as discussed in Section 3-14 and Section 9-14.

Andrade, E. N. da C. *Rutherford and the Nature of the Atom.* Garden City, N.Y.:   **BIBLIOGRAPHY**
    Doubleday, 1964.
Asimov, I. *A Short History of Chemistry.* Garden City, N.Y.: Doubleday, 1965.
Dickson, T. R. *Introduction to Chemistry.* New York: Wiley, 1971.
Masterton, W. L., and E. J. Slowinski. *Chemical Principles.* 3rd ed. Philadelphia:
    W. B. Saunders, 1973.

QUESTIONS
AND
PROBLEMS

1. What are the three states of matter?

2. Give a definition of an element.

3. Give a definition of a compound.

4. Give a statement of the atomic theory.

5. Describe an atom according to the nuclear theory.

6. What are the three basic kinds of subatomic particles?

7. What is the atomic number of an element?

8. Give a definition of the term isotope.

9. Give a statement of the periodic law.

10. What is a group or family of elements in the periodic table? What is a row or period in the periodic table?

11. What are the names of the two distinct classes of elements in the periodic table?

12. What does the electron dot symbol of an element show?

13. Referring to a periodic table, write the electron dot symbols for sodium, magnesium, aluminum, oxygen, and chlorine.

14. What is the meaning of the term atomic weight?

15. Give the symbol for each of the following elements:

    carbon   nitrogen   sulfur   phosphorus   potassium

16. Look up John Dalton in an encyclopedia or history of science book. What is he noted for, and what kind of person was he according to the book?

3-1 CHEMICAL COMPOUNDS AND FORMULAS

Atoms are the fundamental chemical particles. They serve as the building blocks of matter. Some substances consist of aggregations of atoms. For instance, a piece of iron consists of a large aggregation of iron atoms. On the other hand, many substances contain atoms of various elements which have entered into atomic combinations to form compounds. When atoms combine with one another they do not retain their original electronic structures. Rather, they are transformed by the loss, gain, or mutual sharing of electrons into new kinds of chemical particles which compose compounds. These chemical particles, called ions and molecules, are discussed in this chapter.

As was discussed in Chapter 2, a given compound always contains the same elements, and they are always present in the same relative amounts. Water contains two combined atoms of hydrogen for every one combined atom of oxygen. Each pure compound contains a definite proportion of combined atoms of each element present in the compound. Since this is true, compounds can be represented by a symbol indicating the kinds of elements present and the relative number of combined atoms of each element. Such a symbol is called the **formula** of the compound. Elements in the combined state are represented by their usual symbol. Thus, formulas of compounds include the symbol of each element present, and each symbol is followed by a subscript indicating the relative number of combined atoms. For example, the formula for water is H_2O, which indicates that the compound contains twice as many combined atoms of hydrogen as there are combined atoms of oxygen. Note that when the subscript is one, it is not written as the number one, but is omitted. Formulas are often read by just pronouncing the letters and number. Thus, H_2O reads as "H-two-O." Consider another formula and the information it conveys:

$$Na_2CO_3$$

two sodiums   one carbon   three oxygens

The formula reads "N-A-two-C-O-three" and indicates that the compound (named sodium carbonate) contains two combined sodium atoms and one combined carbon atom to every three combined oxygen atoms.

3-2 CHEMICAL BONDING

The forces that hold atoms in combination with one another in compounds are called **chemical bonds.** The ability of an atom to form chemical bonds is related to the distribution of electrons in the atom. Specifically, the combining ability of an atom depends upon the number of outer energy level electrons or valence electrons. The electron dot symbols of the elements can be useful in explaining the formation of chemical bonds. Recall from Section 2-8 that the electron dot symbols indicate the number of outer energy level electrons. Around 1920, W. Kossel and G. Lewis noted that the representative elements tend to enter into chemical combinations involving the loss, gain, or sharing of electrons. In fact, based on their study of numerous compounds, the octet rule was proposed. This rule states that atoms tend to lose, gain, or share electrons so that they attain a total of eight (octet) outer energy level electrons. The inert gases, which form very few compounds, have an outer energy level of eight electrons in the uncombined state. An octet of outer energy level electrons can be shown as eight dots around the elemental symbol. For example, the octet of neon is shown as

$$:\overset{..}{\underset{..}{Ne}}:$$

It has been found that not all elements follow the octet rule. Hydrogen reacts in a manner in which the atoms attain a completed outer energy level of two electrons. Even though there are exceptions to the octet rule, it is observed in a great many cases and is a most useful concept.

There are two fundamental kinds of chemical bonds. The **ionic bond** results when electrons are transferred from one atom to another. The **covalent bond** results when two atoms share electrons. These two kinds of chemical bonds are discussed in the following sections.

3-3 IONIC BONDING AND ION FORMATION

When two elements are mixed, the constituent atoms may interact to form a chemical compound. Such an interaction may result in the outer energy level of electrons in the atoms of one element moving to the outer energy level of the atoms of another element. Such a transfer of electrons allows both kinds of atoms to attain the octet of outer energy level electrons. Atoms which lose electrons have an octet in the next lowest energy level and those that gain electrons gain enough electrons to complete an octet in the outer energy level. When an atom loses electrons this means negative charges are removed (each electron carries a single negative charge), leaving behind a positively charged particle consisting of a nucleus and the remaining electrons. Such a charged particle is called an **ion** or a **positive ion.** (See Figure 3-1.) Atoms which gain electrons take on the negative charges of the electrons. Such a negatively charged particle is also called an ion, but, this time a **negative ion.** (See Figure 3-1.)

Which elements form positive ions and which form negative ions? The elements with few (one, two, or three) outer energy level electrons tend to lose electrons. This includes nearly all the metals. The elements with many outer energy level electrons (five, six, or seven) tend to gain electrons. This includes many of the nonmetals. The Group IA elements (see the periodic table) tend to lose one electron and the Group VIIA elements tend to gain one electron when they react with the atoms of other elements. For example, when sodium (Group IA) and chlorine (Group VIIA) are mixed, the sodium atoms lose electrons to the chlorine atoms to form sodium ions and chloride ions. This process can be represented as shown at the top of the next page.

$$Na\cdot + \cdot\overset{\cdot\cdot}{\underset{\cdot\cdot}{Cl}}: \longrightarrow \quad Na^+ \quad + \quad :\overset{\cdot\cdot}{\underset{\cdot\cdot}{Cl}}:^-$$

sodium ion   chloride ion

Such a transfer of electrons produces positive sodium ions and negative chloride ions. When such negative and positive ions are produced, they aggregate as a result of the attraction between oppositely charged particles. When attraction between oppositely charged ions is strong enough to hold the ions in an aggregation, we say that an **ionic bond** exists. That is, when sodium metal reacts with chlorine gas, the numerous (billions of trillions) ions which are produced aggregate together by ionic bonds to form a three-dimensional stack of ions which usually takes the form of a crystalline solid. Sprinkle a few grains of salt on a piece of paper. Look closely at this crystalline solid. Imagine that such crystals are composed of numerous sodium ions and chloride ions stacked together in space. The formation of sodium chloride or salt from sodium and chlorine is illustrated in Figure 3-2.

It is important to realize that ions are a different kind of chemical particle compared to atoms. They have different properties from the original atoms. We sprinkle salt on our food and eat it. In fact we need it to keep our bodies functioning correctly. No one would dare eat sodium or chlorine. They are highly toxic and dangerous. Chemists and professional people sometimes use terminology which may be misleading to a beginner. For instance, a doctor may refer

Figure 3-1
Positive and negative ions.

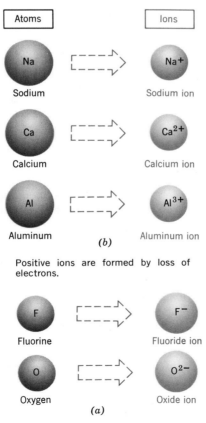

Atoms

Ions

Na
Sodium

Na⁺
Sodium ion

Ca
Calcium

Ca²⁺
Calcium ion

Al
Aluminum

Al³⁺
Aluminum ion

(b)

Positive ions are formed by loss of electrons.

F
Fluorine

F⁻
Fluoride ion

O
Oxygen

O²⁻
Oxide ion

(a)

Negative ions are formed by gain of electrons.

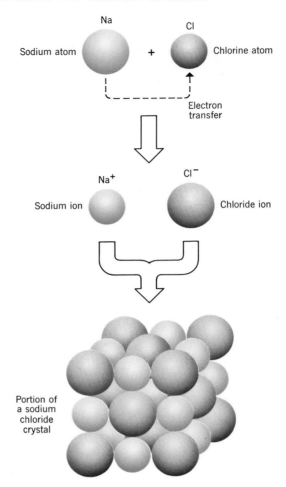

to a low sodium diet or a nutritionist may claim that we need a certain amount of calcium in our diets. These professionals and chemists know that they are referring to sodium ion and calcium ion and not sodium or calcium metal. As you learn more chemistry you will become more aware of this terminology.

An ion is represented by a formula using the symbol of the element from which the ion was formed with the charge of the ion as a superscript following the element symbol. A sodium ion is represented as Na^+ (read "N-A-plus"), and the chloride ion is represented as Cl^- (read "C-L-minus"). The sodium ion has a single positive charge, and the chloride ion has a single negative charge. The formula of the calcium ion, Ca^{2+} (read "C-A-two-plus"), indicates that the calcium ion carries a double positive charge.

For any of the representative metals (Group IA, IIA, and IIIA metals), the number of electrons lost in forming positive ions is given by the group number. (See Table 3-1.) Many of the transition metals (B groups) and Group IVA and VA metals form ions by the loss of one, two, three, and sometimes more electrons. (See Table 3-3.)

The Group VIIA, VIA, and VA nonmetals tend to gain electrons to form negative ions. The number of electrons gained is given by the group number subtracted from eight, as shown in Table 3-2.

Generally, when a metal is mixed with a nonmetal, the transfer of electrons occurs, resulting in the formation of ions which ionically bond to form an

$Na\cdot \longrightarrow$ sodium	Na^+ sodium ion	IA	one electron lost	
$\cdot Ca\cdot \longrightarrow$ calcium	Ca^{2+} calcium ion	IIA	two electrons lost	Table 3-1 Ions Formed by Group IA, IIA, and IIIA Metals
$\cdot \overset{\cdot}{Al}\cdot \longrightarrow$	Al^{3+} aluminum ion	IIIA	three electrons lost	

$:\overset{..}{\underset{\cdot}{O}}\cdot \longrightarrow$	O^{2-} oxide ion	VIA	two electrons gained	Table 3-2 Ions Formed by Group VIA and VIIA Nonmetals
$:\overset{..}{\underset{..}{F}}\cdot \longrightarrow$	F^- fluoride ion	VIIA	one electron gained	

aggregate of ions. Such a compound between a metal and a nonmetal is called an **ionic compound.** Thus, we can say that binary (two-element) compounds involving a metal and a nonmetal are ionic compounds and contain metal ions and nonmetal ions. For every atom which loses electrons, other atoms gain them. Even though ionic compounds are composed of ions, they are neutral. That is, they have the same amount of positive and negative charge. The formula of the compound reflects this. Consider the compound between calcium (Group IIA) and chlorine (Group VIIA). Calcium forms Ca^{2+} ions, and chlorine forms chloride ions, Cl^-. Thus, two chloride ions are formed for every calcium ion which forms. The formula of the compound is

$$Ca^{2+}(Cl^-)_2$$

However, we usually leave out the charges and the parentheses (when not needed) in writing the formula.

$$CaCl_2$$

A few more examples are given in Figure 3-3. A table of ions formed by the common metals and nonmetals is given in Table 3-3.

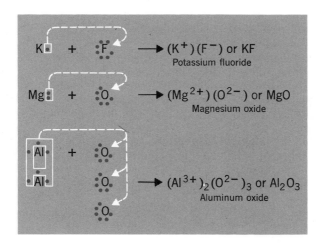

Figure 3-3
The formation of some metal–nonmetal compounds.

Table 3-3

Common Group IA, IIA, and IIIA ions

Li^+	lithium ion	Mg^{2+}	magnesium ion	Al^{3+}	aluminum ion
Na^+	sodium ion	Ca^{2+}	calcium ion		
K^+	potassium ion	Sr^{2+}	strontium ion		
		Ba^{2+}	barium ion		

Common transition and Group IVA metal ions

Cr^{3+}	chromium(III) ion (chromic ion)	Pb^{2+}	lead(II) ion (plumbous ion)
Fe^{2+}	iron(II) ion (ferrous ion)	Sn^{2+}	tin(II) ion (stannous ion)
Fe^{3+}	iron(III) ion (ferric ion)	Sn^{4+}	tin(IV) ion (stannic ion)
Cu^+	copper(I) ion (cuprous ion)		
Cu^{2+}	copper(II) ion (cupric ion)		
Zn^{2+}	zinc ion		
Cd^{2+}	cadmium ion		
Hg^{2+}	mercury(II) ion (mercuric ion)		

Common Group VIA and VIIA negative ions

O^{2-}	oxide ion	F^-	fluoride ion
S^{2-}	sulfide ion	Cl^-	chloride ion
		Br^-	bromide ion
		I^-	iodide ion

3-4 THE COVALENT BOND AND MOLECULES

The elements with many outer level electrons can attain the octet by sharing pairs of electrons. The nonmetals of Groups IV, V, VI, and VII (see the periodic table) are those elements which tend to share electrons. Consider two atoms of chlorine, each with seven outer level electrons,

$$: \overset{..}{\underset{..}{Cl}} \cdot \qquad \cdot \overset{..}{\underset{..}{Cl}} :$$

Each atom needs one more electron to form an outer level octet. Rather than exist as independent atoms, these chlorine atoms interact so that they share a pair of electrons and, thus, both attain the octet. This sharing can be represented as

$$: \overset{..}{\underset{..}{Cl}} \cdot \qquad \cdot \overset{..}{\underset{..}{Cl}} : \longrightarrow : \overset{..}{\underset{..}{Cl}} : \overset{..}{\underset{..}{Cl}} :$$

The sharing of a pair of electrons binds the atoms together to form a new kind of chemical particle called a molecule. The net force of attraction arising when two atoms share electrons is called a **covalent bond**. A **molecule** is a group of covalently bonded atoms. Molecules are stable particles and are the characteristic chemical particles of many compounds. Under certain circumstances molecules can be broken up to form atoms. However, this involves the breaking of the covalent bonds and would represent the decomposition of the compound involving the molecules.

Covalent bonding normally occurs between atoms of nonmetals. Hydrogen can form one covalent bond. The interaction between hydrogen atoms and chlorine atoms can be represented as

$$H \cdot \qquad \cdot \overset{..}{\underset{..}{Cl}} : \longrightarrow \quad H : \overset{..}{\underset{..}{Cl}} :$$

covalent bond

Note that chlorine attains an octet by this sharing, and hydrogen attains two outer level electrons. The molecule consisting of one hydrogen atom and a chlorine atom is called a hydrogen chloride molecule. When hydrogen gas and chlorine gas are chemically combined, numerous hydrogen chloride molecules result. (See Figure 3-4.)

Compounds formed between nonmetals and nonmetals involve the formation of molecules. Such nonmetal–nonmetal compounds are called **covalent** or **molecular compounds.** The number of covalent bonds normally formed by an atom of a nonmetal is given by the group number subtracted from eight. This can also be seen by noting the electron dot symbol of an element. (See Section 2-8.) The elements tend to form as many covalent bonds as there are single (unpaired) electrons in the dot symbol. Oxygen (Group VIA) tends to form two covalent bonds:

$$:\overset{\textstyle..}{\underset{\textstyle.}{O}}\cdot$$

Hydrogen tends to form one:

$$H\cdot$$

The compound formed between oxygen and hydrogen is water, which involves molecules containing two atoms of hydrogen covalently bonded to an oxygen atom.

$$:\overset{..}{\underset{.}{O}}\cdot \quad \cdot H \longrightarrow :\overset{..}{\underset{H}{O}}:H$$
$$H\cdot$$

Such a molecule is often represented in a **structural formula,**

$$\begin{array}{c}\text{covalent} \\ \text{bonds}\end{array} \left\{\begin{array}{c} :\overset{..}{O}\text{---}H \\ | \\ H\end{array}\right.$$

where the lines represent covalent bonds, or simply as the molecular formula H_2O. The formula H_2O indicates that water consists of molecules in which two atoms of hydrogen are bonded to an oxygen atom. Look at a glass of water. Imagine that such a sample of water consists of a vast (about 10^{25}) collection of

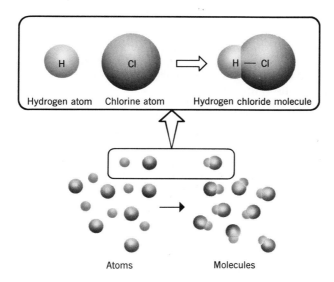

Hydrogen atom   Chlorine atom   Hydrogen chloride molecule

Atoms   Molecules

Figure 3-4
An idealized representation of the reaction between hydrogen and chlorine to form hydrogen chloride molecules.

Figure 3-5
Pictorial
representations of
some common
molecules.

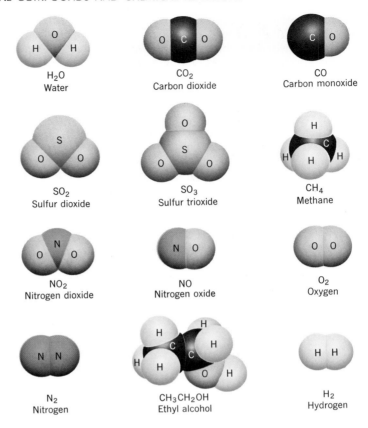

H_2O
Water

CO_2
Carbon dioxide

CO
Carbon monoxide

SO_2
Sulfur dioxide

SO_3
Sulfur trioxide

CH_4
Methane

NO_2
Nitrogen dioxide

NO
Nitrogen oxide

O_2
Oxygen

N_2
Nitrogen

CH_3CH_2OH
Ethyl alcohol

H_2
Hydrogen

H_2O molecules. Covalent compounds may involve one element, two elements, or three or more elements. Often, covalent compounds involve a few atoms bonded together to form simple molecules. However, molecules can consist of several atoms, tens of atoms, hundreds of atoms, or even thousands of covalently bonded atoms. For example, the paper in this page is composed mainly of the compound cellulose, which consists of molecules made up of thousands of covalently bonded carbon, hydrogen, and oxygen atoms. Some covalent compounds are solids, some are liquids, and some are gases. Figure 3-5 illustrates the structures of a few different molecules.

Seven of the nonmetals have such a great tendency to form covalent bonds in the uncombined state (not combined with other elements) that they occur in the form of molecules in which two atoms are covalently bonded. Uncombined oxygen is found in the form of such molecules and, thus, is represented by the formula O_2. Such a molecule is called a **diatomic molecule,** and elements that occur in this form are called diatomic molecular elements. This means that when we have a sample of such an element, the element can be represented by the formula of the diatomic molecule. However, this does not mean that these elements remain in the form of diatomic molecules when they are combined with other elements. The other diatomic elements are hydrogen (H_2), nitrogen (N_2), fluorine (F_2), chlorine (Cl_2), bromine (Br_2), and iodine (I_2).

3-5   POLYATOMIC IONS

In the natural environment the elements are most often found as compounds. A few elements such as the inert gases (helium, neon, argon, krypton, and xenon), sulfur, gold, and copper are often found in the uncombined state. Such elements are just aggregations of atoms. However, the atoms of most elements are found combined as ions in ionic compounds or covalently bonded to other atoms in the molecules of molecular compounds.

Positive and negative ions and molecules are chemical particles of much importance, since it is in these forms that combined atoms are found. Thus, it can be said that the three important chemical particles or **chemical species** are **atoms, ions,** and **molecules.** (See Figure 3-6.) The ions we have referred to are those which result when atoms lose or gain electrons. These are simple ions. When we investigate the combined elements in the natural environment, another kind of ion is found to occur. These ions are composed of two or more covalently bonded atoms which carry an electrical charge as a unit. Such an ion is a **polyatomic ion.** For instance, a piece of limestone or marble is composed of the calcium ion, Ca^{2+}, in combination with the polyatomic ion carbonate ion, CO_3^{2-} (read "C-O-three-two-minus"). The carbonate ion is an ion composed of three oxygen atoms covalently bonded to a carbon atom. The four-atom particle carries a double negative charge. Negative polyatomic ions occur in combination with metallic ions in many common ionic compounds. In such an ionic compound the amount of positive charge and negative charge must be the same

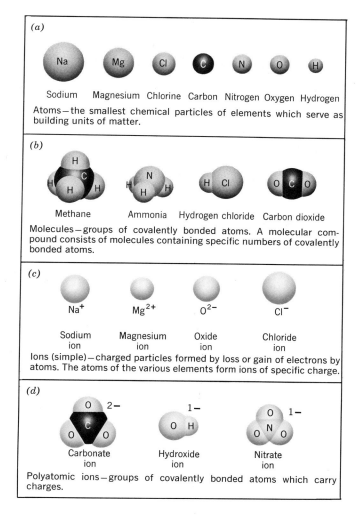

Figure 3-6
The four kinds of chemical species or particles which serve as building units of matter.

(a)

Sodium Magnesium Chlorine Carbon Nitrogen Oxygen Hydrogen

Atoms—the smallest chemical particles of elements which serve as building units of matter.

(b)

Methane Ammonia Hydrogen chloride Carbon dioxide

Molecules—groups of covalently bonded atoms. A molecular compound consists of molecules containing specific numbers of covalently bonded atoms.

(c)

Na^+ Mg^{2+} O^{2-} Cl^-

Sodium ion Magnesium ion Oxide ion Chloride ion

Ions (simple)—charged particles formed by loss or gain of electrons by atoms. The atoms of the various elements form ions of specific charge.

(d)

Carbonate ion Hydroxide ion Nitrate ion

Polyatomic ions—groups of covalently bonded atoms which carry charges.

NH_4^+	ammonium ion	NO_3^-	nitrate ion
$C_2H_3O_2^-$	acetate ion	NO_2^-	nitrite ion
CN^-	cyanide ion	CO_3^{2-}	carbonate ion
HCO_3^-	hydrogen carbonate ion or	SO_4^{2-}	sulfate ion
	bicarbonate ion	PO_4^{3-}	phosphate ion
OH^-	hydroxide ion		

Table 3-4
Some Common
Polyatomic Ions

and the net charge is thus zero. The formula of the ionic compound involving calcium ion and carbonate ion is

$$Ca^{2+}CO_3^{2-}$$

We usually write the formula without the charges as

$$CaCO_3$$

Table 3-4 lists several common polyatomic ions. Only one of the common poly-atomic ions is a positive ion (ammonium ion, NH_4^+), and it is found in combination with negative ions. For instance, ammonium ion can combine with chloride ion, Cl^-, to form ammonium chloride, NH_4Cl, or it can combine with sulfate ion, SO_4^{2-}, to form ammonium sulfate:

$$NH_4^+ \qquad SO_4^{2-}$$

two of these ions
needed for each of these ions,
resulting in the formula

$$(NH_4)_2SO_4$$

Note that, since the charge of ammonium ion is 1+ and the charge of a sulfate ion is 2−, two ammonium ions are needed for every sulfate ion. To denote two ammonium ions, the formula for ammonium ion is enclosed in parentheses with a subscript of two, $(NH_4)_2SO_4$. Figure 3-7 gives a few more examples of such ionic compounds.

3-6 NOMENCLATURE OF CHEMICAL COMPOUNDS

What do we name a chemical compound if we know the formula? Several systems of chemical nomenclature have been developed by chemists. A com-pound can be named according to one of these systems. Such a name is called the systematic name. However, the names of many chemical compounds devel-oped historically before any systems of nomenclature were used. These names are called **common** or **trivial** names and usually do not convey any information concerning the structure of the compound. Some compounds, such as water, H_2O, and ammonia, NH_3, are always referred to by the common names. Other compounds have common names which are used on occasion, but the system-atic names are used by chemists. For example, NaCl is commonly called salt, but is sodium chloride to the chemist, and C_2H_5OH is commonly called alcohol, but is technically ethyl alcohol. Table 3-5 lists the common and systematic names of some common compounds.

3-7 NAMING IONIC COMPOUNDS

The systematic approach to naming a compound depends upon the type of com-pound. Of the kinds of compounds discussed so far there are three distinct

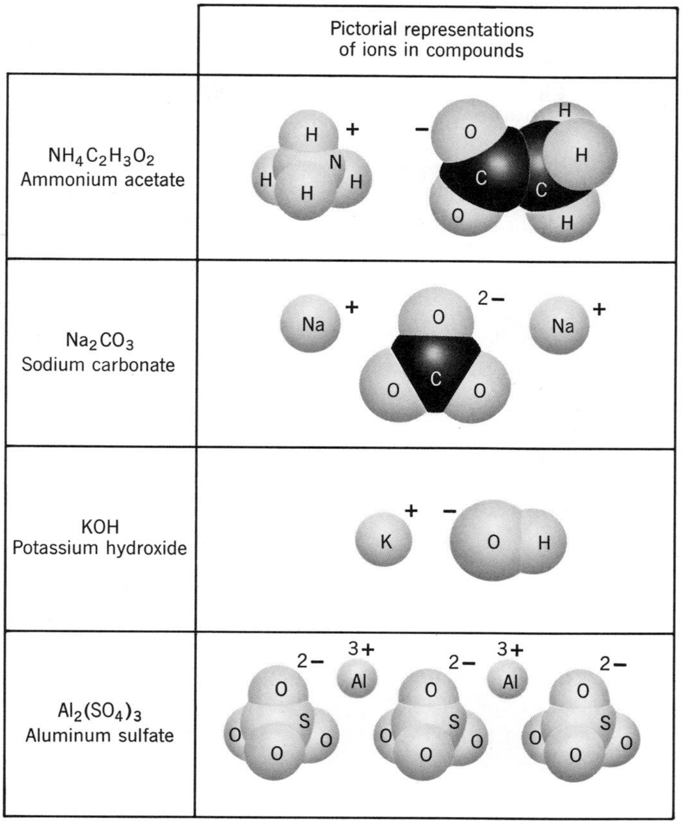

	Pictorial representations of ions in compounds
$NH_4C_2H_3O_2$ Ammonium acetate	
Na_2CO_3 Sodium carbonate	
KOH Potassium hydroxide	
$Al_2(SO_4)_3$ Aluminum sulfate	

Figure 3-7
Some ionic compounds including polyatomic ions.

types. First, there are ionic compounds composed of a positive metal ion and a negative nonmetal ion. These are the **metal–nonmetal** compounds. Second, there are compounds composed of a positive metal ion and a negative polyatomic ion. These are the **metal-polyatomic ion** compounds. Third, there are compounds composed of two nonmetals. The names of metal–nonmetal and metal–polyatomic ion compounds are obtained by giving the name of the metal followed by the name of the negative ion (without the word ion). A few examples are given below.

Al_2O_3	aluminum oxide	(aluminum ion and oxide ion)
KCl	potassium chloride	(potassium ion and chloride ion)
FeS	ferrous sulfide or iron(II) sulfide (read "iron-two-sulfide")	(ferrous ion and sulfide ion)
NaOH	sodium hydroxide	(sodium ion and hydroxide ion)
K_2CO_3	potassium carbonate	(potassium ion and carbonate ion)
$Ca_3(PO_4)_2$	calcium phosphate	(calcium ion and phosphate ion)
$MgSO_4$	magnesium sulfate	(magnesium ion and sulfate ion)

Many combinations of positive and negative ions are possible, resulting in a vast number of ionic compounds. Such ionic compounds are often found in the environment in the rocks and minerals of the earth or in the oceans. These compounds are sometimes grouped according to the negative ion which they share in common. For instance, all compounds which contain the oxide ion, O^{2-}, are called **oxides**. See Table 3-6 for a list of common oxides. Compounds containing

Table 3-5
Common and
Systematic Names for
Some Common
Compounds

Formula	Common name	Systematic name
Al_2O_3	alumina	aluminum oxide
NH_3	ammonia	
$NaHCO_3$	baking soda or bicarbonate of soda	sodium hydrogen carbonate
$Na_2B_4O_7 \cdot 10H_2O$	borax	sodium tetraborate decahydrate
$CaCO_3$	calcite or marble	calcium carbonate
$KHC_4H_4O_6$	cream of tartar	potassium hydrogen tartrate
$MgSO_4 \cdot 7H_2O$	epsom salt	magnesium sulfate heptahydrate
$CaSO_4 \cdot 2H_2O$	gypsum	calcium sulfate dihydrate
C_2H_5OH	grain alcohol or alcohol	ethyl alcohol or ethanol
$Na_2S_2O_3$	hypo	sodium thiosulfate
N_2O	laughing gas or nitrous oxide	dinitrogen oxide
PbO	litharge	lead(II) oxide or plumbous oxide
CaO	lime	calcium oxide
$NaOH$	lye	sodium hydroxide
CH_4	methane	
O_3	ozone	
$2CaSO_4 \cdot H_2O$	plaster of paris	2-calcium sulfate 1-water
K_2CO_3	potash	potassium carbonate
NH_4Cl	sal ammoniac	ammonium chloride
$NaNO_3$	saltpeter	sodium nitrate
$Ca(OH)_2$	slaked lime	calcium hydroxide
$C_{12}H_{22}O_{11}$	sugar	sucrose
$NaCl$	salt	sodium chloride
$Na_2CO_3 \cdot 10H_2O$	washing soda	sodium carbonate decahydrate
H_2O	water	
CH_3OH	wood alcohol	methyl alcohol or methanol

chloride ion, Cl^-, are called **chlorides.** (See again Table 3-6.) Common **hydroxides** (OH^--containing), **carbonates** (CO_3^{2-}-containing), **sulfates** (SO_4^{2-}-containing), **nitrates** (NO_3^--containing), and **phosphates** (PO_4^{3-}-containing) are also listed in Table 3-6.

3-8   NAMING MOLECULAR COMPOUNDS

Compounds comprised of two nonmetals (molecular) are named according to a different method, called the prefix method. According to this method, a compound is named by giving the name of the first element followed by the name of the second element with the ending changed to -ide. Each part of the name is preceded by a Greek or Latin prefix indicating the number of combined atoms of the element in the compound. The prefixes used are mono (one), di (two), tri (three), tetra (four), penta (five), hexa (six), hepta (seven), octa (eight), nona (nine), and deca (ten). As an example, consider the compound CO_2. The name by the prefix method is

carbon dioxide
(the prefix mono is often omitted)

which indicates that the compound consists of molecules containing one carbon atom and two oxygen atoms.

Table 3-6
Common Oxides,
Chlorides,
Hydroxides,
Carbonates, Sulfates,
Nitrates, and
Phosphates (Mineral
Names Given in
Parentheses)

Oxides

CaO	calcium oxide
MgO	magnesium oxide
Al_2O_3	aluminum oxide
MnO_2	manganese dioxide (pyrolusite)
FeO	iron(II) oxide or ferrous oxide
Fe_2O_3	iron(III) oxide or ferric oxide (hematite)
Fe_3O_4	iron(II,III) oxide (magnetite)

Chlorides

$NaCl$	sodium chloride (halite)
KCl	potassium chloride (sylvite)
$CaCl_2$	calcium chloride
$MgCl_2$	magnesium chloride
HCl	hydrogen chloride
$AgCl$	silver chloride (horn silver)

Hydroxides

$NaOH$	sodium hydroxide
KOH	potassium hydroxide
$Mg(OH)_2$	magnesium hydroxide
$Ca(OH)_2$	calcium hydroxide
$Al(OH)_3$	aluminum hydroxide
$Fe(OH)_3$	iron(III) hydroxide or ferric hydroxide

Carbonates

Na_2CO_3	sodium carbonate
$CaCO_3$	calcium carbonate (calcite, marble)
$MgCO_3$	magnesium carbonate (magnesite)
$FeCO_3$	iron(II) carbonate or ferrous carbonate (siderite)
$ZnCO_3$	zinc carbonate (smithsonite)

Sulfates

Na_2SO_4	sodium sulfate (thenardite)
$CaSO_4$	calcium sulfate (anhydrite)
$BaSO_4$	barium sulfate (barite)
$Al_2(SO_4)_3$	aluminum sulfate

Nitrates

$NaNO_3$	sodium nitrate (soda niter, chile saltpeter)
KNO_3	potassium nitrate (niter, saltpeter)
$Ca(NO_3)_2$	calcium nitrate

Phosphates

Na_3PO_4	sodium phosphate or trisodium phosphate
$Ca_3(PO_4)_2$	calcium phosphate (whitlockite, phosphate rock)

Some additional examples of this system of nomenclature are:

SO_2	sulfur dioxide	PCl_3	phosphorus trichloride
N_2O_5	dinitrogen pentoxide	CO	carbon monoxide
NO_2	nitrogen dioxide		

Table 3-7 lists some more typical binary molecular compounds named by this method.

3-9 NAMING HYDRATES

Water is a compound that is widely distributed in nature. It is the major component of the natural waters of the earth, and it is intimately involved in the cells

Table 3-7
Some Common
Molecular
Compounds Named
by the Prefix Method

HF	hydrogen fluoride
HCl	hydrogen chloride
NO	nitrogen oxide (commonly called nitric oxide)
N_2O	dinitrogen oxide (commonly called nitrous oxide)
SO_3	sulfur trioxide
H_2S	dihydrogen sulfide (commonly called hydrogen sulfide)
P_4S_3	tetraphosphorus trisulfide (used in "strike anywhere" matches)
Sb_2S_3	diantimony trisulfide (used in safety matches)

and fluids of animals and plants. It is even found incorporated in the minerals and rocks of the crust of the earth. In fact, many ionic compounds found in the earth are also found to contain water. Such ionic compounds in association with water are called **hydrates.** Actually, this association of ionic compounds and water is such that the hydrates obey the Law of Constant Composition and are actually chemical compounds that can be represented by a special formula. For example, the mineral gypsum is a hydrate containing calcium sulfate. This hydrate contains two parts of water for every one part of calcium sulfate. Thus, the hydrate can be represented by the formula

$$CaSO_4 \cdot 2H_2O$$

This is the standard way to write a formula for a hydrate. The formula of the ionic compound is given followed by the formula for water preceded by a number indicating the relative amount of water in the hydrate. The two formulas are separated by a dot. A hydrate is named by stating the name of the ionic compound followed by the word hydrate preceded by a prefix (the same prefixes as used in the prefix nomenclature) indicating the amount of water. Thus, the name of the hydrate gypsum is

calcium sulfate dihydrate

Table 3-8 lists the names and formulas of a few typical hydrates.

Table 3-8
Some Common
Hydrates

$Na_2B_4O_7 \cdot 10H_2O$	sodium tetraborate decahydrate (borax)
$MgSO_4 \cdot 7H_2O$	magnesium sulfate heptahydrate (epsomite)
$CuSO_4 \cdot 5H_2O$	copper sulfate pentahydrate (blue vitriol, cahalcanthite)
$CaCl_2 \cdot 2H_2O$	calcium chloride dihydrate
$Al_2O_3 \cdot 3H_2O$	aluminum oxide trihydrate
$Na_2SO_4 \cdot 10H_2O$	sodium sulfate decahydrate (glauber salt)
$Na_2CO_3 \cdot 10H_2O$	sodium carbonate decahydrate (washing soda)

3-10 SOME COMMON ACIDS

One other type of compound deserves mentioning at this time. These compounds contain hydrogen combined with other nonmetals and display certain chemical similarities. The obvious similarities are that they taste sour (i.e., vinegar), can dissolve many metals, and can burn the skin. The hydrogen-containing compounds which display these characteristics are called **acids.** Not all hydrogen–nonmetal compounds are acids. The names of many of these acids have developed historically. For instance, the acid H_2SO_4 is called sulfuric acid, a name which indicates the presence of sulfur but not oxygen. Table 3-9 lists the names and formulas of some common acids.

				Table 3-9
HCl	hydrochloric acid	H_3PO_4	phosphoric acid	Some Common Acids
H_2SO_4	sulfuric acid	H_2CO_3	carbonic acid	
$HC_2H_3O_2$	acetic acid	HNO_3	nitric acid	

3-11 CHEMICAL REACTIONS AND EQUATIONS

The elements can combine to form compounds. Similarly, many compounds can be decomposed to form the constituent elements. These changes occur as chemical processes. Any chemical process in which certain substances are converted into other substances is called a **chemical reaction.** In a chemical reaction the initial substances, called **reactants,** are said to react chemically to form new substances called **products.** A chemical reaction involves the breaking of certain chemical bonds and the forming of new chemical bonds. Thus, the chemical particles (atoms, ions, or molecules) comprising the reactants are transformed into new chemical compounds which are the products. Chemical reactions are occurring in and around us continuously. The digestion and metabolism of food involves chemical reactions. The rusting of metal and the burning of fuel are chemical reactions. Some reactions occur quickly and some quite slowly. The explosion of dynamite is a fast chemical reaction, while the formation of a fossil involves extremely slow chemical processes. Chemical reactions account for many of the changes we see around us. A flame is the indication of a relatively simple chemical reaction, whereas the growth of a plant involves a complex series of chemical reactions.

Chemical reactions can be described by indicating the reactants and products. For instance, when we burn charcoal (e.g., charcoal briquettes in the barbeque), the main chemical reactants are carbon (C, the main component of charcoal) and oxygen (O_2, remember oxygen occurs as diatomic molecules) gas in the air. They react to form carbon dioxide as a product. Thus we say that carbon reacts with oxygen to form carbon dioxide. Such a chemical reaction can be represented symbolically by a **chemical equation.** A chemical equation must include the proper formulas for the reactants and the products. Furthermore, the reactant formulas are separated by plus signs which are not intended to mean a mathematical plus but rather should be interpreted as meaning "and." The product formulas are also separated by plus signs. The reactants and products are separated by an arrow ($\rightarrow$) which is a symbol that means "react to produce yield or give." For the reaction between carbon and oxygen mentioned above, the chemical equation is written as follows:

$$C + O_2 \longrightarrow CO_2$$

This equation says that carbon and oxygen react to produce carbon dioxide, or carbon plus oxygen react to yield carbon dioxide. Notice that the symbol used for solid carbon is the atomic symbol C and the symbol used for gaseous oxygen is the formula for the diatomic molecule O_2.

The combined atoms comprising the substances are not created or destroyed in a chemical reaction, but chemical bonds are broken and new combinations result. Consequently, the total number of combined atoms of each element involved remains constant. Since atoms are not destroyed or created in a chemical reaction, the total mass of the reactants must equal the total mass of the products. Let us consider what happens when we use a flashbulb in a camera. The bulb contains some magnesium metal and some oxygen gas. A small electrical current passing through the magnesium initiates a violent chemical reaction between the magnesium and the oxygen. Even though this reaction is accompanied by the release of energy in the form of heat and light, no detectable mass is lost or gained in the reaction. That is, the mass of the flashbulb is the same

Figure 3-8
From *Introduction to Chemistry* by T. R. Dickson, John Wiley & Sons, New York, 1971.

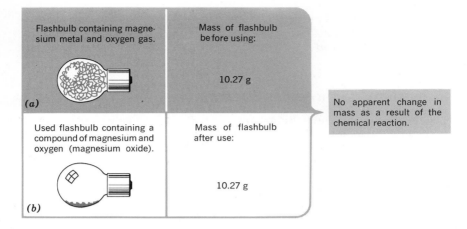

before and after the reaction. (See Figure 3-8.) This observation is generally true for all chemical reactions. The fact that mass is conserved in a chemical reaction is called the Law of Conservation of Matter. Recall that this idea was mentioned in Section 1-6.

3-12 BALANCING CHEMICAL EQUATIONS

The Law of Conservation of Matter affects the way in which an equation is written. The number of atoms of each element in a reaction must be the same on either side of the equation. Consider the reaction which occurs when methane, CH_4 (the main component of natural gas), reacts with oxygen gas (i.e., burns in the air) to produce carbon dioxide, CO_2, and water, H_2O. The equation can be written showing the formulas of the reactants and products separated by an arrow:

$$CH_4 + O_2 \longrightarrow CO_2 + H_2O$$

However, we must be sure that the number of atoms of each element is the same on both sides of the equation, since atoms are not created or destroyed in a reaction. We cannot change the formulas of the compounds in order to make the number of atoms of each element the same on each side of the arrow, since this would not be representative of the proper compounds. We can, however, change the number of each species that appears in the equation. We do this by placing a numerical coefficient in front of the formula of any species to refer to a specific number of such species. As illustrated in Figure 3-9, to have four hydrogens on each side of the above equation, two water molecules must be produced; and to have four oxygen atoms on each side, two oxygen molecules must react. This is denoted by placing the necessary number, called a **coefficient**, in front of the formula of the appropriate species. Thus, the correctly written equation is

$$CH_4 + 2\,O_2 \longrightarrow CO_2 + 2\,H_2O$$

This equation can be read as: methane reacts with oxygen to produce carbon dioxide and water; or more precisely, one molecule of methane and two molecules of oxygen react to produce one molecule of carbon dioxide and two molecules of water.

In any equation representing a chemical reaction, the coefficients must be adjusted to satisfy the requirement that the number of atoms of each element must be the same in the reactants and the products. An equation in which the

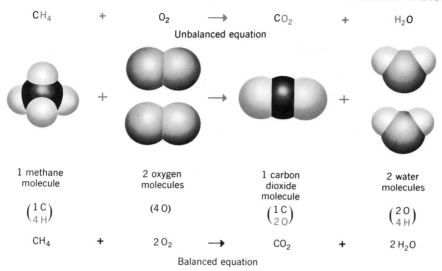

Figure 3-9
Balancing of the
equation for the
reaction between
methane and oxygen
to give carbon dioxide
and water.

number of atoms of each element is the same on both sides of the arrow is called a **balanced** equation. The process of adjusting the coefficients is called **balancing** the equation. Often such balancing is carried out by trial and error; that is, by trying various coefficients.

Chemical equations are used in chemistry to show a chemical change in a shorthand fashion. All chemical reactions could be stated in words, but the equation provides for the precise description of a chemical process in convenient symbolism. You will see many other chemical equations in this book.

Sometimes special conditions are needed to have a chemical reaction occur. Often these conditions are indicated above or below the arrow. For example, nitrogen, N_2, and hydrogen, H_2, react to form ammonia, NH_3, under high pressure. The equation

$$N_2 + 3\,H_2 \xrightarrow{\text{pressure}} 2\,NH_3$$

indicates that pressure is needed in order for the reaction to proceed. Some chemical reactions occur readily only in the presence of another chemical substance which is neither a reactant nor a product. Such a substance, the presence of which is necessary for the reaction to occur but is not permanently altered chemically in the reaction, is called a **catalyst.** The nitrogen and hydrogen reaction mentioned above only occurs readily in the presence of a catalyst such as iron, Fe. To denote this, we include iron above or below the arrow:

$$N_2 + 3\,H_2 \xrightarrow[\text{Fe}]{\text{pressure}} 2\,NH_3$$

This indicates that nitrogen and hydrogen under pressure and in the presence of the catalyst iron will react to form ammonia. Whenever needed, additional information concerning a chemical reaction is usually written above or below the arrow.

3-13   CHEMICAL ENERGY

As you have undoubtedly noticed, some chemical reactions are accompanied by energy changes. A burning candle gives off heat and light (two forms of energy). A camera flashbulb gives off a brilliant flash of light along with some heat. Some chemical reactions, such as those occurring in green plants during photosynthesis, actually absorb energy (energy from sunlight, in the case of photosynthesis).

Chemical reactions involve the breaking and forming of chemical bonds. Whenever these processes occur, energy exchanges are involved. When a chemical reaction occurs, a corresponding energy change occurs. Some chemical reactions release energy and are called **exergonic** reactions. Other reactions require energy and are called **endergonic** reactions. The energy changes that accompany reactions often take the form of heat energy. A reaction which releases heat to the surroundings is called **exothermic** and a reaction which absorbs heat from the surroundings is called **endothermic.** Many chemical reactions are carried out for the sole purpose of releasing energy. We burn gasoline and oil (react them with oxygen in the air) in our cars and trucks and convert the chemical energy of these fuels to useful mechanical work. Natural gas, coal, and oil are burned to produce energy for many industrial processes. Today our major source of energy comes from the combustion or burning of gasoline, oils, coal, and natural gas.

The sources of the energy changes in chemical reactions are the chemical species involved in the reactions. Each species has a **chemical energy** and, when a chemical reaction does occur, there is an overall change in the chemical energy. If the total residual chemical energy content of the products is greater than the total initial chemical energy of the reactants, an external source of energy is needed so that the reaction can occur. This is an endergonic reaction. If the total residual chemical energy content of the products is less than the total initial chemical energy content of the reactants, the excess energy is released. This is an exergonic reaction.

To denote whether a chemical reaction releases or absorbs energy, the word energy or heat can be written as a reactant or product. The reaction of methane and oxygen gas can be written as

$$CH_4 + 2\ O_2 \longrightarrow CO_2 + 2\ H_2O + heat$$

indicating that the reaction is exothermic. The reaction between nitrogen gas, N_2, and oxygen gas, O_2, to produce NO gas is endothermic and can be written as

$$N_2 + O_2 + energy \longrightarrow 2\ NO$$

Sometimes an endergonic reaction is denoted by writing the word energy above the arrow.

$$N_2 + O_2 \xrightarrow{\text{energy}} 2\ NO$$

The idea of chemical energy is very important in chemistry and especially in the chemistry of life processes. Through endergonic reactions which involve the formation of chemical bonds, energy can be stored in the form of chemical compounds. Since this energy has the potential of being released in an exergonic reaction, we call it chemical energy. When the proper reaction takes place, the energy is released. In this way many chemicals act as sources of energy or storehouses of energy and are involved in many of the energy changes which occur in the environment. Many other energy changes, such as tides, winds, flow of water, and earthquakes, do not involve chemical reactions.

3-14 FORMULA WEIGHTS

Since atoms have mass, and molecules and ionic compounds are composed of atoms, it is sometimes convenient to refer to the relative masses of compounds. The formula of a compound indicates which elements are present and the proportion of each element. The relative masses of compounds can be expressed as formula weights. The **formula weight** of a compound is the sum of the product of the atomic weights of each element present multiplied by the subscript in the formula. For example, the formula weight of water, H_2O, is (see the periodic table for atomic weights)

H	$2(1.01 \text{ amu}) = 2.02 \text{ amu}$
O	$16.00 \text{ amu} = \underline{16.00 \text{ amu}}$
H_2O	18.02 amu

We can interpret this formula weight as indicating that since an atom of hydrogen has a mass of 1.01 amu and an atom of oxygen has a mass of 16.00 amu, then a water molecule composed of a combination of two hydrogen atoms and an oxygen atom will have a mass of 18.02 amu. In other words, the formula weight of water is the mass of a molecule of water. Sometimes formula weights of molecular substances are called **molecular weights.**

Now let us consider the formula weight of an ionic compound such as sodium sulfate, Na_2SO_4. We calculate the formula weight, according to the definition given above, as follows:

Na	$2(23.0 \text{ amu}) = 46.0 \text{ amu}$
S	$32.1 \text{ amu} = 32.1 \text{ amu}$
O	$4(16.0 \text{ amu}) = \underline{64.0 \text{ amu}}$
Na_2SO_4	142.1 amu

This formula weight indicates that a combination of two sodium ions and a sulfate ion (a formula unit) will have a mass of 142.1 amu. Formula weights are quite useful in certain chemical calculations. See sections 9-14 and 9-15.

3-15 PARTS PER MILLION AND PARTS PER BILLION

The percentage (see Section 1-12) is used to express what fraction one quantity is of the whole. Percentage is an expression of parts per hundred. In some mixtures certain components are present in very small amounts and the use of percentage to express the amount is not convenient. To express the amount of such components of a mixture other terms are used. One such term is called **parts per million** (abbreviated **ppm**). It is an expression of how many parts of the component there are in a million parts of the whole. Parts per million is often placed on a mass basis. Since there are 1 million milligrams in 1 kilogram (1,000 mg = 1 g and 1,000 g = 1 kg), the parts per million of a component in a mixture can be expressed as the number of milligrams in a kilogram of the whole. For example, it is found that a sample of fish contains 5 milligrams of mercury in a kilogram of the fish. Thus, parts per million of mercury in the fish is

$$\left(\frac{5 \text{ mg}}{1 \text{ kg}}\right) \quad \text{or} \quad 5 \text{ ppm}$$

In water solutions the parts per million term is used to express the presence of small amounts of components. Since 1 liter of water has a mass of about 1 kilogram, the parts per million of a component in a water solution can be expressed as the number of milligrams of the component per liter of solution. Suppose 1 liter of tap water is evaporated and found to leave behind 400 milligrams of solids which were dissolved in the water. The parts per million dissolved solids in the tap water is

$$\left(\frac{400 \text{ mg}}{1 \text{ } \ell}\right) = 400 \text{ ppm}$$

One part per million is a small amount. Some comparisons which illustrate the concept of parts per million are given on the next page.

1 ppm is 220 people out of the entire population of the United States
1 ppm is 1 in. in 15.8 miles or 1 mm in 1 km
1 ppm is 1 penny in $10,000.00
1 ppm is a 1 g needle in a stack of hay weighing about 1 ton
1 ppm is one drop in about 61 ℓ or 16 gal

Even though 1 part per million is a small amount of the whole, in certain cases the presence of a few parts per million of a substance mixed with another substance can be significant. Fish with too many parts per million of mercury can be dangerous to eat. A few parts per million of an impurity in a transistor prevents the transistor from functioning. The presence of certain impurities in water at the parts per million level renders the water undrinkable.

Another term used to refer to very small portions of the whole is **parts per billion (ppb).** It is an expression of how many parts out of a billion (10^9) parts a component is of the whole. Since a microgram (μg) is 1 billionth part of a kilogram, parts per billion can be expressed as the number of micrograms per kilogram or micrograms per liter for solutions. One part per billion is indeed a small portion. Some comparisons which illustrate the concept of parts per billion are:

1 ppb is 4 people out of the entire population of the world
1 ppb is 1 in. in 15,783 miles or 1 mm in 1,000 km
1 ppb is 1 penny in $10,000,000.00
1 ppb is a 1 mg hair in a stack of hay weighing about 1 ton
1 ppb is one drop in about 60,500 ℓ or 16,000 gal

The presence of certain substances at parts per billion levels can be significant. For instance, the amount of lead in the air that we breath can be expressed in terms of parts per billion, but this amount of lead is significant to our health.

BIBLIOGRAPHY   BOOKS

Benfey, O. T. *Classics in the Theory of Chemical Combination.* New York: Dover, 1963.
Griswold, E. *Chemical Bonding and Structure.* Boston: D. C. Heath, 1968.

QUESTIONS AND PROBLEMS

1. The formula of sodium bicarbonate is $NaHCO_3$. What information does this formula convey about the compound?

2. Give a definition of a chemical bond.

3. What is the octet rule?

4. Explain how atoms form ions.

5. Give the formulas for the simple ions you would expect the following elements to form by loss or gain of electrons:

   sodium   magnesium   aluminum   oxygen   chlorine

6. Describe the ionic bond.

7. Describe the covalent bond.

8. What is a molecule?

9. Which of the nonmetals occur in the form of diatomic elements? Give the formulas for the molecules of each of these elements (e.g., hydrogen H_2).

10. What is a polyatomic ion?

11. Give the common names of the following compounds:

    H_2O NH_3 CH_4

12. Give the names of the following ionic compounds:
    Na_2CO_3 $MgSO_4$ Al_2O_3 $Ca_3(PO_4)_2$

13. Give the names of the following nonmetal–nonmetal compounds:
    CO_2 NO NO_2 SO_2 SO_3

14. Give the names of the following hydrates:

    $CaCl_2 \cdot 3H_2O$ $FeSO_4 \cdot 7H_2O$ $Na_2C_2H_3O_2 \cdot 3H_2O$

15. Give a definition of a chemical reaction.

16. What is a chemical equation?

17. Which of the following equations are balanced correctly?

    (a) $2\ Al_2O_3 \rightarrow 4\ Al + 3\ O_2$
    (b) $C_4H_{10} + 9\ O_2 \rightarrow 4\ CO_2 + 5\ H_2O$
    (c) $H_2 + Cl_2 \rightarrow 2\ HCl$

18. Balance the following unbalanced equations:

    (a) $H_2 + O_2 \rightarrow H_2O$ (d) $P + Cl_2 \rightarrow PCl_3$
    (b) $C_3H_8 + O_2 \rightarrow CO_2 + H_2O$ (e) $CH_4 + O_2 \rightarrow CO + H_2O$
    (c) $K + Cl_2 \rightarrow KCl$

19. Explain what is meant by the following terms:

    (a)   exergonic reaction   (c)   exothermic reaction
    (b)   endergonic reaction   (d)   endothermic reaction

20. Determine the formula weights of the following compounds. Use the periodic table inside the back cover of this book for atomic weights.

    (a) CO_2 (b) SO_3 (c) Na_3PO_4 (d) $CaCO_3$

21. A 2.0 kilogram sample of tuna fish is found to contain 0.80 milligram of mercury. What is the concentration of mercury in parts per million?

22. Ocean water contains about 4 micrograms of lead (as lead ion) per liter. What is the parts per billion concentration of lead in ocean water?

CHEMICAL
ELEMENTS
IN THE
ENVIRONMENT

4

4-1 CHEMICAL ELEMENTS IN THE ECOSPHERE

There are four distinct regions of our earth environment. The rocky and mountainous crust of the earth is called the **lithosphere** (Greek: litho = stone). The envelope of gases surrounding the earth is the **atmosphere** (Greek: *atmos* = vapor). The **hydrosphere** (Greek: *hydro* = water) includes the vast amount of water in lakes, rivers, oceans, and underground deposits; the water contained in the ice and snow of the earth; and the water which makes up the clouds and moisture in the atmosphere. Within the atmosphere and hydrosphere and upon the lithosphere dwell the various plants and animals which constitute the **biosphere** (Greek: *bio* = life). Since these four realms are interrelated and form our normal environment, they are referred to as a whole as the **ecosphere** (Greek: *oîkos* = house). The components of the lithosphere, hydrosphere, and atmosphere are intermingled to a certain extent. The atmosphere contains water vapor and dust particles. The hydrosphere contains dissolved gases and particles of the lithosphere. Waters of the hydrosphere cover and permeate portions of the lithosphere. This harmonious interfacing of the three portions of the ecosphere provide for the existence and maintenance of the biosphere. Certain activities of man are affecting the very compositions and the natural balances of the subspheres.

The four portions of our environment contain elements in various states of combinations. Table 4-1 gives the percentages by mass of the eighteen elements which make up about 99.5 percent of the lithosphere, atmosphere, hydrosphere, and biosphere. The lithosphere, which is made up mainly of rock and some soil, contains combined silicon, oxygen, aluminum, and many other metals. Most of the metals are combined in chemical compounds, and a few, such as gold, are found only occasionally as the free uncombined metal. Metals are distributed within the lithosphere in very small percentages. Sometimes, as a result of geological processes, certain chemical substances are localized in the lithosphere. These accumulations of substances are called **mineral deposits.** When such mineral deposits contain elements which man needs and mining is feasible, the mineral is called an **ore.** It is estimated that usable mineral deposits make up far less than 1 percent of the crust of the earth. Let us consider an example of an ore. Aluminum content in most rocks is about 8 percent. However, deposits of the mineral bauxite contain around 25 percent aluminum. Thus, such

Table 4-1
The Percentages by
Mass of the Elements
in the Ecosphere

Element	Percentage	Element	Percentage
Oxygen	49.20	Chlorine	0.19
Silicon	25.67	Phosphorus	0.11
Aluminum	7.50	Manganese	0.09
Iron	4.71	Carbon	0.09
Calcium	3.39	Sulfur	0.06
Sodium	2.63	Barium	0.04
Potassium	2.40	Fluorine	0.03
Magnesium	1.93	Nitrogen	0.03
Hydrogen	0.87		
Titanium	0.58	Other elements	0.47

bauxite deposits are aluminum ore from which aluminum metal can be extracted by chemical processes. The minerals which are used the most by industrial societies are iron ore, copper ore, aluminum ore, and minerals which find use as fertilizers (phosphorus-, nitrogen-, and potassium-containing minerals).

The hydrosphere consists mainly of hydrogen and oxygen combined as water. However, many elements in various dissolved forms are found in seawater. These elements are present in small percentages. If we evaporated the water from 1 liter of typical seawater, about 35 grams of chemical substances would remain. Much salt (sodium chloride, NaCl) is obtained from seawater. Significant amounts of magnesium and bromine are also extracted from seawater by chemical processes.

The atmosphere is the layer of air surrounding the earth. The atmosphere is most dense near the surface of the earth and decreases in density as it extends into space. The portion of the atmosphere extending out about 12 kilometers is called the **troposphere.** This is where cloud formation and movement of air (wind) occurs. The portion of the atmosphere extending out from 12 kilometers to about 50 kilometers is called the **stratosphere.** The stratosphere is free of clouds and winds. The portion of the atmosphere above about 50 kilometers is called the **ionosphere.**

Air is a homogeneous mixture of gases. The major components of dry air, in percent by volume, are

nitrogen	N_2	78.1 percent		neon	Ne	0.0018 percent
oxygen	O_2	21.0 percent		helium	He	0.00053 percent
argon	Ar	0.93 percent		krypton	Kr	0.0001 percent
carbon dioxide	CO_2	0.03 percent				

The atmosphere also contains variable amounts of water vapor and dust. The amount of water vapor varies from small amounts in desert regions to large amounts in tropical regions. The water vapor condenses to form clouds, rain, and snow.

The biosphere includes all plant and animal life. It centers near the interface between the lithosphere and hydrosphere, but in the extremes extends to thousands of feet into the atmosphere (birds) and thousands of feet into the ocean depths (fishes). The major elements which structure living matter are carbon, oxygen, and hydrogen. These elements are combined in the chemical substances of life so that the percentage by mass composition of the biosphere is 52 percent oxygen, 39 percent carbon, and 6.7 percent hydrogen. In addition, there are over ten elements which are found in the biosphere in small amounts. The most important of these elements are nitrogen, sulfur, and phosphorus.

4-2 NATURAL RESOURCES

Throughout the centuries man has sought to isolate usable substances from the environment. These have become known as **natural resources.** Great civilizations of the past developed in the vicinity of adequate water supplies, since water is a fundamental natural resource. The natural resources include such things as metals, minerals, coal, oil, and wood. Man developed various methods and techniques to extract and isolate these substances. Today we have an elaborate extractive technology by which we locate and exploit the natural resources. Most often these extractive technologies are directed toward virgin deposits of natural resources. Once the resource has been tapped and converted to useful products, new sources are sought. Some resources, such as oil, can be used only once, while others, such as iron, can be recycled and used again. Unfortunately, in most cases it has been economically advantageous to exploit virgin sources of resources rather than to attempt recycling. There are limits to the amounts of certain resources and new sources will become increasingly more difficult to find. Consequently, most resources are deemed **nonrenewable resources.** In many cases, the development of recycling technologies will become necessary as the sources of the resource dwindle. In other cases, resource management is needed. Water and forest management have been successful to varying degrees. However, the pressure of growing populations and inordinate population distribution could easily counteract effective resource management.

An obvious question which arises concerning nonrenewable resources is: "When will the resources of the world be exhausted?" This question is not easily answered, since it involves the nature of the resources, economics, politics, technology, scientific development, geological exploration, population pressures, and pollution, among other factors. Projections based upon current trends indicate that virgin sources of many resources will be depleted within a few hundred years. Such projections are tentative and subject to change, but it is obvious that recycling techniques and resource management must be developed and expanded.

4-3 CYCLES OF NATURE

Many elements in the ecosphere are exchanged between two or more of the four subspheres. These exchanges result from physical transfer of substances and chemical reactions in which the atoms of an element are changed from one state of combination to another. These exchanges are often cyclic in the sense that atoms which cross the boundaries between subspheres often return to the subsphere in which they previously existed. Such cyclic exchanges are called natural cycles. As we shall see, humans on occasion can affect these cycles by their endeavors.

4-4 OXYGEN CYCLE

In the lithosphere, oxygen is found in combination with silicon and aluminum in aluminum silicates (rocks). Some is found in combination with metals such as metallic oxides, carbonates, sulfates, nitrates, and phosphates. Oxygen in the lithosphere is not readily exchanged with the other subspheres except by the process in which water carries dissolved minerals including carbonates, sulfates, nitrates, and phosphates to the oceans. The atmosphere contains oxygen mainly in the form of molecular oxygen (O_2) and carbon dioxide (CO_2). The hydrosphere contains oxygen in the combined state with hydrogen in water (H_2O). In addition, oxygen is present in small amounts in the form of dissolved minerals and some dissolved oxygen and carbon dioxide. In the biosphere, oxygen is an

Figure 4-1
The natural oxygen
cycle.

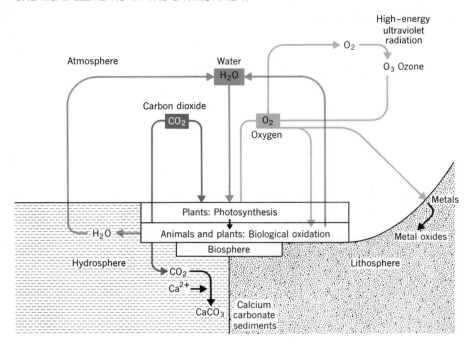

essential element in most biologically important chemical compounds. In fact, about one-quarter of all atoms in animals and plants are oxygen atoms.

The **oxygen cycle** is illustrated in Figure 4-1. Processes which occur in the biosphere stimulate the exchange of oxygen between the various spheres of the environment. A very profound and essential process which occurs in the biosphere is photosynthesis. **Photosynthesis** is the chemical process by which green plants convert carbon dioxide and water into molecular oxygen and molecules which structure the plants. The energy needed to accomplish the chemical reactions (synthesis) which occur is obtained from sunlight (solar or radiant energy). Thus, exposure of green plants to sunlight stimulates the natural process of photosynthesis. The term "photo" refers to light, and the term "synthesis" refers to the manufacture of molecules. The energy of the sunlight is transformed by photosynthesis into chemical energy contained in the plant molecules. Most of the molecules produced by plants contain carbon, hydrogen, and oxygen, and are called **carbohydrates.** Such carbohydrates are complex molecules, and they usually contain one atom of combined carbon and oxygen for every two atoms of combined hydrogen. Thus, we can represent such molecules by the simple formula CH_2O. Keep in mind that there is a large variety of different plant molecules. Glucose, $C_6H_{12}O_6$, is one of the most common, but there are many others. The formula CH_2O does not represent a real molecule, but is merely a simple symbolic way to represent plant carbohydrates. The structures of carbohydrates are discussed in Chapter 13. Using the CH_2O notation, we can represent photosynthesis by the equation

$$\text{sunlight (energy)} + CO_2 + H_2O \longrightarrow CH_2O + O_2$$

Note that molecular oxygen is a product and, as it is produced in the biosphere, it is released to the atmosphere and some which is produced in the photic zone of the hydrosphere remains dissolved in the waters of the hydrosphere. The photic zone is the layer of the hydrosphere that is penetrated by sunlight and in which plant life abounds. The atmosphere contains large amounts of molecular oxygen. Geological evidence suggests that the vast majority of the oxygen has been produced by plants over many millions of years. The oxygen content of the atmosphere is vast and virtually unaltered by the activities of man or any natural processes. We do not have to worry about our oxygen supply. It is essentially inexhaustible.

The biosphere, in addition to producing oxygen, uses oxygen from the atmosphere and hydrosphere. Animals and plants utilize molecular oxygen in **biological oxidation** in which food molecules are converted to carbon dioxide and water. This biological oxidation releases energy which is utilized by the living material for life processes called metabolism. In other words, the solar energy absorbed by the plants in photosynthesis is released when the animals and plants use the products of photosynthesis in metabolism. Even though metabolism is a complex process, the overall result of the biological oxidation of carbohydrates can be represented by the equation

$$CH_2O + O_2 \longrightarrow CO_2 + H_2O + energy$$

Note that this process is essentially the reverse of the process of photosynthesis. Biological oxidation by plants and animals produces carbon dioxide and water which are given off to the atmosphere and hydrosphere. It is interesting to note that the plants of the ocean and land produce 1.3×10^{17} grams of oxygen each year by photosynthesis. However, animals (including bacteria) and plants use up, in biological oxidation, an amount of oxygen nearly the same as that produced by plants. Thus, there is virtually no net change in the oxygen content of the atmosphere.

The three substances, water, carbon dioxide, and molecular oxygen, along with the molecules of living systems, are the major substances involved in the oxygen cycle. Figure 4-1, which illustrates the oxygen cycle, shows other processes which need explanation. Some molecular oxygen in the higher regions of the atmosphere is converted from diatomic molecules to a triatomic molecular form called **ozone,** O_3. This conversion is stimulated by high-energy ultraviolet solar energy and is called a photochemical reaction:

$$3\ O_2 \xrightarrow[\text{radiation}]{\text{ultraviolet}} 2\ O_3\ (\text{ozone})$$

The ozone soon decomposes to reform diatomic oxygen. Since the upper atmosphere is continuously subjected to sunlight, a relatively constant amount of ozone exists in a portion of the upper atmosphere called the **ozonosphere** (about 50 kilometer altitude). The significance of this photochemical production of ozone is that it serves as a chemical barrier which screens out the high-energy ultraviolet radiation from the surface of the earth. Such radiation would have an extremely detrimental effect on the components of the biosphere. In fact, some ultraviolet radiation does reach the surface of the earth—it can injure plants in certain cases and cause sunburn in humans. Incidentally, sunburn preventatives are merely chemicals which screen out ultraviolet radiation.

Most of the carbon dioxide dissolved in the hydrosphere is used by aquatic plants in photosynthesis. However, some of the dissolved carbon dioxide can be converted to carbonate ion (CO_3^{2-}). When calcium ion (Ca^{2+}) is also present in the water, the solid calcium carbonate (limestone) can be formed and settle out to be deposited as part of the crust of the earth. This process occurs to a certain extent and removes some oxygen from the oxygen cycle. However, this does not deplete much oxygen and, in fact, the oxygen may return if the limestone is ever redissolved to form the carbonate ion. A small amount of oxygen is removed by reaction with metals to form metallic oxides. We are familiar with examples of these corrosion reactions, such as the rusting of iron, but they do not account for the loss of very much oxygen.

4-5 CARBON CYCLE

Carbon, the essential element of the biosphere, is involved in a cyclic exchange in the ecosphere. Before considering the carbon cycle, let us consider the chemical combinations in which carbon is found. In the atmosphere, carbon occurs in carbon dioxide, CO_2. In the lithosphere, carbon occurs in metallic carbonates such as calcium carbonate ($CaCO_3$). Carbon occurs in the hydrosphere as dissolved carbon dioxide and as carbonate ions (CO_3^{2-}) or hydrogen carbonate ions (HCO_3^-). In the biosphere, carbon is a major component of virtually all the molecules of living matter. Nearly one-quarter of all atoms in the biosphere are carbon atoms. The most prevalent atomic combination in the biosphere occurs in carbohydrates. Thus, it is useful to represent the carbon in the biosphere by the simple formula CH_2O. Since carbon dioxide is involved in the process of photosynthesis, the exchange of carbon between the spheres of nature is closely related to the oxygen cycle. The **carbon cycle** is illustrated in Figure 4-2. Carbon dioxide from the atmosphere and that dissolved in the hydrosphere is absorbed into the biosphere by photosynthesis of green plants on the land and aquatic plants. As a result of photosynthesis, the carbon becomes part of the biological molecules of life. Plants obtain energy by the biological oxidation of molecules made by photosynthesis and, thus, plants also release some carbon dioxide back into the atmosphere or hydrosphere. Within the biosphere, animals directly or indirectly use plants as food. This food is biologically oxidized to produce carbon dioxide and water. Thus, land and sea animals return the carbon dioxide to the atmosphere and hydrosphere. Another biological oxidative process that goes on in the soils of the lithosphere and certain regions of the hydrosphere involves the decomposition of dead plant and animal material. Generally, this decomposition is accomplished by microorganisms and bacteria, and most often produces carbon dioxide and water. Some of the carbon in the oceans is essentially removed from the main carbon cycle by the formation of calcium carbonate sediments. This sediment becomes

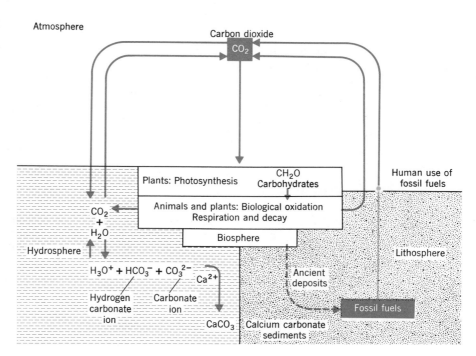

Figure 4-2 The natural carbon cycle and alteration by use of fossil fuels.

part of the vast amount of metallic carbonates which exist in the lithosphere. The fact is that the majority of the carbon of the earth is in the lithosphere mainly as carbonates. Less than 1 percent of the carbon of the earth is involved in the carbon cycle.

As a result of geological phenomena which occurred millions of years ago, some collections of living organisms became trapped within the lithosphere and have been transformed into deposits of coal, oil, and natural gas. These deposits represent a portion of an ancient biosphere which was removed from the carbon cycle and transformed by heat and pressure into a variety of carbon-containing compounds. After discovering some of these deposits, man found that useful energy could be obtained by the burning of these **fossil fuels.** Thus, the fossil fuels have become the major source of energy in our industrial society. (See Chapter 5 for a discussion of energy.) One of the major products of the combustion of fossil fuels is carbon dioxide. Around 20 billion tons of carbon dioxide are produced each year by the burning of fossil fuels. As shown in Figure 4-2, this carbon dioxide enters the atmosphere and becomes part of the carbon cycle. There is evidence which indicates that the burning of fossil fuels by man is increasing the amount of carbon dioxide in the atmosphere. This increase can affect the carbon cycle and could have some influence on the climate of the earth. As a result of the increased use in the amount of fossil fuels, the carbon dioxide content of the atmosphere is expected to increase. However, it is difficult to predict the effects which the increase in carbon dioxide will produce.

4-6 NITROGEN CYCLE

Nitrogen, next to carbon, hydrogen, and oxygen, is the fourth most abundant element in the biosphere. Since nitrogen is an important component of the amino acids and proteins, it is one of the critical nutrients of plants and animals. Consequently, the interchange of nitrogen in the ecosphere is an important cycle. The **nitrogen cycle** involves the transfer of nitrogen between the biosphere, lithosphere, atmosphere, and hydrosphere in various chemical forms. Within the atmosphere, nitrogen exists in the form of diatomic molecules, N_2. This form of nitrogen is quite stable and relatively inert. A small amount of nitrogen in the atmosphere is in combination with oxygen in the nitrogen oxides. Within the lithosphere, nitrogen exists mainly in nitrate ion, NO_3^-, and, to a small extent, in the form of nitrite ion, NO_2^-, and ammonium ion, NH_4^+. Within the hydrosphere, nitrogen exists mainly as dissolved diatomic nitrogen, N_2, and dissolved nitrate ion, NO_3^-. The biosphere contains combined nitrogen in plant and animal proteins. **Proteins** are complex molecules of living organisms containing carbon, hydrogen, oxygen, and nitrogen, along with small amounts of sulfur and other elements. See Section 13-4 for a discussion of proteins.

The nitrogen cycle is illustrated in Figure 4-3. As can be seen, the overall cycle includes an outer cycle involving the atmosphere, lithosphere, or hydrosphere linked by the biosphere and an inner cycle involving the biosphere, lithosphere, or hydrosphere. The outer nitrogen cycle includes the conversion of atmospheric nitrogen to nitrate ion and/or ammonium ion. The conversion of molecular nitrogen to these forms is called **nitrogen fixation.** One mode of nitrogen fixation is the process in which molecular nitrogen is converted to nitrogen–oxygen (nitrogen oxides) compounds by the high energy of lightning in the atmosphere. Only a small amount of nitrogen is fixed in this fashion. This fixed nitrogen is carried to the surface of the earth by rain and enters the nitrate portion of the cycle. Another more important mode of fixation is that in which microorganisms (often those closely related to certain plants) convert molecular nitrogen to forms (ammonium ion, nitrite ion, and nitrate ion) in which it

Figure 4-3
The natural nitrogen
cycle and human
intervention in the
cycle.

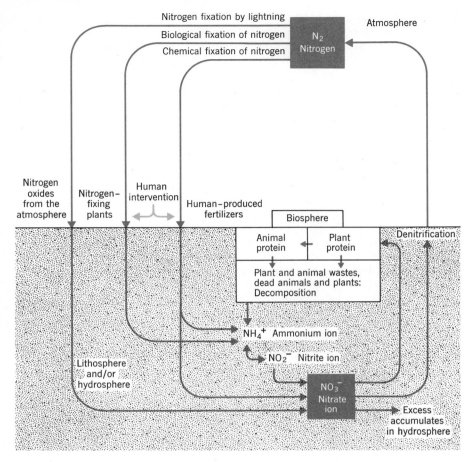

Figure 4-3 The natural nitrogen cycle and human intervention in the cycle.

becomes available to the nitrogen inner cycle. This process of fixation is called **biological fixation,** and the microorganisms involved are called **nitrogen-fixing bacteria.**

Within the inner cycle, nitrate ion serves as a source of nitrogen for most aquatic and terrestrial plant life. The plants incorporate the nitrogen into plant protein. Many of the plants are eaten by animals which convert the plant protein to animal protein. Smaller animals pass the nitrogen along to larger animals in the food chain. The inner cycle is completed by the death and decay of the plants or animals. When these systems die or give off wastes (e.g., animal excrement), protein decomposition produces ammonium ion. Certain microorganisms in soil and in the hydrosphere use the ammonium ion and convert it ultimately to the form of nitrate ion deposited in the soil or dissolved in water. Nitrate ion is exchanged between the soil and the hydrosphere by the process in which dissolved nitrate ion is carried by ground waters. Other microorganisms in the soil and hydrosphere utilize nitrate ion in a process called denitrification. **Denitrification** is a biological process in which certain bacteria convert nitrate ion to molecular nitrogen, N_2. The molecular nitrogen produced by the **denitrifying bacteria** becomes dissolved nitrogen or atmospheric nitrogen. The entry of the molecular nitrogen into the atmosphere completes the nitrogen cycle.

As mentioned previously, nitrogen as nitrate ion serves as an essential nutrient in plant growth. Of course, plants serve as the fundamental foods of man. In order to produce enough plant foods using modern agricultural methods, man has found it necessary to alter the nitrogen cycle by fixing more nitrogen than would occur in the absence of the intervention of man. Man has encouraged more biological fixation by the intentional cultivation of crops which are associated with nitrogen-fixing bacteria. The most common of these crops are the legumes such as alfalfa. Cultivation of such crops provides fixed nitrogen for the cultivation of other plants. In addition to the designed increase in biological fixation, man has developed chemical methods by which nitrogen can be fixed. This chemically fixed nitrogen is incorporated in nitrogen-containing fertilizers which find wide use in agriculture. The method used to chemically fix nitrogen is called the **Haber process** and is accomplished by reacting hydrogen gas and nitrogen gas to produce ammonia. This process can be represented by the chemical equation

$$N_2 + 3 H_2 \longrightarrow 2 NH_3 \text{ (ammonia)}$$

Once molecular nitrogen is converted to ammonia, the ammonia can be chemically converted to the various forms in which it is useful in fertilizer. As fertilizers, the nitrogen is introduced into the nitrogen cycle. See Section 15-6 for a discussion of fertilizers.

Man has been introducing large amounts of fixed nitrogen into the cycle for about 50 years. Since the denitrification part of the cycle (see Figure 4-3) converts only a certain fraction of fixed nitrogen back to molecular nitrogen, man's effect on the nitrogen cycle is resulting in the accumulation of fixed nitrogen. Much of this fixed nitrogen remains as nitrate ion in the various bodies of water which are part of the hydrosphere. Too much nitrate ion in drinking water can be dangerous. Human babies and grazing animals have become sick and in some cases died after drinking water containing too much nitrate ion. Nitrate ion is converted to nitrite ion in the stomachs of babies which, when it enters the bloodstream is capable of causing methemoglobinemia, a blood disorder (blue babies). Well water in agricultural areas should be checked often for nitrate ion. The U.S. Public Health Service recommends a limit of 45 parts per million of nitrate ion in drinking water.

4-7 MINING AND REFINING

In an industrial country like the United States, the most important nonfuel resources are iron ore, aluminum ore, copper ore, and minerals used in fertilizers. In addition, large amounts of cement, gravel, stone, and asphalt are used. To support the industrial society of the United States, about 5 billion tons of materials and ore are extracted from the earth each year. Once an ore deposit is located, much energy is expended in the mining operations and the transportation of the ore. To make these operations of physical extraction and transportation worthwhile, the ores must contain enough of the desirable compound or element. Some low-grade ore deposits are not usable due to the economics of the mining process.

The materials extracted from the earth have to be processed for use. The physical processes of screening, separating, washing, and grinding produce wastes (some are called mine tailings) and dust (fine solid particles which become suspended in the air). Often, the desirable element in an ore is in an undesirable chemical form and must be converted to a desirable one. For instance, aluminum in aluminum ore is in the form of aluminum ion, Al^{3+}, combined with oxide ion, O^{2-}. To be useful, the aluminum ion has to be converted to aluminum metal, Al. In such cases the ore must be processed to extract the desirable element. Chemical or electrical processing of an ore to extract an element is called the **refining** of the ore. The refining process normally involves a chemical reaction or a series of chemical reactions which produce the desired form of the element or compound sought from the ore. These chemical reactions usually require large amounts of energy, and the cooling and washing processes involved utilize large amounts of water. In the following sections, the refining and processing of some important ores and materials are discussed.

4-8 IRON AND STEEL

Iron ores are usually a mixture of iron oxides and silicate rocks. The principal oxides are Fe_2O_3 (hematite ore) and Fe_3O_4 (magnetite ore). These iron oxides are converted to iron by allowing the ore to react with carbon (in a form called coke), limestone, and air in a blast furnace at temperatures exceeding 1,300°C. The **blast furnace** is illustrated in Figure 4-4. The main chemical reactions which take place in the refining process are

$$2\,C + O_2 \longrightarrow 2\,CO$$
$$3\,CO + Fe_2O_3 \longrightarrow 3\,CO_2 + 2\,Fe\ (\text{molten iron})$$

The molten iron is tapped from the blast furnace and cast into ingots called pig iron or cast iron. Most of this iron is used to manufacture steels. **Steels** are alloys of iron—iron mixed with small amounts of carbon and other metals such as manganese, nickel, chromium, vanadium, and/or tungsten. Stainless steel is iron containing some carbon, 15–20 percent chromium, and about 10 percent nickel.

Figure 4-4
A blast furnace used in the refining of iron from iron ore.

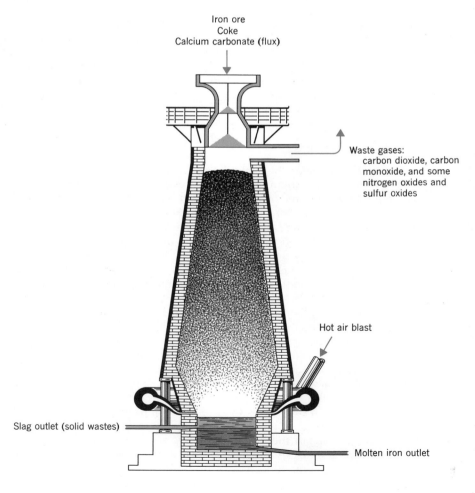

Iron ore
Coke
Calcium carbonate (flux)

Waste gases:
carbon dioxide, carbon monoxide, and some nitrogen oxides and sulfur oxides

Hot air blast

Slag outlet (solid wastes)

Molten iron outlet

Steel is manufactured by melting pig iron and mixing in the alloying metals. The preparation of steel is carried out by use of devices called Bessemer converters, open-hearth furnaces, or electric arc furnaces.

Steel is used widely in manufacturing to make pipes, wires, beams, rails, steel sheets, and bars. In the United States, about 10,000 kilograms of steel are in use for each person.* Most of this steel is in structural materials and in motor vehicles. Nearly 500 kilograms of steel per person is added to the environment each year, but about 330 kilograms per person is lost each year as a result of junking, dumping trash, and corrosion. The presence of acids in the smoggy air of heavily urbanized, industrialized areas hastens the corrosion of exposed iron. Of this discarded steel, about 40 percent is recycled to the steel furnaces and the rest is scrapped or buried. The buried and scrapped iron may never be utilized again. In other words, we throw away nearly 200 kilograms of iron per person each year. The uses of iron and the potential sources of pollution in the steel-making process are illustrated in Figure 4-5.

The use of steel by a country is a direct indicator of the extent of industrialization and the economic development. Each year the United States uses about 600 kilograms per person, Japan about 500 kilograms per person, Russia about 400 kilograms per person, and India about 10 kilograms per person. Large amounts of energy are expended in the mining, manufacturing, and distribution of steel. In the United States, energy equivalent to the burning of about 15 tons of coal is needed to put 1 ton of steel into use.† The manufacture of 1 ton of steel requires over 40,000 gallons of water.

4-9   ALUMINUM

Large amounts of aluminum occur in the form of aluminum compounds in the lithosphere. Aluminum is the most abundant metal in the crust of the earth. Combined aluminum is found in clays and rocks. **Bauxite** is **aluminum ore** consisting of a mixture of aluminum compounds and some iron oxides. Aluminum oxide, Al_2O_3, can be extracted from the bauxite. The refining of aluminum oxide to aluminum is more difficult than the production of iron from iron ore. The problem is that aluminum metal cannot be produced from the oxide by a simple chemical process such as the reaction with carbon, as is done in the refining of iron. To refine aluminum, a process called **electrolysis,** in which electric current converts the aluminum ion to aluminum metal is used. All aluminum is made electrolytically by the **Hall process** (named after Charles M. Hall, who first developed the process in 1886). In this process (see Figure 4-6), the aluminum oxide is dissolved in a large vat of the molten (around 1000°C) mineral cryolite (Na_3AlF_6). When the aluminum oxide dissolves in the cryolite, it forms the aluminum ion (Al^{3+}) and oxide ion (O^{2-}). Immersed in the vat are large pieces of carbon which serve as the anode (positive electrode). The carbon lining of the iron vat serves as the cathode (negative electrode). When electrical current flows, the aluminum ion is attracted to the cathode, where it is converted to molten aluminum metal by gaining electrons. The oxide ion is attracted to the anode, where it loses electrons and combines with carbon to form carbon

* Throughout this book environmental topics will be discussed and many numerical references will be made. Most often these references involve percentages and tonnages of specific chemical compounds. An attempt has been made to have these references as accurate as possible, however, in many cases, numerical references may fluctuate depending upon economic, technological, and other factors. As a result, many numerical values quoted are not to be considered precise and invariant, but should be regarded as good approximations.

† In this book, ton refers to a metric ton. One metric ton is 1,000 kilograms and corresponds to 1.1 tons or 2,200 pounds. A metric ton and an English ton differ only slightly.

Figure 4-5
The uses of iron and
the potential sources
of pollution in the
refining of iron ore
and the manufacture
and processing of
steel. From "Pollution
Control in the Steel
Industry," by
Henry C. Bramer,
*Environmental Science
and Technology* **5**,
(10), 1004 (Oct. 1971).

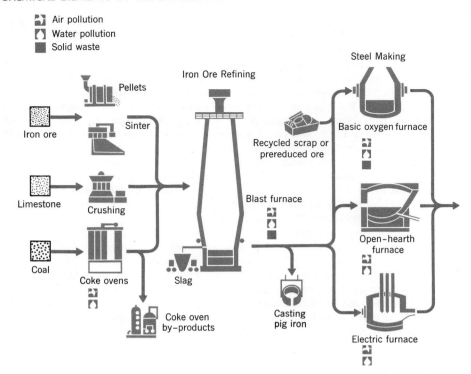

dioxide. Aluminum produced by this process is called primary aluminum. Primary aluminum contains some impurities which can be removed by subjecting the aluminum to a second electrolytic process. In this second process, the impure aluminum is converted to a highly pure form of aluminum. The production of aluminum by the Hall process requires large amounts of electricity and water for cooling and washing. Consequently, large aluminum plants are usually located near convenient sources of water in regions where electricity from hydroelectric sources is available.

In the United States, well over 200 million tons of aluminum are produced each year. In addition to large amounts of electrical energy, about 360,000 gallons of water are used in the production of 1 ton of aluminum. As shown in Figure 4-6, carbon dioxide and other waste gases are given off during the production of aluminum. The generation of fluorine-containing compounds (from cryolite) as air pollutants has created a very serious environmental problem around many aluminum refineries. The fluorine compounds settle out around the refinery and kill trees and other plant life. Animals which ingest these compounds suffer from fluorosis (decay of bones and teeth) which leads to maiming or death. Large amounts of solid wastes result when aluminum oxide is extracted from the bauxite ore.

Aluminum is a lightweight, nontoxic metal which can be easily formed or cast into a variety of shapes ranging from structural beams to sheets and foils. It is corrosion-resistant, has a high heat conductivity, and is a good conductor of electricity. Aluminum alloyed with small amounts of other metals is used for such items as kitchen utensils, building decorations, and structural components; aircraft construction; electrical transmission lines; mirrors; automobile parts; foils; and food and drink cans. Aluminum used in buildings and other structural components may remain in use for long periods of time, but much of

Figure 4-5

Steel Processing   Steel Products

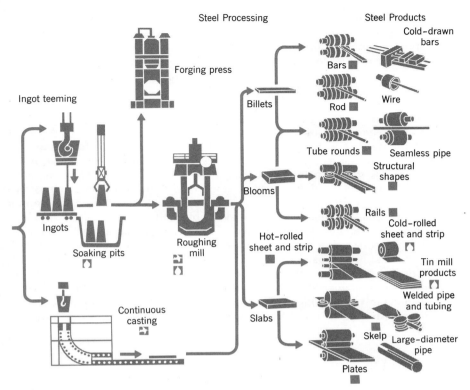

the aluminum we use is thrown away in a short time. Such aluminum is widely distributed in junked cars and refuse disposal dumps. Since aluminum is corrosion-resistant, junked aluminum oxidizes back to aluminum oxide very slowly. Some of the junk aluminum we dispose of today (i.e., aluminum cans) may be around for hundreds of years as metallic aluminum. Of course, there is an abundant supply of aluminum ores in the lithosphere and there is little prospect of running out. However, the refining of 1 ton of aluminum requires five times the energy needed to refine 1 ton of steel. Thus, it is an energetically expensive

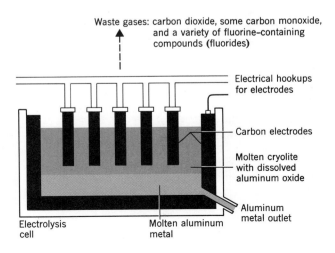

Figure 4-6

A typical Hall process electrolysis cell used in the refining of aluminum ore.

metal. The recycling of aluminum is possible, and some effort has been directed to recycling aluminum in cans. Unfortunately, only about 20 percent of the annual production of aluminum returns to be used again. Over 400,000 tons of aluminum are used each year to make cans, can tops, and lids.

4-10   TOXIC METALS IN THE ENVIRONMENT

Most of the elements of the earth are widely distributed in the lithosphere, hydrosphere, and atmosphere. We are exposed to these elements when we eat food, drink water, and breathe air. Some of the elements are beneficial and essential to our life processes, and some are very toxic. Fortunately, most of the toxic elements found in food, water, and air are present in extremely small amounts. Nevertheless, there are trace amounts of many different elements entering our bodies each day. Since these trace amounts result from the natural distribution of elements, they are unavoidable and are referred to as background levels. Sometimes, the endeavors of man and, on occasion, a geological accident, introduce undesirable amounts of certain toxic elements into the environment. Most notorious of these are the **heavy metals** mercury (Hg), lead (Pb), and cadmium (Cd). Note the position of these metals in the periodic table. However, other elements of concern are arsenic (As), beryllium (Be), antimony (Sb), vanadium (V), and nickel (Ni). The heavy metals are of greatest concern, since they are used in great amounts and distributed widely in our industrial society.

4-11   MERCURY

Mercury is a very useful element, since it is the only metal which is a liquid at normal temperatures and has a high electrical conductivity. Unfortunately, mercury compounds are poisonous to all living systems. Most mercury is obtained from ore containing cinnabar, HgS, by heating in air. The heating converts the HgS to mercury metal and sulfur dioxide:

$$HgS + O_2 \xrightarrow{heat} Hg + SO_2$$

Over 3,000 tons of mercury are used in the United States each year. Mercury is used in a variety of industrial processes and in various products such as paints, fungicides, electrical apparatus, and thermometers. Table 4-2 lists the common

Use	Percentage of total tonnage used
Chlor–alkali industry	26.0
Electrical apparatus (including batteries)	22.9
Paint (fungicides)	12.2
Scientific instrumentation	6.5
Catalysts	3.7
Dental preparations (amalgams)	3.5
Agriculture (pesticides and fungicides)	3.4
General laboratory uses	2.1
Pharmaceuticals	0.9
Paper and pulp processing (fungicides)	0.7
Industrial amalgamation	0.2
Miscellaneous	17.9

Table 4-2
Mercury Uses in the United States (Percentage of Total Tonnage Used), 1969

Adapted from "Effects of Mercury on Man and the Environment" (1970), Senate Subcommittee on Energy, Natural Resources, and the Environment Hearings.

uses of mercury in the United States. The greatest single use of mercury is in the electrolytic manufacture of chlorine and sodium hydroxide, which is discussed in Section 10-6. The second greatest use of mercury is in electrical apparatus such as mercury vapor lamps, electronic tubes and switches, and mercury batteries.

Mercury enters the environment in the elemental form as loss from industrial processes and scrapped equipment, and in the form of mercury compounds from industrial and agricultural endeavors. Mercury also enters the environment from unexpected sources. Significant amounts of mercury are found in sewage resulting from use of small amounts of mercury-containing chemicals, pharmaceuticals, and paints, by large numbers of people. Another mercury source appears to be the combustion of coal, oil, and gasoline. These fossil fuels contain small amounts of mercury. However, since fossil fuels are used in such large amounts, significant quantities of mercury enter the atmosphere by fossil fuel combustion. It is estimated that about 3,000 tons of mercury enter the atmosphere from this source each year.

Mercury is incorporated into the food we eat by agricultural uses and the water supply. Mercury is toxic in the metallic form and also in the combined form. The two general forms of combined mercury are called inorganic mercury and organic mercury. Within the environment, the various forms of mercury are interconverted. In fact, it appears that metallic and inorganic mercury are converted to methylated mercury by biological processes occurring in water in which mercury wastes are found. Furthermore, **methylated mercury (dimethyl mercury,** $Hg(CH_3)_2$, and **methyl mercury ion,** $HgCH_3{}^+$) are absorbed in the tissue of living organisms. Once absorbed, these forms of mercury can remain in an organism for long periods. As one animal eats another, the mercury can be incorporated into the **food chain.** This can result in the biological concentration of mercury within the food chain. For example, algae ingest mercury from the water, a small fish eats the algae, a larger fish eats the small fish. This biological concentration results in animals at the top of the food chain having more than normal (above background) amounts of mercury incorporated in the tissue. Plants can incorporate mercury compounds from the soil and mercury-containing seed coatings. Mercury enters our bodies through the plants and animals we eat and the water we drink and, thus, becomes incorporated into our body tissue. The U.S. Food and Drug Administration (FDA) has established maximum concentrations of 0.5 part per million of mercury in fish and 0.005 part per million of mercury in water. The FDA periodically checks samples of food for mercury content. In 1971, the FDA discovered that commercial swordfish was highly contaminated with mercury, and all sales were banned. In the same year, certain canned tuna samples were found to have intolerable mercury content. As a result, large numbers of cans were recalled from retail stores.

Too much mercury in our systems can produce **mercury poisoning,** which can be deadly or cause permanent brain damage. In Japan, illness, deaths, and birth defects have been directly attributed to mercury-containing seafoods. Around 100 persons living in the Minimata Bay area of Japan became afflicted with a mysterious illness, and many died. It was found that the victims' main diet was mercury-contaminated fish from the bay. The mercury had been dumped into the bay in the waste water of a plastics plant. The Japanese government soon imposed strict requirements on mercury disposal. The Swedes, who also rely on a seafood diet, have strictly limited the use of mercury and mercury compounds in their country. The symptoms of acute mercury poisoning are loss of appetite, numbness of extremities, metallic taste, diarrhea, vision problems, lack of coordination, speech and hearing difficulties, and mental instability. The damage to the body by mercury poisoning is usually permanent. Not much is known about the effects of small amounts of mercury in the body, since the symptoms are not specific and cannot be distinguished from other diseases.

Since mercury is used in such great amounts and can be converted to highly toxic forms which are biologically concentrated in the environment, it is necessary that we exercise greater control over the intended or unintended discharge of mercury. Alternatives to industrial discharges should be enforced. The use of mercury compounds in agriculture should be restricted or eliminated if possible. Finally, the content of mercury in the food, water, and air should be closely watched.

4-12   LEAD

Lead, a heavy metal like mercury, is a toxic metal which accumulates in the body as it is inhaled from the air or ingested from food and water. Lead has been used for centuries to fabricate devices used by man. Lead is mentioned in the Old Testament of the Bible. The Romans used lead for water pipes and for cooking utensils. Actually, lead is somewhat of a rare element in the lithosphere. The percentage of lead in the crust of the earth is about 0.00002 percent. However, lead ore deposits consisting of the ore galena, PbS, are used as sources of lead. Lead has a low melting point and is quite a soft metal. This malleability allows lead to be cast and formed easily. Lead can be mixed with other metals to form useful alloys. Nearly 1.5 million tons of lead are used in the United States each year. About one-half of this is recycled lead, and the rest is virgin lead. As shown in Table 4-3, lead is used in a variety of products including storage batteries, gasoline antiknock chemicals, paint pigments, and ceramic glazes. Most lead is used in storage batteries, and most of this is salvaged so that the lead can be reused. The environmental threat comes from lead used in chemicals. About 200,000–300,000 tons of lead are used each year to manufacture **lead alkyl** compounds (lead tetraethyl and lead tetramethyl) which are used as **antiknock** additives in gasoline. This lead becomes widely distributed in the environment as it is discharged in automobile exhaust. This is illustrated in Figure 4-7. Much of the lead compounds given off in automobile exhaust is dispersed into the atmosphere as gaseous substances or tiny bits of solids called particulates. This atmospheric lead can be inhaled by people, settle out on plants and soil, or be absorbed by water. The lead from gasoline is supplemented by lead compounds produced in the burning of coal and oil, and the manufacturing of lead and lead products.

Investigations show that the amounts of lead in the environment have increased greatly since the beginning of the Industrial Revolution, and the amounts are currently increasing at a rapid rate. This is illustrated in Figure 4-8. Many urban areas have significant amounts of airborne lead. Recent studies revealed an average lead content of 2.3 $\mu g/m^3$ (micrograms per cubic meter, see Section 7-7) of air in California metropolitan areas. A typical person

	Use	Percentage of total tonnage used
Table 4-3 Uses of Lead in the United States (Percentage of Total Tonnage Used), 1969	Storage batteries	43.1
	Metal products (ammunition, cable covering, pipes and plumbing uses, solder, and type metal for printing)	26.2
	Lead alkyl (lead tetraethyl and lead tetra- methyl antiknock compounds)	20.0
	Paint pigments	7.5
	Miscellaneous (chemicals and pottery glazes)	3.2

From *Minerals Yearbook—1969.* U.S. Department of the Interior.

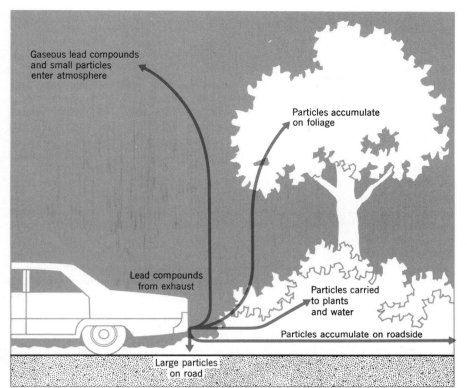

Figure 4-7
The distribution of lead from automobile exhaust. Lead compounds from the exhaust of automobiles using leaded gasoline are distributed in the environment. About one-third of the lead from exhaust enters the atmosphere as gaseous compounds or as tiny solids called particulates. Around two-thirds of the lead is deposited on the highways and nearby landscape as larger particles of lead-containing compounds.

breathes 15 cubic meters of air per day. Lead compounds are present in most food we eat and some are in the water we drink. Fresh orange juice contains about 0.15 microgram per milliliter, and potatoes contain about 0.3 microgram per gram. Of course, not all of the lead ingested and inhaled is absorbed into the body. Each day, the average American absorbs about 15 micrograms of lead from food and water and about 15 micrograms from air. This represents a small amount of lead but, unfortunately, lead is a cumulative poison. Most of the lead which remains in the body localizes in the bones, since it is chemically similar to calcium, a main constituent of bones. As a person grows older, more and more lead accumulates in the body. Not all lead that enters the body remains, but enough does to result in about 0.2 gram of lead in an adult American. Actually, the average lead excretion by an American is around 30 micro-

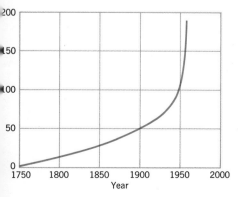

Figure 4-8
The increase in the lead content of the environment. Lead content of arctic snow over a period of years. The snow fall in some arctic regions is slight, and the snow remains packed in layers for centuries. These snow layers can be dated and serve as good sources of evidence of the lead content of the atmosphere in years past. As the graph indicates, the atmospheric lead content has apparently increased steadily over the last 200 years. Further, a very significant increase has occurred since 1950, corresponding to the wide use of lead alkyls in gasoline and the increase in the automobile population. The graph is based on average data collected by C. Patterson of the California Institute of Technology, *Scientist and Citizen* 66 (Apr. 1968).

grams per day. This is just about the same as the amount of lead taken in each day, and there is little danger of lead poisoning unless a person is exposed to high lead levels in food and air. Lead localized in the bones can enter the fluids of the body during illness or treatment with certain medicines. The major toxic effect of lead appears to be that it interferes with the production of red blood cells and other bodily functions. **Lead poisoning** is a possibility if the concentration of lead in the blood is 0.8 part per million for adults or 0.4 part per million for children. Currently, the lead concentration in the blood of a typical American is about 0.2 part per million. The symptoms of mild lead poisoning are loss of appetite, fatigue, headaches, and anemia. Lead poisoning is especially dangerous and prevalent in children. Severe lead poisoning can cause permanent brain damage in children.

Another source of lead in the environment is from lead-containing paint pigments. Such pigments are not supposed to be used in interior paints today but were used before World War II. In fact, U.S. Food and Drug Administration regulations call for a limit of 0.06 percent lead in paint pigments by 1974. Tragically, many older buildings, such as those found in ghettos, contain layers of paint with high lead content. Children suffering from malnutrition tend to nibble on bits of dirt and paint. In this way significant amounts of lead are ingested by children. It is estimated that about 400,000 children in the United States are afflicted with various degrees of lead poisoning each year. Some 200 of these children die and about 800 suffer from permanent brain damage.

Still another source of lead in the diet comes from the lead compounds used in some glazes and ceramics. Food in contact with such glazed ceramics can sometimes extract lead in significant amounts. This does not occur with all ceramic glazes. However, it is good practice not to store carbonated beverages, fruit juices, wines, any food containing vinegar, sauerkraut, or fruit products in ceramic vessels. Furthermore, it is best to avoid using ceramic vessels with cracked, chipped, or faulty glazes.

Since lead is a cumulative poison, it is important to continuously investigate the amounts of lead which are entering the environment. This is especially true in the United States, which uses nearly 60 percent of the lead consumed in the entire world. Of this 60 percent, about one-half will never be used again, and will become distributed in the environment. The greatest waste of lead is in the use of lead alkyls as gasoline additives. About 20 percent of the world consumption of lead is lost as nonrecoverable lead alkyls. Since this is a waste of a valuable resource and, more importantly, a threat to human health, the use of lead alkyls in gasolines will have to be stopped. (See Section 5-7.) In any case, the content of lead in food, water, air, and the bodies of Americans should be closely watched.

4-13   CADMIUM

Cadmium is a toxic metal which is significant, since it also is becoming widely distributed in the environment by man's endeavors. About 7,000 tons of cadmium are produced in the United States each year. Virtually all the cadmium is produced as a by-product in the refining of zinc from zinc ore which contains some cadmium. Since this process is a plentiful source of cadmium, only a small amount of cadmium is recycled. As shown in Table 4-4, the uses of cadmium include electroplating or iron alloys, pigments, alloys, batteries, and plastic additives. Cadmium from plating operations enters the waste water environment. Airborne cadmium comes about as a by-product in the processing of scrap steel, the reclaiming of copper, and the refining of lead, copper, and zinc. Phosphate fertilizers contain some cadmium which can be absorbed by plants. Some cadmium enters the water supply from water pipes. There appears to be only small

Use	Percentage of total used
Electroplating (cadmium plating)	45.1
Pigments (paints and coloring agents)	21.1
Plastics (plastic stabilizers)	15.0
Alloys (steels and other metals)	7.5
Batteries (nickel–cadmium batteries)	3.0
Other uses (television tubes, fungicides, nuclear reactors, curing agents for rubbers)	8.3

Table 4-4
The Uses of Cadmium

Based on 1968 use. Data from *Environmental Science and Technology* 755 (Sept. 1971).

amounts of cadmium in air, water, and food. However, cadmium can be absorbed by the body and can remain in the system for long periods of time. Excessive amounts of cadmium can cause liver, kidney, and spleen damage. Furthermore, there is evidence that cadmium is related to hypertension (high blood pressure), which can cause heart ailments. The problem with cadmium is that the amounts in the body that can cause problems are not known. Thus, it is important to be aware that increased cadmium levels in the environment could present a significant health hazard.

4-14 SOLID WASTES

The wastes of a society take the form of gases, liquids, and solids. Gaseous wastes are released into the atmosphere; liquids and some solids are washed away in sewage water. Most of the **solid wastes** have to be moved from one place to another and disposed of. Such disposal usually takes the form of dumping on the ground, dumping in the ocean, and sometimes burning in a dump or an incinerator. Solid wastes have been a problem since ancient times. Many ancient cities were smothered with wastes, and new cities were built on the rubble or on new locations. The ruins of the ancient city of Troy were found in Turkey under a vast mound of solid wastes. The problem of solid wastes in the United States revolves around the large-scale industrial and mining endeavors, high population densities in urban areas, the use of excessive amounts of packaging materials and the throw-away-nonrecycling economy. The solid waste dilemma which exists today results from utilizing virgin resources, distributing the products of these resources widely, and then collecting the used products for disposal on valuable land.

4-15 URBAN WASTES

About 4 billion tons of solid wastes are generated in the United States each year. This includes municipal, industrial, agricultural, animal, and mining wastes. It seems reasonable that some wastes could be put to use. The problems are finding economic use of some wastes and separating the conglomeration of materials which accumulate. There are three distinct types of solid wastes, each of which present different problems. The most familiar type of solid waste is **urban refuse** from domestic, commercial, municipal, and industrial sources. This refuse amounts to about 400 million tons annually and includes such things as cans, bottles, paper, plastic, junked automobiles, rubble, and worn out consumer products. Table 4-5 lists the composition of typical urban wastes. Each year the population of the United States generates enough trash to cover 20,000 acres of ground with a 7 foot deep pile. The cost of handling solid wastes amounts to billions of dollars each year. Most of the cost (about 75 percent) involves the

	Percentage by mass		Percentage by mass
Kitchen garbage		Garden trash	
Vegetable food wastes	2.5	Wood	2.5
Meat scraps	2.5	Tree leaves	2.5
Fried fats	2.5	Flowers and garden	
Citrus peelings	1.7	plants	1.7
Subtotal	9.2	Lawn grass	1.7
Paper and paper products		Evergreens and shrub	
Cardboard	25.7	clippings	1.7
Newspapers	10.3	Subtotal	10.1
Magazines	7.5	Miscellaneous	
Brown paper	6.2	Glass, ceramics, ashes	8.6
Mail	3.0	Various metals	7.6
Paper food cartons	2.3	Dirt	1.7
Tissue paper	2.2	Vacuum cleaner wastes	0.8
Plastic coated paper	0.8	Plastics	0.8
Wax paper	0.8	Rags	0.8
Subtotal	58.8	Paints and oils	0.8
		Rubber products	0.4
		Leather products	0.4
		Subtotal	21.9
		Total	100.0

From I. Remson, A. A. Fungaroli, and A. Lawrence. "Water Movement in an Unsaturated Sanitary Landfill." *Journal of The Sanitary Engineering Division, Proceedings of the American Society of Civil Engineers* **94,** SA 2 (Apr. 1968).

collection and transportation of the wastes. Once the wastes are collected, they are usually dumped on land somewhat removed from the urban areas. About 75 percent of all urban wastes is disposed of in open dumps, sometimes to be burned, but more often it is left in mounds. The best method of trash burial is the sanitary landfill technique which minimizes the health and water pollution hazards. Figure 4-9 illustrates the **sanitary landfill** technique. Only about 10 percent of the trash dumps practice sanitary landfill techniques. About 12 percent of the urban trash is burned in municipal **incinerators.** This method consumes the combustible materials in the trash, but, unfortunately, contributes to air pollution. In fact, about 5 percent of all air pollutants come from the incineration of refuse. Once refuse is burned, the noncombustible ash which remains must be buried or dumped into the ocean. One of the biggest problems of urban trash disposal is finding enough land for disposal purposes. Sanitary landfill sites can be reclaimed if the landfill is carried out properly. Urban trash represents a potential source of resources if the mixture can be separated into the usable parts. Processes such as the one illustrated in Figure 4-10 are being tested to accomplish the profitable separation of the components of urban wastes.

Another aspect of urban solid wastes is that of **junked automobiles.** The motor vehicle population in the United States is well over 100 million (over 90 million automobiles) and is increasing at a rate which is greater than the increase in human population. Large amounts of our natural resources are included in the manufacturing of cars; however, some of these resources are recycled. Needless to say, a significant number of junked cars are never reclaimed and represent a waste of resources. According to the U.S. Bureau of Mines, a typical 1,600 kilogram junk automobile contains 1,140 kilograms of steel, 230 kilograms of cast iron, 15 kilograms of copper, 25 kilograms of zinc, 23 kilograms of aluminum, and 9 kilograms of lead. The remainder includes plastics and other materials.

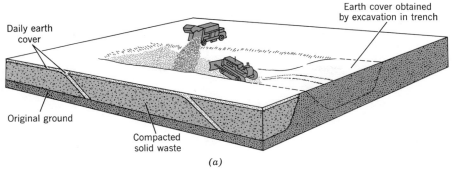

Earth cover obtained
by excavation in trench

Daily earth
cover

Original ground

Compacted
solid waste

(a)

Trench landfill method.

Figure 4-9
Sanitary landfill
methods. From "Sanitary Landfill: Alternative to Open
Dumping," *Environmental Science and
Technology* **6** (5), 408
(May 1972).

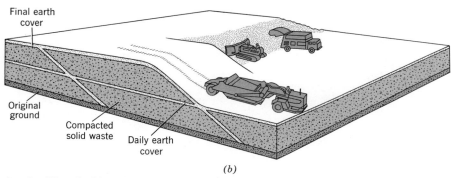

Final earth
cover

Original
ground

Compacted
solid waste   Daily earth
cover

(b)

Area landfill method (mounding rather than excavations).

Methods of salvaging junked automobiles involve stripping of the cars and shredding the hulks to yield scrap steel. The problems of reclaiming the metals in cars are (*1*) separating the various kinds of metals and (*2*) collecting the junked cars for processing.

4-16   INDUSTRIAL WASTES

Another category of solid waste includes **industrial** and **mining** wastes. Industrial wastes involve a wide variety of materials which differ from urban wastes. Often, an industry will utilize all economically valuable and usable wastes. Consequently, most industrial solid wastes are materials which are not usable by current technological methods. Most industrial wastes of this type are incinerated, carted to private dumps, or disposed of by ocean dumping. Since the United States utilizes great quantities of natural resources obtained through mining and refining, large amounts of solid wastes have accumulated from these operations. These wastes include mine tailings (residues and scraps), rejected material from washing plants, and slags from mining and processing. It is not uncommon for this material to be accumulated in vast piles near the mines and processing plants. Some of these wastes have potential economic value, while others have no foreseeable value.

4-17   AGRICULTURAL WASTES

Agricultural endeavors and animal products are still another source of solid wastes. Table 4-6 lists some specific sources of these wastes. The animal population of the United States produces about twenty times as much body wastes as

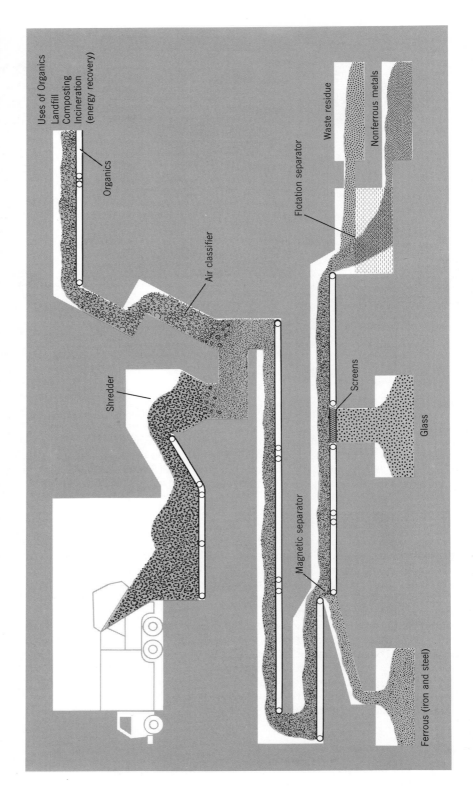

Figure 4-10
Separation and recycling of domestic trash. Possible scheme for the separation and recycling of domestic trash. The trash is shredded and subjected to an air stream which separates the lighter organic items (plastics, papers, and garbage) from the heavier items (metals, glass, etc.). The organics are collected and disposed of or put to some use as shown. The nonorganic items are subjected to a magnetic separator to separate the ferrous (iron-containing) scrap from the other materials. Passing these materials over a screen separates out the glass and finally a flotation separator is used to separate the nonferrous metals from lighter waste materials. Adapted from "Tackling Resource Recovery on a National Scale," by Richard L. Lesher, *Environmental Science and Technology* **6** (13) (Dec. 1972).

Livestock	Population (millions)	Annual production of solid wastes (million tons)	Annual production of liquid wastes (million tons)
Cattle	107	1,004.0	390.0
Horses[a]	3	17.5	4.4
Hogs	53	57.3	33.9
Sheep	26	11.8	7.1
Chickens	375	27.4	
Turkeys	104	19.0	
Ducks	11	1.6	
Totals		1,138.6	435.4

Table 4-6
Production of Wastes by Livestock in the United States, 1965

From "Cleaning Our Environment—The Chemical Basis for Action." American Chemical Society, 1969.
[a] Horses and mules on farms as work stock.

the human population. Unfortunately, not all of these wastes are disposed of properly. Land spreading is the common method of disposal. This presents health hazards in populated areas, uses valuable land, and is a potential source of water pollution. Animal wastes contain nitrates and phosphates which can be useful when used correctly (fertilizers) and can be detrimental when they end up in water sources. In some small-scale operations, animal wastes have been subjected to a bacterial decomposition, called **anaerobic** (in absence of oxygen) degradation, to produce methane, CH_4, gas. Methane can be used as a fuel.

Solid wastes are a dilemma of our highly industrialized and growing society. Solid waste disposal is costly, creates a cluttered landscape, uses valuable land, encourages the growth of insects and rodents, and causes some air and water pollution. We know that many solid wastes contain valuable resources. It is now necessary to view solid wastes as valuable and exert a greater effort to reclaim and recycle our wasted resources.

BOOKS

BIBLIOGRAPHY

American Chemical Society. *Solid Wastes*. Washington, D.C.: American Chemical Society, 1971. 96 pp.

Lee, D. H. K. *Metallic Contaminants and Human Health*. New York: Academic Press, 1972. 241 pp.

National Academy of Sciences. *Lead: Airborne Lead in Perspective*. Washington, D.C.: National Academy of Sciences, 1972. 330 pp.

U.S. Bureau of Mines. *Automobile Disposal: A National Problem*. Washington, D.C.: U.S. Government Printing Office, 1967. 569 pp.

ARTICLES AND PAMPHLETS

Aaronson, T. "Mercury in the Environment." *Environment* **13** (4), 16–23 (1971).

Bramer, H. C. "Pollution Control in the Steel Industry" *Environmental Science and Technology* (Oct. 1971).

Chisholm, J. J., Jr. "Lead Poisoning." *Scientific American* **224**, 15–23 (1971).

Goldwater, L. "Mercury in the Environment." *Scientific American* **224**, 15–21 (1971).

Grinstead, R. R. "Machinery for Trash Mining." *Environment* **14** (4) (1972).

Hall, S. "Lead Pollution and Poisoning," *Environmental Science and Technology* (Jan. 1972).

Hannon, B. M. "Bottles, Cans, Energy." *Environment* **14** (2) (1972).

Hershaft, A. "Solid Waste Treatment Technology." *Environmental Science and Technology* (May 1972).

Lesher, R. "Tackling Resource Recovery on a National Scale." *Environmental Science and Technology* (Dec. 1972).

Medeiros, R. W. "Lead from Automobile Exhaust." *Chemistry* (Nov. 1971).

Scientific American **223** (Sept. 1970). The entire issue is devoted to the biosphere and cycles of nature.

U.S. Department of Commerce. *Animal Wastes.* Staff Report, National Industrial Pollution Control Council, Washington, D.C.: U.S. Department of Commerce, 1971.

Woodwell, G. M. "Toxic Substances and Ecological Cycles." *Scientific American* **216,** 24–31 (1967).

QUESTIONS AND PROBLEMS

1. What are the four components of the ecosphere?

2. Which three elements are most abundant in the lithosphere?

3. Which two elements are most common to the hydrosphere?

4. Which two elements are most abundant in the atmosphere?

5. Which three elements make up most of the biosphere?

6. What is an ore?

7. Which minerals are used the most in an industrialized society?

8. Briefly describe the oxygen cycle by indicating the important compounds of oxygen involved in the cycle and the process by which oxygen is exchanged between the parts of the ecosphere.

9. Briefly describe the carbon cycle by indicating the important compounds of carbon involved in the cycle and the processes by which carbon is exchanged between the parts of the ecosphere.

10. How have human endeavors interfered with the carbon cycle?

11. Briefly describe the nitrogen cycle by indicating the important compounds of nitrogen involved in the cycle and the processes by which nitrogen is exchanged between the parts of the ecosphere.

12. How have human endeavors interfered with the nitrogen cycle?

13. Explain how iron is refined from iron ore.

14. Explain how aluminum metal is obtained from aluminum ore.

15. How many tons of aluminum are produced per person in the United States each year? (Assume about 200 million people in the United States.) About how much of this aluminum is recycled each year?

16. Which three heavy metals are of greatest environmental concern? Why?

17. What are the two main uses of mercury in the United States?

18. How does mercury become concentrated in certain fish?

19. What are the three main uses of lead in the United States?

20. What use of lead is of greatest environmental concern? Why?

21. A typical person breathes 15 cubic meters of air per day. If the average lead content of the air in an urban area was 2.3 micrograms per cubic meter, how much lead would a person breathe in each day. (Of course, a person may not breathe air with the lead content for 24 hours and much of the inhaled lead is exhaled.)

22. Why are some children in ghetto areas especially prone to lead poisoning?

23. What are the five major sources of solid wastes in the United States?

24. On the average, each person in the United States discards about five cans per week. Assuming that there are about 200 million people in the country, determine the number of cans discarded each year.

25. Make a list of all the objects you throw away in 1 week at your home. Include any trash generated for food preparation.

ENERGY
AND THE
ENVIRONMENT

5

5-1 ENERGY AND SOCIETY

The United States uses more than one-third of all the energy produced in the world. If current trends continue, the energy needs of the United States will more than double by the turn of the century, and the world needs may triple. You might wonder what this energy use means and what it has to do with the environmental dilemma. Energy, the capacity to do work, is fundamental to an industrialized society. In fact, industrialized systems rely on a continuous source of cheap energy: energy to produce electricity, energy for industry, energy for transportation, and energy for commercial and household use. An industrialized nation like the United States is said to be **energy-subsidized.** A growing economy, successful industries, agriculture, and commercial endeavors require the input of energy. As an example of this energy-subsidizing, consider a hamburger purchased from a fast-food service drive-in restaurant. The energy content or food value of the hamburger is about one-tenth of the energy used to get it from the farm to the customer. In a like manner modern agriculture is highly energy-subsidized by the input of energy into the manufacturing of farm machinery, pesticides and fertilizers, and irrigation practices. Added to this is the energy expended in the transportation of these materials to the farms and of the produce from the farms to urban areas.

Any society needs energy. A primitive society uses the energy of working humans and domestic animals. Historically, energy from sources other than humans and animals was not available in significant quantities until the waterwheel was developed around the fourth century BC. The development of the windmill around the twelfth century added to the available energy. In the seventeenth and eighteenth centuries, steam engines came into use. The use of steam as a supplement to the waterwheel and windmill gave impetus to the Industrial Revolution. Steam engines became the basic industrial energy source and a significant transportation source, giving rise to the development of railroads. Ultimately, water and steam turbines were developed to provide industrial energy and the generation of electricity. These, supplemented by the internal combustion engine as a mobile energy source, are fundamental to the industrial systems as they exist today.

5-2 ENERGY

Energy is simply defined as the capacity for doing work. When work is done, energy is expended. Table 5-1 lists some common forms of energy. Recall from Section 1-5 that energy can be stored and transformed from one form to another. For instance, the kinetic energy or energy of motion of an automobile comes from the mechanical energy output by the engine, which uses the chemical energy of burning gasoline. Before the sources and uses of energy are discussed, it is important to consider how to express amounts of energy. A truck can do more work than a car; a match gives off less heat energy than a fire. Just as amounts of mass can be measured in grams, kilograms, pounds, or tons, various terms have been established for the measurement of energy. We often associate energy with heat. An automobile engine becomes hot; a burning match is hot; an electric lightbulb becomes hot as it lights; a pan of water becomes hot on the stove. Heat is not the only form of energy, but since it is a familiar and easily observed form, it serves as a means to define a unit for the measurement of energy. When a sample of water is heated, it absorbs the heat energy which changes the temperature of the water. The amount of heat absorbed is directly related to the amount of water and the increase in the temperature of the water. It is possible to measure the amount of heat absorbed by measuring the increase in temperature of a given amount of water.

A unit of heat or energy commonly used is the **calorie (cal)**, which is defined as the amount of heat required to raise the temperature of 1 gram of water by 1°C. It would take about 75,000 calories to heat 1 quart of water from room temperature to the boiling point. The calorie is used to express amounts of energy in various forms other than heat, since that energy could be converted to heat.

Another unit of energy used in the United States is the **British thermal unit (BTU)**. The British thermal unit is defined as the amount of heat needed to raise the temperature of 1 pound of water by 1°F (1 pound of water is about 0.12 gallons). One British thermal unit is equivalent to 252 calories. Thus, about 300 British thermal units are needed to heat 1 quart of water from room temperature to the boiling point.

Physicists often use another unit of energy called the joule. The **joule** is the defined unit of energy in the metric system of measurement. The magnitude of the joule can be understood if you consider that you would expend approximately 2 joules of energy if you lifted a 1 pound object from the floor to waist height. One calorie is equivalent to 4.185 joules. Consequently, it requires about 320,000 joules to heat 1 quart of water to the boiling point.

Each of these three units of energy are used to express amounts of energy.

5-3 POWER

There is still another aspect to the expression of amounts of energy. When energy is being expended to do work, it is desirable to express the amount of energy utilized in a given amount of time. **Power** is the term used to refer to the amount of energy or work per unit time or the time rate of work. When work is performed at the rate of 1 joule per second, the power involved is 1 watt. A **watt** is defined as a joule per second. A 100 watt lightbulb uses 100 joules of energy each second it is burning. Many household appliances have the wattage require-

Table 5-1 Kinds and Forms of Energy		
Solar or radiant energy	Heat energy
Mechanical energy	Nuclear energy
Electrical energy	Kinetic energy—energy of motion
Chemical energy	Potential energy—energy of position

Table 5-2
Units of Energy
and Power

Energy
   Calorie (cal)—amount of energy required to raise temperature of 1 gram of water by 1°C:
      1 calorie = 4.185 joules
   British thermal unit (BTU)—amount of energy required to raise the temperature of 1 pound of water by 1°F:
      1 BTU = 252 calories
   Joule—SI (or metric) unit of energy:
      1 joule = 1 newton-meter;
      1 newton = 1 kilogram-meter per square second
Power
   Watt—SI (or metric) unit of power:
      1 watt = 1 joule per second
   Horsepower:
      1 horsepower = 746 watts
Electrical energy—sometimes expressed in units of kilowatt-hours

ments listed on them. In the United States, another unit of power is the horsepower—a term which is still used even though the name indicates that it arose in the days in which horses were a major source of power. One **horsepower** is equivalent to about 746 watts. This means, for instance, that a 100 horsepower automobile engine could expend 74,600 joules for each second of operation. Note that the difference between energy and power is that power is an expression of the amount of energy used per unit of time. Table 5-2 summarizes the various units of energy and power. An electric utility company charges for the amount of electrical energy used. As a consequence, amounts of electrical energy are often expressed in terms of kilowatt-hours. Your electric bill shows how many kilowatt-hours of electricity you used as recorded by your home electric meter. Incidentally, a study by the Federal Power Commission in 1972 revealed the following average rates: residential areas, 2.22 cents per kilowatt-hour; commercial areas, 2.08 cents per kilowatt-hour; industrial areas, 1.02 cents per kilowatt-hour. Senator Lee Metcalf of Montana claims that this study shows that poorer people living in congested areas where cost of service should be low pay three times as much per kilowatt-hour as industry and twice as much as suburban homeowners.

5-4 SOLAR ENERGY AND PHOTOSYNTHESIS

As a result of intricate cosmic phenomena occurring in the sun, this ball of fire at the center of our solar system is continuously giving off tremendous amounts of energy in the form of **radiant** or **solar energy.** As the earth rotates about the sun and revolves about its axis, various regions of the earth are subjected to solar energy. (See Figure 5-1.) Some of this solar energy (about 30 percent) is reflected back into space by the clouds and dust-like particles in the atmosphere. A certain portion of the solar radiation is absorbed by the atmosphere, increasing the air temperature. Likewise, the solar energy is absorbed by the rocks and soils of the lithosphere and the waters of the hydrosphere. This absorbed energy affects the movement of winds, weather conditions, and ocean currents. At night, in the absence of sunlight, the earth cools down and much of the daily dose of solar radiation is reradiated as heat to outer space. A most significant event which occurs during the daily dose of solar radiation is the process by which green plants of the earth capture a fraction (less than 0.1 percent) of the solar radiation. This, of course, is **photosynthesis.** (Refer to Section

Figure 5-1
Solar radiation striking
the Earth.

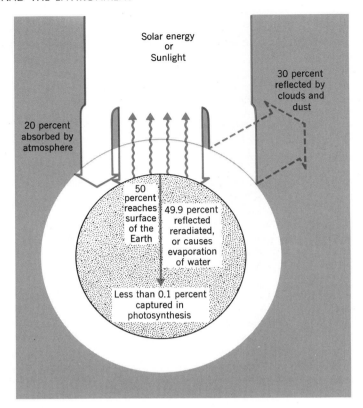

4-4 for a discussion of photosynthesis.) Photosynthesis takes place in the plants of forests, oceans, fresh waterways, deserts, and mountains of the earth, and in agricultural areas. Agriculture takes place on about 9 percent of the surface of the earth in which photosynthesis occurs. Recall that photosynthesis is capable of converting solar energy into chemical energy contained in plant molecules. These plant molecules, as potential sources of energy, serve as energy reservoirs or energy reserves. Such molecules or compounds made from them by further chemical reactions in plants can be eaten by man or animals for energy. Furthermore, wood as fuel is certainly an important energy source in the world.

5-5   HYDROCARBONS AND FOSSIL FUELS

Today our major source of energy for our industrial society comes from the combustion or burning of gasoline, oils, coal, and natural gas. These substances, originating from prehistoric plants and animals which became entrapped in the lithosphere by geological processes, are called **fossil fuels.** Fossil fuels are composed of hundreds of different carbon-containing compounds. Compounds of carbon are called **organic compounds.** There are so many organic compounds that an entire branch of chemistry called **organic chemistry** is devoted to their study. See Chapter 12 for a discussion of organic chemistry. The majority of the organic compounds found in fossil fuels contain only carbon and hydrogen. Such hydrogen and carbon compounds are called **hydrocarbons.**

Before discussing the various kinds of fossil fuels, let us consider some common hydrocarbons. Carbon forms four covalent bonds in organic compounds. However, carbon atoms have the ability to form chemical bonds with one another as well as the atoms of a variety of other elements. Because of this, a vast number of carbon-containing compounds are possible. Hydrocarbons range from those made up of molecules containing one carbon atom to those composed of tens or hundreds of carbon atoms bonded as carbon chains or as a ring of carbon atoms. A few examples will illustrate the variety of hydrocarbons.

Methane, the simplest of hydrocarbons, is composed of molecules containing one carbon bonded to four hydrogens (recall from Section 3-4 that lines in a formula represent covalent bonds or pairs of shared electrons):

$$
\begin{array}{c}
\text{H} \\
| \\
\text{H}-\text{C}-\text{H} \\
| \\
\text{H}
\end{array}
$$

Methane is usually represented by the molecular formula CH_4. Butane, C_4H_{10}, is a hydrocarbon composed of molecules which have four carbons bonded in a sequence or chain:

$$
\begin{array}{c}
\text{H}\quad\text{H}\quad\text{H}\quad\text{H} \\
|\quad\ |\quad\ |\quad\ | \\
\text{H}-\text{C}-\text{C}-\text{C}-\text{C}-\text{H} \\
|\quad\ |\quad\ |\quad\ | \\
\text{H}\quad\text{H}\quad\text{H}\quad\text{H}
\end{array}
$$

The hydrocarbon commonly called isooctane (2,2,4-trimethylpentane) can be represented by the formula

$$
\begin{array}{c}
\text{H}\qquad\qquad\text{H} \\
|\qquad\qquad\ | \\
\text{H}\quad\text{H}-\text{C}-\text{H}\quad\text{H}\quad\text{H}-\text{C}-\text{H}\quad\text{H} \\
|\qquad|\qquad\qquad|\qquad|\qquad\qquad| \\
\text{H}-\text{C}\text{——}\text{C}\text{——}\text{C}\text{——}\text{C}\text{——}\text{C}-\text{H} \\
|\qquad|\qquad\qquad|\qquad|\qquad\qquad| \\
\text{H}\quad\text{H}-\text{C}-\text{H}\quad\text{H}\qquad\text{H}\qquad\text{H} \\
|\\
\text{H}
\end{array}
$$

Note that isooctane involves a carbon chain with other carbons bonded to the carbons of this main chain. Such a bonding sequence is called branching and is a characteristic of **branched hydrocarbons** like isooctane.

Ethylene, C_2H_4, is composed of molecules involving two carbon atoms mutually sharing two pairs of electrons.

$$
\begin{array}{c}
\text{H}\qquad\quad\text{H} \\
\diagdown\qquad\diagup \\
\text{C}=\text{C} \\
\diagup\qquad\diagdown \\
\text{H}\qquad\quad\text{H}
\end{array}
$$

The sharing of two pairs of electrons in this manner is called a **double bond.** Any hydrocarbon which involves a carbon sequence including such a double bond is known as an **unsaturated hydrocarbon.** They are also known as **alkenes** or **olefins.**

An example of a hydrocarbon involving a carbon ring or cyclic carbon sequence is cyclohexane, C_6H_{12}. The ring can be represented by the formula

$$
\begin{array}{c}
\text{H}\ \ \text{H} \\
\diagdown\diagup \\
\text{C} \\
\diagup\qquad\diagdown \\
\text{H}\diagdown\text{C}\qquad\qquad\text{C}\diagup\text{H} \\
\text{H}\diagup\qquad\qquad\diagdown\text{H} \\
\\
\text{H}\diagdown\qquad\qquad\diagup\text{H} \\
\text{C}\qquad\qquad\text{C} \\
\text{H}\diagup\qquad\qquad\diagdown\text{H} \\
\diagdown\qquad\diagup \\
\text{C} \\
\diagup\diagdown \\
\text{H}\ \ \text{H}
\end{array}
$$

Any hydrocarbon made up of molecules composed of such carbon rings is called a **cyclic hydrocarbon.** Cyclic hydrocarbons can involve rings of three or more carbons.

One especially unique carbon ring sequence is that found in **benzene,** C_6H_6. Benzene molecules can be represented as

The hydrocarbons which include this cyclic unsaturated benzene sequence are collectively called **aromatic hydrocarbons.** Table 5-3 lists common hydrocarbons, some of which we make reference to in the following discussion.

Alkanes (aliphatics)		Alkenes (olefins)	
Methane	CH_4	Ethylene	C_2H_4
Ethane	C_2H_6	Propylene	C_3H_6
Propane	C_3H_8	Butylene	C_4H_8
Butane	C_4H_{10}		
Pentane	C_5H_{12}		
Hexane	C_6H_{14}		
Heptane	C_7H_{16}		
Octane	C_8H_{18}		
Aromatics		Cyclic hydrocarbons	
Benzene	C_6H_6	Cyclopentane	C_5H_{10}
Toluene	C_7H_8	Cyclohexane	C_6H_{12}
Xylene	C_8H_{10}		
Styrene	C_8H_8		
Naphthalene	$C_{10}H_8$		

5-6 PETROLEUM

Petroleum (Latin: *petra* = rock and *oleum* = oil) or **oil** is a complex liquid made up of numerous organic compounds. It is not known exactly how petroleum deposits came about in the lithosphere. It is thought that petroleum was formed millions of years ago when plants and animals living in shallow waters died and became buried in muddy sediments. Certain microorganisms caused these components of the biosphere to undergo partial decay. The sediment layers built up into sedimentary rocks. As a result of the forces of microorganism decay, and pressure and heat from geological phenomena, the plant and animal molecules were converted to petroleum and natural gas deposits. **Crude oil,** as it is extracted from underground deposits, is a viscous liquid. Crude oil is made up chiefly of hydrocarbon (94–99 percent) with some organic compounds containing sulfur, nitrogen, or oxygen. Typical crude oil contains over 500 different compounds. Crude oil composition varies depending upon the region of the lithosphere from which it is obtained. Of the hydrocarbons in crude oil, most are alkanes of varying structure, some are cyclic and aromatic compounds.

The **refining** of petroleum is the process in which the crude oil is separated

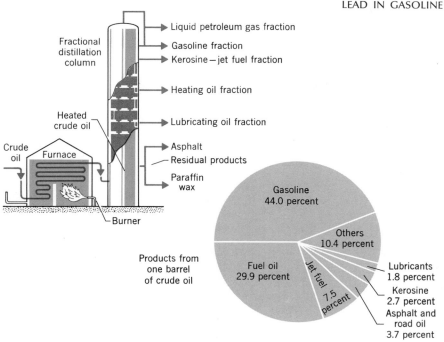

Figure 5-2
Petroleum refining
tower.

into useful components by distilling fractions of various boiling point ranges. The refining process is illustrated in Figure 5-2. The fraction of crude oil ranging in boiling point from 0 to 200°C is gasoline. The other fractions are kerosine and jet fuel (boiling point 175–275°C), fuel oil and diesel oil (boiling point 250–400°C), and lubricating oils (boiling point over 400°C). The solids separated from the liquids are paraffin wax and asphalt. Lubricating oils are purified and mixed with various additives before use.

5-7 LEAD IN GASOLINE

The **gasoline** fraction is, of course, used to make gasoline for motor vehicles. However, the gasoline has to be chemically altered to form usable gasoline. To be usable in an automobile engine, gasoline has to have the correct octane rating. **Octane rating** is related to the engine knock or ping in an internal combustion engine. Such knocking results from uneven burning of gasoline in the cylinders of the engine. Gasolines can be rated by comparing the knocking tendency to standard octane rated fuels. The hydrocarbon heptane, C_7H_{16}, is designated an octane rating of 0 and 2,2,4-trimethylpentane (isooctane) is designated an octane rating of 100. Standard fuels of various octane ratings can be made by mixing the two standards. A gasoline can be rated by comparing the knocking tendency to standard fuels in a special internal combustion engine. Octane rating of gasolines indicates that higher octane ratings result from the presence of more branched hydrocarbons, unsaturated hydrocarbons, and aromatic hydrocarbons in the gasoline. Gasoline octane rating can be increased by special additional refining processes called cracking, reforming, and alkylation. These processes are shown in Figure 5-3.

In 1923 it was found that the addition of **lead tetraethyl,** $Pb(C_2H_5)_4$, increased the octane rating of gasoline. Since then "ethyl" has become a common additive used in regular and premium gasoline. The organic lead compounds known as **lead alkyls** (lead tetraethyl and lead tetramethyl) retard the premature burning process in the internal combustion engine and, thus, are called **antiknock addi-**

Figure 5-3
Methods of increasing
the octane rating of
gasoline.

Cracking — long-chain, high-boiling-point hydrocarbons are heated in the presence of hydrogen and a catalyst to form shorter-chain gasoline hydrocarbons. Example:

$$C_{15}H_{32} + H_2 \xrightarrow[\text{heat}]{\text{catalyst}} C_8H_{18} + C_7H_{16}$$
pentadecane   octane   heptane

Reforming — normal and cyclic hydrocarbons are heated in the presence of a catalyst to form aromatic hydrocarbons. Example:

$$C_6H_{12} \xrightarrow[\text{heat}]{\text{catalyst}} C_6H_6 + 3 H_2$$
cyclohexane   benzene

$$C_7H_{16} \xrightarrow[\text{heat}]{\text{catalyst}} C_6H_5CH_3 + 3 H_2$$
heptane   toluene

Alkylation — butane and butene hydrocarbons are reacted to form octanes. Example:

$$C_4H_{10} + C_4H_8 \xrightarrow{H_2SO_4} C_8H_{18}$$
butane   butene   octane

tives. Commercial antiknock additives consist of about 60 percent lead tetraethyl (or methyl) and 12 percent ethylene bromide, 25 percent ethylene chloride, and 3 percent kerosine and dye. The ethylene compounds are called scavengers and function by reacting with the lead to form inorganic lead compounds which leave the engine in the exhaust gases. High-octane, leaded gasolines contain about 3 grams of lead (as lead tetraethyl or methyl) per gallon. **Low-lead** gasolines (most are 91 octane) contain about 0.5 gram of lead per gallon. As mentioned in Section 4-12, tremendous amounts of lead are introduced into the environment from the use of lead alkyls in gasoline. This is easily seen if we consider the fact that nearly 100 billion gallons of gasoline are burned each year in the United States. According to a U.S. Environmental Protection Agency (EPA) regulation, most service stations must offer a 91 octane lead-free gasoline by mid-1974. Another pending regulation will require the phasing out of all lead in gasoline. High-octane lead-free gasolines require higher proportions of aromatic and unsaturated hydrocarbons. Such gasolines will be more expensive and will require the building of new refining facilities by petroleum companies.

5-8 NATURAL GAS AND COAL

Natural gas consists of gaseous hydrocarbons produced from fossil fuels which have accumulated in pockets in the lithosphere. Natural gas is withdrawn from gas wells, processed, and piped long distances for use as fuel. Natural gas varies in composition, but is made up chiefly of methane gas, CH_4, with some ethane, C_2H_6, propane, C_3H_8, and butane, C_4H_{10}. Before natural gas is used, most of the hydrocarbons other than methane are removed for use in the manufacture of other organic chemicals. The production of synthetic natural gas is discussed in Section 5-14.

Coal is solidified plant material which has been deposited in rock layers, has undergone partial decay, and has been subjected to geological heat and pressure. Most coal is thought to have been derived from peat bogs. In fact, the various forms of coal, lignite, bituminous (soft) coal, and anthracite (hard) coal are ancient peat in various stages of decay and compaction. Coal contains carbon and a variety of hydrocarbons. Coal can be burned in air and used as a fuel. However, coal does contain some compounds containing sulfur, oxygen, and nitrogen. When soft coal is heated in the absence of air, hydrocarbon substances are given off, leaving a residue of impure carbon called **coke.** Vast amounts of coke are used in the manufacture of steel. The hydrocarbon substances given off during the coking of coal are cooled down and separated into a fraction which condenses to a liquid, called **coal tar,** and a fraction which remains gaseous, called **coal gas.** Coal gas consists of hydrogen, H_2, and methane, CH_4, along with a variety of other hydrocarbons and gases. Coal gas is used as a fuel. Coal tars contain a wide variety of organic compounds used in the manufacture of other organic compounds.

5-9   ENERGY USE AND ENERGY SOURCES

Each year the United States uses about 70 quadrillion (70×10^{15}) British thermal units of energy. This amounts to around 2.4 trillion (2.4×10^{12}) watts. As an indication of how much energy this is, consider each of the over 200 million Americans burning 120 100 watt lightbulbs for 24 hours each day every day of the year. The energy consumption per person by a country is directly related to the level of industrialization and affluence. The United States, Canada, and Western European countries have high per capita energy consumption, and underdeveloped countries, such as India and China, have very low energy consumption. In fact, as mentioned previously, the United States, with about 6 percent of the people in the world, consumes over one-third of all the energy used throughout the world each year. The use of energy in the United States is increasing each year, as can be seen from Figure 5-4, which shows the increase in energy use over a period of 100 years along with the per capita increase. The per capita energy use of the world is increasing at a rate somewhat greater than that of the United States.

Figure 5-5 shows the sources and uses of energy in the United States. As can be seen, fossil fuels provide over 95 percent of the energy. Coal provides 20 per-

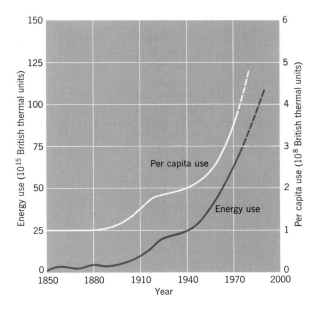

Figure 5-4
The increase in energy use and the per capita increase in the United States.

Figure 5-5
Energy sources and
energy uses in the
United States.

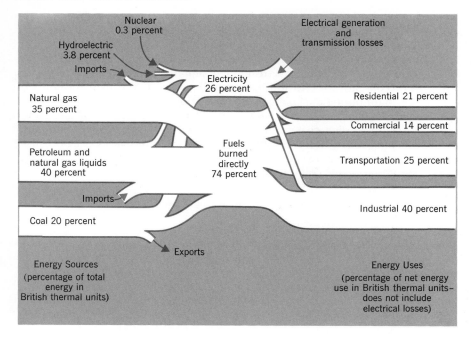

cent of our energy. Somewhat more than half of the coal is used to produce electricity, and the remainder is used by industry. Natural gas accounts for about 35 percent of our energy. Only about one-sixth of this gas is used to produce electricity, and the rest is used directly as a heating fuel. Petroleum and natural gas liquids (butane and propane) contribute about 40 percent of our energy. Of this, only about one-tenth is used to produce electricity, and the rest is used directly as fuel. A small portion of our energy comes from hydroelectric sources (around 3.8 percent) and nuclear fuels (around 0.3 percent). Both of these sources are used entirely in the production of electricity.

Overall, about 26 percent of our energy sources is used to produce electricity, and 74 percent is consumed directly—mainly as fuel, but some in nonenergy uses such as lubricating oils, coke (from coal), and petrochemical products. Consumption of electricity is growing at a faster rate than any other use of energy. Consequently, it is expected that an increasing fraction of our energy sources will be used to produce electricity. However, it is thought that nuclear energy sources will provide much of the energy for electricity in the future. See Chapter 6 for a discussion of nuclear energy.

Referring to Figure 5-5, let us now consider the uses of energy in the United States. Industry uses over 40 percent of the energy in producing consumer goods. The largest fraction of this energy is used in blast furnaces and metal refining to produce metals such as iron and aluminum. Other large industrial energy consumers include oil refineries, mines, chemical plants, glass factories, food processing plants, and paper plants. Commercial establishments (stores, offices, hotels, etc.) account for about 14 percent of energy use. Half of the energy used commercially is for space heating (heaters) and air conditioning. Transportation uses 25 percent of our energy. Most of this comes from the burning of gasoline and diesel oil in cars, buses, and trucks. Trains, ships, and airplanes are additional energy-consuming transportation modes.

Domestic uses consume about 20 percent of our energy. Around half of this energy goes for space heating using natural gas and oil heaters. Other domestic uses include electricity, cooking, and the heating of water. The rapid growth in

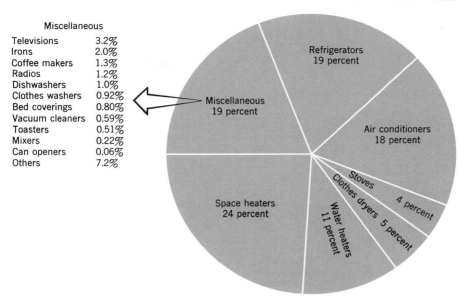

Miscellaneous

Televisions	3.2%
Irons	2.0%
Coffee makers	1.3%
Radios	1.2%
Dishwashers	1.0%
Clothes washers	0.92%
Bed coverings	0.80%
Vacuum cleaners	0.59%
Toasters	0.51%
Mixers	0.22%
Can openers	0.06%
Others	7.2%

Figure 5-6
Contributors to residential use of electricity (average electrical uses in American households).

residential energy use which has taken place over the last few decades is due to increased use of electricity. Increased electricity consumption has resulted from the increasing number of home appliances. The various contributors to the use of residential electricity are shown in Figure 5-6.

5-10 LIMITATIONS OF FOSSIL FUELS

Currently, our industrial society is totally dependent upon fossil fuels. Fossil fuels were deposited in the earth over a period of some 600 million years. Man is extracting these fuels at an increasing rate. There are limits to the amounts of fuels which can be extracted. In other words, the fossil fuels are nonrenewable resources, since they are formed very slowly, and once they are used they can never be used again.

The problems are: "When will the fossil fuel supply run out?" and "What other sources of energy are available?" There is a time in the not too distant future when new sources of fuels will be harder to find and harder to extract from the earth. For each fossil fuel there is a peak of maximum use which will drop off as the fuel becomes more difficult to find and extract. It is not easy to predict the exact time at which a fuel will become so difficult to extract that it will be of no more use. Nevertheless, there is such a time for each of the fossil fuels. Actually, it is possible to determine an approximate lifetime for the fossil fuels based upon technological and geological knowledge. Figure 5-7 shows the projected expected production of coal and oil in the world from the beginning of fossil fuel use to the time at which supplies will be depleted. Coal and lignite, which make up nearly 90 percent of the fossil fuel supply, will reach peak production around the year 2100 and will diminish in supply around 2400–2500. Coal is our most plentiful fossil fuel and will find more use in the future. Unfortunately, it is difficult to mine and transport, and it is the most polluting of the fossil fuels. (See Section 8-14 concerning air pollution from coal.) Petroleum is expected to reach peak production around the year 2000 and to diminish in supply by 2050. Currently, somewhat less than 30 percent of our petroleum is imported. However, the majority of the untapped petroleum sources of the world are outside the United States. Thus, it is expected that by 1985 57 percent of our petroleum

Figure 5-7
Adapted from
Resources and Man,
by Committee on
Resources and Man,
National Academy of
Sciences–National Re-
search Council, W. H.
Freeman, San Fran-
cisco, 1969.

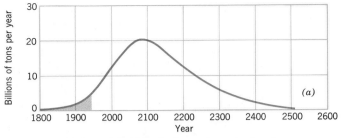

Estimated world production and supply of coal.

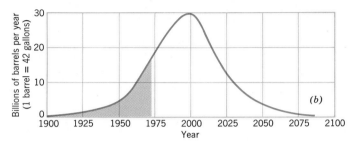

Estimated world production and supply of petroleum.

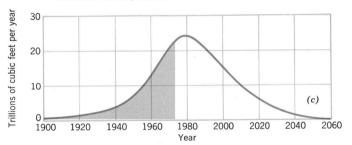

Estimated production and supply of natural gas in the United States.

needs will be imported. This petroleum (mostly from North Africa and the Near East) will constitute one-fourth of our total energy consumption. In the United States, natural gas supplies are thought to be currently near peak production and may diminish in supply by 2000. (See Figure 5-7.) Natural gas is our most limited fossil fuel. However, since it is so useful, additional sources are being sought, and in the future synthetic natural gas (see Section 5-14) may be used.

Although the demand for energy will undoubtedly increase in the future, it may be possible to institute certain **energy conservation** measures to help control the rate of growth. Several methods of energy conservation have been suggested. Better thermal insulation of homes and buildings could save significant amounts of energy used for space heating. It may be possible to develop energy-conserving architectural designs for new buildings. The use of gas instead of electricity for home and commercial heating would be more efficient. The use of solar energy for heating of homes is being investigated. More efficient air conditioners are possible, and reduced commercial lighting levels are feasible. Railroad freight transportation is energetically less costly than trucking. Likewise, intercity transportation by rail or bus is more efficient than air travel. Mass transit systems can save energy when compared to individual automobile use. More efficient automobile engines are feasible, but the increased use of smaller cars with greater fuel economy would be the easiest way to conserve fuels. Considering the fact that over half of our automobile trips are less than 5 miles, more walking or bicycling could save some energy. Finally, it is possible to design more efficient industrial processes, and recycling of metals and glass might help.

5-11 COMBUSTION AND ELECTRICAL POWER

Since the combustion of fossil fuels is our main source of energy, let us look at combustion more closely. **Combustion** refers to the reaction of an organic compound (hydrocarbon) with oxygen gas of the atmosphere to produce carbon dioxide and water. The chemical energies of carbon dioxide and oxygen are less than the chemical energies of the organic compound and oxygen. Thus, various amounts of energy are released in combustion, depending upon the organic compound involved. Combustion is exothermic, and if it is carried out in a controlled manner, useful work can be done by the energy released.

Let us consider a few typical combustion reactions. The combustion of natural gas involves the burning of methane, CH_4, as represented by the chemical equation

$$CH_4 + 2\ O_2 \longrightarrow CO_2 + 2\ H_2O + energy$$

Energy is denoted as a product of the reaction in order to emphasize that the reaction is exothermic. Note that the equation is balanced (see Section 3-12). Gasoline is a complex mixture of hydrocarbons. Consequently, combustion of gasoline involves a variety of reactions. As an example, consider the combustion of octane, C_8H_{18}:

$$2\ C_8H_{18} + 25\ O_2 \longrightarrow 16\ CO_2 + 18\ H_2O + energy$$

As with gasoline, the combustion of jet fuel, diesel fuel, fuel oil, and coal involve a variety of reactions, all of which are exothermic. The combustion of coke or the carbon in coal can be represented by the equation

$$C + O_2 \longrightarrow CO_2 + energy$$

This simple exothermic reaction is essentially the same reaction that occurs when charcoal (coked wood) is burned.

The products of combustion are ideally carbon dioxide and water. These products are released as hot gases into the atmosphere. Unfortunately, as a result of **incomplete combustion** and the combustion of compounds found in fossil fuels as impurities, other gases are released into the atmosphere. Some of these become air pollutants. See Chapter 8 for a discussion of air pollution.

About 26 percent of the combustion takes place in transportation vehicles, about 21 percent for household and commercial purposes, about 30 percent by industrial endeavors, and about 23 percent for production of electricity. A portion of our electricity (around 17 percent) is produced by hydroelectric and nuclear sources. The major source of electricity comes from combustion in the numerous electrical generating plants located across the United States. The main components of a **steam–electric power** plant are shown in Figure 5-8. The fossil fuel is burned to produce heat in a **boiler** which converts water to steam which is heated to a high temperature (around 550°C). This steam jets into a large **turbine** which turns a shaft attached to the electricity **generator.** The rotating shaft of the generator turns the rotor, a huge electromagnet. The spinning of the rotor causes electrical current to be generated. The current is sent to a transformer which increases the voltage for cross-country transmission to substations where the voltage is reduced for use. The steam which turns the turbine is cooled down to reform water by passing it through a heat exchanger cooled by vast amounts of **cooling water.** The fossil fuel steam plants produce some air

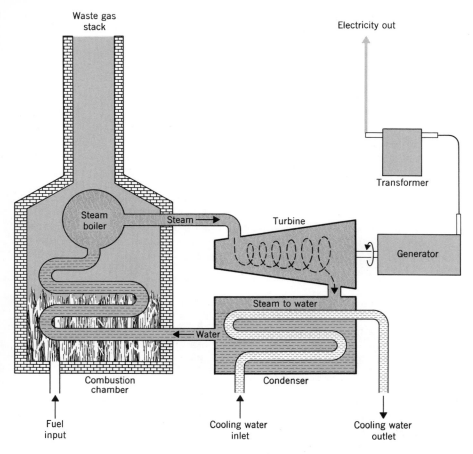

Figure 5-8
Steam–electric power
plant.

pollutants in the combustion process. These pollutants are dispersed into the atmosphere through the waste gas stack.

It is estimated that future electric power demands in the United States will double every 10 years. As shown in Figure 5-9, most of this will be steam-generated. However, as can be seen in the figure, it is projected that by the turn of the century over 50 percent of the energy will come from nuclear-fueled steam plants. (See Section 6-11 for a discussion of nuclear power plants.) The rapid growth in the demand for electricity is expected to produce environmental dilemmas in many parts of the country. For instance, the five major utility companies in California have projected electrical power needs of the state up to the 1990s. If these projections are accurate, California will need 130 new power plants of 1,200 Megawatt capacity between now and the year 2000. If all these plants are located on the coast to use ocean water for cooling, there will be one plant every 8 miles along the coastline. Fortunately for California, projections by independent researchers (Rand Corporation) indicate that power demands can be curtailed and that other sources, such as geothermal energy, may limit the number of new power plants needed.

5-12 THERMAL POLLUTION

The ultimate form which energy takes when used by man is heat. Heat is given off by electrical generating plants and by the machines, appliances, and lights in

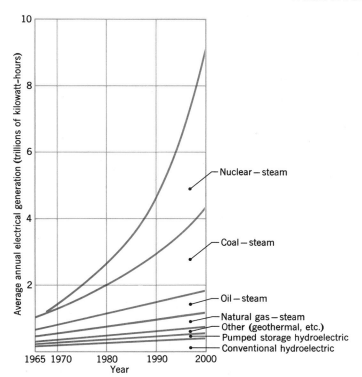

which the electricity is used. Heat is given off in industrial endeavors, heating of buildings, and in the hot exhaust and hot engines of motor vehicles. The production of waste heat is an unavoidable consequence of energy use.

In any engine designed to convert heat into work, a portion of the heat is converted to work, but the rest is rejected by the engine as waste heat. This fact is always true no matter what kind of engine is used. This is not a violation of the Law of Conservation of Energy (see Section 1-6), but it is a fact involved in energy conversion. This fact is related to a physical law called the **Second Law of Thermodynamics.** According to this law, when a heat engine (e.g., a steam turbine) using a heated fluid (steam) is used to perform work (turn a generator), only a certain portion of the energy of the fluid is converted to work and the rest of the energy is given off as waste heat (hot water and hot engine parts). Thus, some heat energy is always wasted and cannot be converted to work. This waste of energy is called **inefficiency.** For example, in a typical electrical power generating plant, only about 40 percent of the energy obtained from the combustion of fossil fuels is converted to electricity. Such an energy conversion is said to be 40 percent efficient. Most of the wasted heat energy is absorbed by the cooling water used by the generating plant. Typically, the temperature of cooling water is increased by about 11°C (20°F). Once this cooling water passes through the generating plant, it normally is released into the environment from which it was obtained. When this warm water disrupts the animal and plant life around the generating plant, it is called **thermal pollution.** The increased temperature of the water decreases the amount of dissolved oxygen in the water. This can affect the organisms and animal life in the water. The increased temperature tends to stimulate the animals to be more active. Greater activity increases the need for oxygen which has been diminished by the higher temperature.

Thermal pollution is potentially a serious problem, but studies of numerous power plants have revealed that at the current levels of waste heat production no significant ecological disruptions are apparent. In fact, in certain cases the warmer waters have stimulated plant and animal growth. However, the long-term effects of the warmed water on plant and animal life are not known.

Currently, around 40 percent of all the water used in the United States is steam–electric cooling water. By the year 2000 it is expected that 60 percent of the water will be used for this purpose. Thermal pollution could become a serious ecological problem as greater amounts of waste heat are produced by larger power plants. An alternative to such thermal pollution is to cool the water in cooling ponds or cooling towers before returning it to the source. This results in evaporation of some water, which utilizes the waste heat. However, this approach can greatly increase the amount of water vapor in the air, which could have an effect on the local weather conditions.

It is important to keep close watch on the thermal effects of cooling waters around large generating plants. Furthermore, more research is needed to determine the maximum amount of waste heat that can be released in specific aquatic environments.

5-13 THE ENERGY CRISIS

National concern has developed over what has been termed the **"energy crisis."** The problem revolves around the diminishing supply of domestic energy sources and the increasing demand for energy. The problem of supply and demand has not reached a critical point at this time, with the possible exception of natural gas. However, a potential crisis is predicted, based on the demand for energy, which is rising at a rate of over 4 percent each year. This rate of increase is more than twice the rate of the population growth in the United States.

Some experts indicate that a national energy policy is needed to define future energy needs and to establish means of reaching these needs. Perhaps on the federal level a Department of National Resources, a Council on Energy Policy, and a Federal Energy Commission should be established.

Three possible approaches to the energy dilemma are apparent: (1) Energy demand can be curtailed by increased energy conservation practices and by enforced limitation of energy use. (2) Energy sources needed to meet energy demands can be imported. It is estimated that by 1985 about 30 percent of our energy sources will be imported. (3) Domestic energy sources can be developed to meet energy needs. Exploitation of domestic resources may include the large-scale use of public lands for strip mining of coal, the mining of uranium ores, and the extraction of oil and natural gas. Furthermore, more off-shore oil leases will be needed, and wildlife sanctuaries may be encroached upon. Alaskan oil and gas reserves will have to be rapidly developed, and numerous fossil fuel and nuclear power plants will be needed.

The national approach to energy use and future energy sources will have profound political, social, economic, and environmental manifestations in the years to come. In fact, decisions concerning energy use and sources will fundamentally affect the environmental dilemma.

5-14 FUTURE ENERGY SOURCES

Let us now discuss the possibilities of some **future energy sources.** (See Figure 5-10.) Fossil fuels will remain the major energy sources throughout this century. By 1985, some 57 percent of our petroleum and 28 percent of our natural gas may have to be imported. It is expected that in 1985 around 8 percent of imported natural gas will be shipped as **liquid natural gas (LNG)** in special tankers. Domestic coal production may double between now and 1985. More efficient use of fossil fuels is being developed, including new combustion methods and

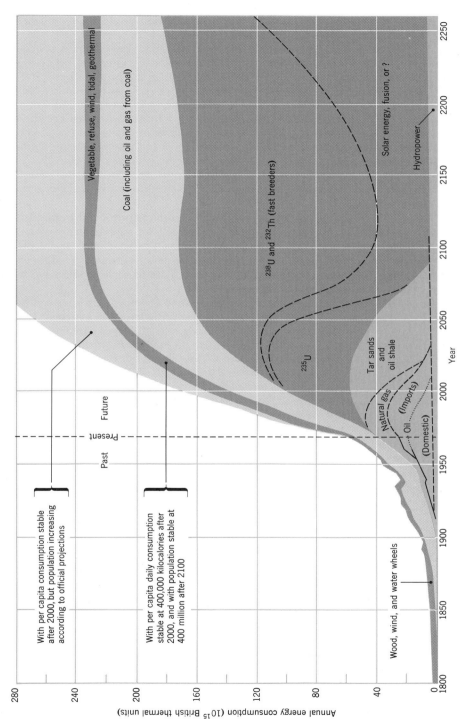

Figure 5-10
Possible future energy sources in the United States. From Earl Cook, *Chemical and Engineering News* 28 (Jan. 10, 1972).

antipollution techniques. One promising approach which is now being re-searched is the use of **magnetohydrodynamic (MHD)** generators that convert heat from combustion reactions directly into electricity. Magnetohydrodynamics appears to be a low-pollution alternative to the generation of electricity and has the added advantage of using coal as a fuel.

In the late 1970s it is expected that diminishing natural gas supplies will be supplemented by **synthetic natural gas (SNG)** prepared by chemical processes from petroleum or coal. Figure 5-11 gives a schematic of the production of SNG from petroleum or coal. As can be seen, petroleum or coal gasification involves mixing these carbon-containing substances with steam and converting to a mixture of carbon monoxide (CO) and hydrogen (H_2). These two gases are then combined in the presence of a special catalyst to form methane gas (SNG) and water. One attractive feature of SNG production is that sulfur-containing compounds responsible for some air pollution problems can be removed (this is termed **desulfurization**). Petroleum gasification is most easily accomplished and will be the first source of SNG. However, coal gasification is a promising long-term source of SNG.

Other long-term sources of oil are petroleum products produced from coal, tar sands, and oil shale. Large **oil shale** (shale rock formations impregnated with petroleum) deposits are located in the western United States. These deposits represent a large supply of oil, but the petroleum products are difficult to remove from the shale. Furthermore, large amounts of shale would have to be processed and the spent shale disposed of after use.

Hydroelectric power sources are limited and, since few new dam sites can be developed, this will be a source of decreasing importance. It is expected that hydroelectric power will decrease from the current 15 percent of electricity generated in the United States to around 8 percent in 1985. **Geothermal** sources are now beginning to be exploited. Certain regions of the earth (especially those of recent volcanic activity) have geological formations in which ground water comes into contact with hot rocks and produces hot water or steam which rises to the surface. In certain cases, the hot water or steam can be contained and used to turn a turbine to generate electricity. Geothermal sources occur in many regions of the western United States and can be used to supplement the energy needs of certain western states. However, geothermal energy is not expected to contribute much to the total energy needs of the nation.

Other potential energy sources include tides and winds. **Tidal energy,** which uses the tidal forces to generate power, and **wind energy,** obtained from wind-

Figure 5-11
Schematic of synthetic natural gas production.

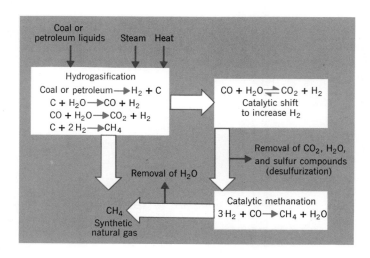

mills, could be used in certain regions of the country. They are not expected to contribute significantly to the national energy needs.

It is expected that **fission energy** will supply around half of our electricity by 1985 if development of nuclear power plants continues without delay. The continued use of fission, however, will depend upon the development of commercial breeder reactors.

Two possible long-range future energy sources are nuclear fusion and solar energy conversion. **Nuclear fusion** is promising but is just now in the basic research stage, and neither the technological nor commercial feasibility has been demonstrated. See Section 6-10 for a discussion of fusion.

5-15 SOLAR ENERGY

Solar energy represents a large energy source. In fact, the solar energy incident on about 0.1 percent of the land area of the United States would meet all our current energy needs. Some experts feel that solar energy will have to be our future energy source. It is a continuous, environmentally sound source. However, sunlight is diffuse, is interrupted by night, and depends upon weather conditions. Use of solar energy will require large land areas for collection and some means of energy storage.

The methods of capturing solar energy and converting it to other energy forms are currently being researched. One method uses **photovoltaic cells** to convert sunlight to electricity. Solar conversion cells seem to work well to supply energy for spacecraft, but the technology of solar conversion is not developed to the extent that large amounts of energy could be produced by this means. Nevertheless, photovoltaic conversion remains a long-range possibility which would involve energy plants in which several square kilometers of land area would be covered with solar cells. Other methods of solar energy conversion are being investigated which involve the use of mirrors or lenses to concentrate the sunlight. The concentrated sunlight would then be used to heat circulating fluids which could then be used to generate electricity.

Even though solar conversion as a large-scale energy source appears to be unlikely in the near future, some scientists claim that solar energy for domestic use will be commercially available by 1980. This will involve the use of **solar energy collection systems** on the roofs of homes. The captured solar energy can then be used for heating and cooling of the homes. Of course, this source of energy could only be used in certain regions of the country and would have to be backed up by normal fossil fuel supplies.

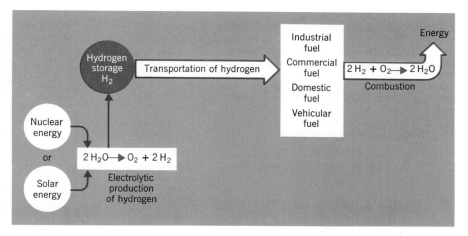

Figure 5-12
The use of hydrogen gas as an energy source.

One of the problems of obtaining energy from solar sources or from large-scale nuclear sources is storing energy. That is, since solar sources are intermittent and nuclear sources generate electricity, some means of storing energy for future use would be needed. Several **energy storage** methods are possible, but one involving the use of hydrogen gas is being considered as a very likely possibility. As shown in Figure 5-12, the idea is to use electrical energy to decompose water into hydrogen and oxygen. The hydrogen then would be stored or transported through gas lines, such as existing natural gas lines, for use. Hydrogen is an excellent and environmentally sound fuel. Combustion of hydrogen produces only water as a product. Hydrogen gas could be used as a source of industrial energy as well as a fuel for domestic use. It is even possible to use hydrogen as a fuel for automobiles. Hydrogen is highly explosive, but with proper handling it should be no more dangerous than natural gas or gasoline.

BIBLIOGRAPHY   BOOKS

Fabricant, N., and R. M. Hallman. *Toward a Rational Power Policy: Energy, Politics, and Pollution.* New York: George Braziller, 1971. 292 pp.
Scientific American. *Energy and Power: A Scientific American Book.* San Francisco: W. H. Freeman, 1971. 144 pp.

ARTICLES AND PAMPHLETS

Clark, J. R. "Thermal Pollution and Aquatic Life." *Scientific American* **220,** 19 (May 1969).
Lessing, L. P. "Coal." *Scientific American* **195,** 58 (Oct. 1956).
Levin, A., et al. "Thermal Discharges from Electrical Utilities." *Environmental Science and Technology* (Mar. 1972).
Lincoln, G. "Energy Conservation." *Science* **180,** 155 (1973).
Makhijani, A. B., and A. J. Lichtenberg. "Energy and Well Being." *Environment* **14** (5) (June 1972).
Stein, R. G. "A Matter of Design." *Environment* **14** (8) (1972).
Williams, R. H. "When the Well Runs Dry." *Environment* **14** (5) (1972).

QUESTIONS AND PROBLEMS

1. Give a brief definition or description of energy.

2. Define the following energy terms:
    (a) calorie   (b) BTU

3. What is power?

4. What are the three kinds of fossil fuels?

5. Describe the following terms:

    (a) hydrocarbon   (d) cyclic hydrocarbon
    (b) branched hydrocarbon   (e) aromatic hydrocarbon
    (c) unsaturated hydrocarbon

6. What is the meaning of the octane rating of gasoline?

7. Why is lead tetraethyl added to gasoline?

8. What is the chief component of natural gas?

9. The United States uses 70×10^{15} British thermal units of energy each year. Assuming that there are about 2.0×10^8 people in this country, (*a*) determine the amount of energy used per person each year and (*b*) determine the amount of energy used per person each day of a year.

10. What portion of the energy produced in the world each year is used by the United States? If the United States uses 70×10^{15} British thermal units of energy each year, how much energy is used in the world each year?

11. Considering that 70×10^{15} British thermal units of energy are used in the United States each year and the information below, determine the number of tons of coal, the number of barrels of oil, and the number of cubic feet of natural gas used in the United States each year.

    1 ton coal $= 26 \times 10^6$ British thermal units
    1 barrel oil $= 5.8 \times 10^6$ British thermal units
    1 cubic foot of natural gas $= 1,000$ British thermal units

    Coal provides 20 percent of the energy in the United States
    Petroleum provides 40 percent of the energy in the United States
    Natural gas provides 35 percent of the energy in the United States

12. What are the four major uses of energy in the United States?

13. List some possible ways in which energy can be conserved.

14. What is combustion?

15. Describe a steam–electric power plant.

16. What is thermal pollution?

17. What is the "energy crisis"?

18. List some possible future energy sources.

19. Why is hydrogen a good fuel and why might it be used extensively in the future?

20. The decision concerning the location or siting of new electrical power plants in the United States is generally made at the county or regional level. Some persons feel that this decision should be made at the state or federal level. What is your opinion? Why?

21. Some utility companies charge lower rates for electricity to industries and commercial establishments that use larger amounts of electricity. What is your opinion of this practice? Why?

22. Some utility companies advertise in a way which is designed to encourage use of more electricity. Do you think such advertising should be controlled? Why?

23. Roughly 100 billion gallons of gasoline are used in the United States each year in approximately 100 million gasoline-powered vehicles. Using this information, deduce the following:
    (*a*) How much gasoline is used in an average vehicle each year? How much per day?
    (*b*) Assuming an average gasoline mileage of 20 miles per gallon, how many miles are driven in the United States each year in gasoline-powered vehicles?
    (*c*) Assuming a typical lead content of gasoline of 1.5 grams per gallon

of gasoline, how many grams of lead are released to the environment each year?

(d) If smog control devices on cars decrease gas mileage from 20 to 15 miles per gallon, how will this affect the amount of gasoline used each year?

(e) It is estimated that the Alaskan oil deposit will provide around 420 billion gallons of crude oil. If all the oil is used to make gasoline, how long would this gasoline last in the United States?

(f) If each vehicle in the United States was not used one day each week, how many gallons of gasoline would be saved in 1 year?

6-1 NUCLEAR CHEMISTRY AND
NUCLEAR ENERGY

The structure and properties of the nuclei of atoms is the concern of nuclear chemists and nuclear physicists. The utilization of atomic nuclei in constructive and potentially destructive manners is important to all people. The same nuclear energy that is involved in nuclear bombs may become, through the use of nuclear reactors, the main source of useful energy available to mankind. Furthermore, the same nuclear radiation that can be destructive to human life is utilized as an effective medical tool.

This chapter serves as an introduction to the field of nuclear science. The structure and properties of atomic nuclei will be discussed. The use of nuclear reactors as energy sources will be considered.

6-2 ATOMIC NUCLEI

Theories of nuclear structure are not as well-established as the theory of atomic structure. Nuclear structure is a very active area of research by nuclear scientists. For our purposes we can consider that atomic nuclei are aggregates of the nuclear particles, **protons** and **neutrons.** The proton and the neutron are called **nucleons.** Other particles, such as the **electron,** the **alpha particle,** the **neutrino,** and the **gamma photon,** are associated with the properties of nuclei. The symbols and properties of all these nuclear particles are summarized in Table 6-1.

Atomic nuclei are, of course, the very massive centers of atoms about which the electrons are in motion. When discussing the properties of the nuclei we will consider them without consideration of the electrons. However, keep in mind that, with the exception of few important cases, nuclei are always parts of atoms and not separate particles.

Recall from Chapter 2 that the nuclei of atoms of a given element always have the same number of protons and that this number of protons is the **atomic number** of the element. For most elements there is more than one possible combination of neutrons and protons that can make up the nuclei of the atoms. However, only certain specific combinations are observed. Atoms of a given element must contain the same number of protons, but when different numbers of neutrons are present the species are called **isotopes** of that element. The nuclei

Table 6-1
Nuclear Particles

Name	Mass (amu)	Charge	Symbol
Proton	1.007825	+1	$_1^1H$, p
Neutron	1.008665	0	$_0^1n$, n
Electron	0.000549	−1	$_{-1}^0e$, e
Alpha particle	4.00260	+2	$_2^4He$, α
Gamma particle (photon)	0	0	γ, $h\nu$
Neutrino	0	0	ν

From *Introduction to Chemistry* by T. R. Dickson, John Wiley & Sons, New York, 1971.

of the various isotopes of the elements are called **nuclides.** Numerous nuclides occur in nature, and others can be produced synthetically. (See Section 6-7.)

Special symbolism is used to represent a nuclide. The number of protons in a nuclide is the atomic number. The **mass number** of a nuclide is defined as the sum of the number of protons and neutrons. A nuclide is represented as

$$_Z^MX$$

where X is the symbol of the element corresponding to the nuclide, M is the mass number, and Z is the atomic number. (The number of neutrons in a nuclide can be determined by subtracting the atomic number from the mass number; $M - Z$). A few examples are

$$_1^1H \qquad _8^{16}O \qquad _6^{12}C \qquad _{92}^{238}U$$

These symbols are often read as the element name followed by the mass number (i.e., $_1^1H$, hydrogen-1; $_6^{12}C$, carbon-12; $_8^{16}O$, oxygen-16; $_{92}^{238}U$, uranium-238). A tabulation of all the observed nuclides has been compiled and is called the nuclide table. Table 6-2 lists the nuclides for a few elements.

6-3 RADIOACTIVITY

Most naturally occurring nuclides are stable and retain their structure indefinitely. Some nuclides, however, are not stable and are said to be radioactive. **Radioactivity** can be defined as spontaneous decay of a nucleus to form another nucleus and a nuclear particle. Often in this decay a more stable nucleus is formed from a less stable nucleus. The nuclear particles produced during radioactive decay are ejected from the original nucleus with large amounts of kinetic energy (high speeds). These energetic particles make radioactive substances dangerous (see Section 6-6) but sometimes, in certain controlled situations, useful.

Radioactive decay occurs only in certain ways. These are called **modes of radioactive decay.** The three most important decay modes are described below.

Table 6-2
Nuclides of Some
Elements (Naturally
Occurring)

Hydrogen	$_1^1H$ (proton)	$_1^2H$ (deuterium)	$_1^3H$ (tritium)
Helium	$_2^3He$	$_2^4He$	
Carbon	$_6^{12}C$	$_6^{13}C$	$_6^{14}C$
Oxygen	$_8^{16}O$	$_8^{17}O$	$_8^{18}O$
Uranium	$_{92}^{234}U$	$_{92}^{235}U$	$_{92}^{238}U$

From *Introduction to Chemistry* by T. R. Dickson, John Wiley & Sons, New York, 1971.

ALPHA PARTICLE DECAY A nucleus (called the parent) decays by emitting a high-speed helium-4 nucleus called an **alpha particle** (α), to form a new nucleus (called the daughter) with a mass number that is four less than the original nucleus and an atomic number that is two less. The decay can be represented by the general nuclear equation

$$\underset{\text{parent nucleus}}{{}^{M}_{Z}X} \longrightarrow \underset{\text{daughter nucleus}}{{}^{M-4}_{Z-2}Y} + \underset{\text{alpha particle}}{{}^{4}_{2}He}$$

where X is the parent element and Y is the newly formed element. Some examples of alpha decay are

$$^{238}_{92}U \longrightarrow {}^{234}_{90}Th + {}^{4}_{2}He$$

uranium-238 decays to form thorium-234, and

$$^{226}_{88}Ra \longrightarrow {}^{222}_{86}Rn + {}^{4}_{2}He$$

radium-226 decays to form radon-222. The alpha particles emitted from a radioactive substance are called **alpha radiation.**

One characteristic of a nuclear equation is that the sums of the mass numbers on each side of the arrow are equal, and the sums of the atomic numbers are equal.

BETA PARTICLE DECAY A nucleus decays by emitting a high-speed electron called a **beta particle** (β), to form a new nucleus with the same mass number as the original nucleus and an atomic number that is one greater. Also, a particle called an antineutrino is emitted to conserve energy:

$$^{M}_{Z}X \longrightarrow {}^{M}_{Z+1}Y + {}^{0}_{-1}e + \nu \text{ (antineutrino)}$$

Note that the mass number M does not change, but the atomic number does. Some examples of beta decay are

$$^{238}_{93}Np \longrightarrow {}^{238}_{94}Pu + {}^{0}_{-1}e + \nu$$

neptunium-238 decays to form plutonium-238, and

$$^{241}_{94}Pu \longrightarrow {}^{241}_{95}Am + {}^{0}_{-1}e + \nu$$

plutonium-241 decays to form americium-241. The beta particles emitted from a radioactive substance are called **beta radiation.**

GAMMA RAY (PHOTON) EMISSION Sometimes the daughter nucleus formed in an alpha decay or a beta decay will be in an energetically excited state. When this nucleus drops to a lower energy state, it emits a photon of electromagnetic energy called a **gamma ray,** γ. The emission of **gamma radiation** by a radioactive substance often accompanies the alpha radiation or beta radiation produced by the decay of the nuclides in the substance.

In summary, there are three major types of radiations: alpha radiation, beta radiation, and gamma radiation. Nuclides that are **alpha emitters** produce alpha radiation (α) accompanied, in some cases, by gamma radiation (γ). Nuclides that are **beta emitters** produce beta radiation (β) accompanied, in some cases, by gamma radiation (γ).

Radioactive decay is spontaneous and cannot be prevented. A sample of a radioactive substance will continue to decay by emission of radiation until all of the sample has decayed. Not all nuclides decay at the same rate. One of the characteristics of a given radioactive nuclide is the rate at which it decays. A common way to express the rate of decay of a radioactive element is in terms of its half-life. **Half-life** is defined as the time required for the decay of one-half of a sample of a radioactive substance. Half-life is illustrated in Figure 6-1. The half-

Figure 6-1
Starting with a given
amount of radioactive
nuclide (N), one-half
the amount will
remain after one
half-life has passed.
After a time equal to a
second half-life has
passed, one-fourth of
the original amount
remains. After the
passage of another
half-life, one-eighth of
the original amount
remains. This
continues until all the
nuclides have
decayed. From
*Introduction to
Chemistry* by T. R.
Dickson, John Wiley &
Sons, New York, 1971.

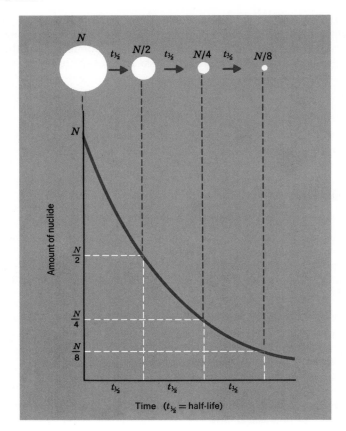

lives of nuclides vary widely, ranging from fractions of seconds to billions of
years. A few typical half-lives are given in Table 6-3.

When working with a radioactive sample, it is important to know how radioactive it is (how "hot" the sample is). The curie is one unit that is used to express
intensity of radioactivity. A **curie** is a unit of activity equal to 37 billion disintegrations per second. The actual amount in grams of a radioactive material that will
have an activity of 1 curie can be determined by knowing the rate of decay of a
radioactive element. For example, 1 gram of radium-226 is equivalent to 1 curie.
A curie is a large amount of radioactivity. Normally, for safety, **millicurie** (0.001
curie) and **microcurie** (0.000001 or 10^{-6} curie) amounts of radioactive substances
are used in laboratory work.

Table 6-3
Some Half-Lives

Nuclide	Half-life	Nuclide	Half-life
$^{238}_{92}$U, uranium-238	4.5×10^9 years	$^{42}_{19}$K, potassium-42	12.4 hours
$^{14}_{6}$C, carbon-14	5,680 years	$^{183}_{78}$Pt, platinum-183	6 minutes
$^{226}_{88}$Ra, radium-226	1,620 years	$^{17}_{9}$F, fluorine-17	66 seconds
$^{22}_{11}$Na, sodium-22	2.60 years	$^{223}_{90}$Th, thorium-223	0.9 second

From *Introduction to Chemistry* by T. R. Dickson, John Wiley & Sons, New York,
1971.

6-4 RADIOACTIVE ELEMENTS

About 330 different nuclides are found in nature. Some elements have only one naturally occurring isotope (i.e., fluorine, $^{19}_{9}F$) while other elements have numerous natural isotopes (i.e., tin, $^{112}_{50}Sn$, $^{114}_{50}Sn$, $^{115}_{50}Sn$, $^{116}_{50}Sn$, $^{117}_{50}Sn$, $^{118}_{50}Sn$, $^{119}_{50}Sn$, $^{120}_{50}Sn$, $^{122}_{50}Sn$, $^{124}_{50}Sn$). With some elements, all of their nuclides are radioactive. All isotopes of the 22 elements of atomic weight greater than bismuth (atomic number 83) are radioactive. A few nuclides of the other elements are naturally radioactive (i.e., $^{204}_{82}Pb$ and $^{40}_{19}K$).

Radioactive nuclides are unstable and undergo decay to form stable nuclides and/or other radioactive nuclides. Studies of the naturally occurring elements have revealed that radioactive nuclides of atomic number greater than 82 belong to one or another of several **decay chains** or **series**. In other words, these elements decay by alpha or beta emission according to definite patterns. The radioactive decay series of uranium-238 is given in Figure 6-2. In this series, uranium-238 decays into thorium-234, which decays into another nuclide, and so on, until stable lead-206 is formed. Other nuclides belong to other series.

Knowledge of radioactive decay series provides a way to estimate the age of the earth. Assuming that uranium-238 was formed when the earth was formed, then a sample of undisturbed rock containing uranium would contain a certain amount of lead-206. (The end product of the uranium-238 decay chain.) The amount of lead-206 would depend on how long the uranium had been decaying. Since the half-life of uranium-238 is 4.5×10^9 years, a measurement of the amounts of uranium-238 and lead-206 indicates the age of the rock. (In one half-life, 1 gram of $^{238}_{92}U$ would decay to about 0.4 gram of $^{206}_{82}Pb$, and 0.5 gram of $^{238}_{92}U$

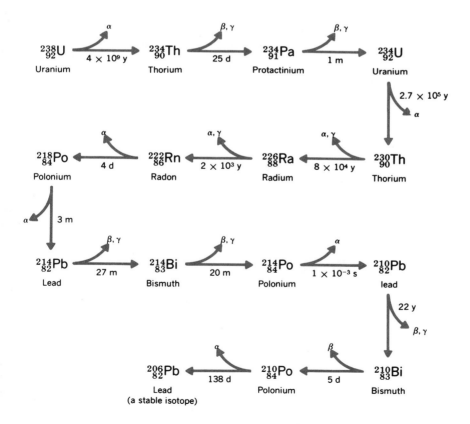

Figure 6-2
Uranium-238 radioactive disintegration series. The number beneath each arrow signifies the half-life of the preceding isotope: y = years, d = days, m = minutes, s = seconds. The small arrow that curves away from each main arrow indicates the kind (or kinds) of radiation emitted by the preceding isotope. From *Elements of General and Biological Chemistry*, 3rd ed., by John Holum, John Wiley & Sons, New York, 1972.

would remain.) Using this method of rock dating and other similar methods, the age of the Earth is estimated to be around 4.5 billion years.

6-5 RADIATION DETECTION

Radioactivity is characterized by emission of radiation during the decay process. This radiation is used to detect and identify radioactive substances, and it can be very dangerous to human beings.

The particles given off during alpha or beta decay are very high-speed, high-energy particles. These particles are ejected from the decaying nucleus and travel outward into the material surrounding the radioactive substance. The alpha and beta particles lose their energy by interacting with matter. This interaction results in an alpha or beta particle, causing an electron to be lost by an atom or molecule as illustrated in Figure 6-3. In other words, these particles lose their energies by causing atoms and molecules to ionize. The electron and positive ion produced by this interaction are called an **ion pair.** As an alpha or beta particle passes through matter, it will cause the formation of many ion pairs. The more massive alpha particles ($_2^4$He) are less penetrating than beta particles ($_{-1}^0$e), but they produce more ion pairs. The typical alpha particle could travel about 6 centimeters in air and produce about 40,000 ion pairs, while a typical beta particle would travel 1,000 centimeters in air and produce about 2,000 ion pairs. Gamma radiation consists of gamma photons (γ) which are photons of electromagnetic radiation similar to x rays. Gamma particles also interact with matter to form ions, and they are very penetrating. Because radiation causes the formation of ion pairs in matter it passes through, it is called **ionizing radiation.**

The detection of radiation is based on its ability to form ions. A typical device used to detect radiation is the **Geiger-Müller tube,** which is illustrated in Figure 6-4. The tube consists of a metal cylinder with a thin (mica) window on the end. The tube is filled with a special gas and has a thin metal rod in the center. Using an external source of power, a voltage difference is maintained between the rod and the cylinder. When a particle of ionizing radiation enters the tube, some ion pairs are formed. The electrons formed in the ionization are strongly attracted to the center rod. They move toward the rod with great speed, and in the process

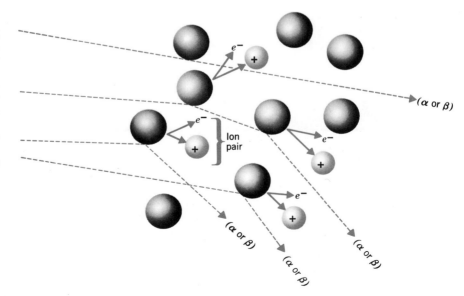

Figure 6-3
When ionizing radiation passes through matter, some of the atoms and molecules are ionized, producing an ion pair involving an electron and a positive ion. From *Introduction to Chemistry* by T. R. Dickson, John Wiley & Sons, New York, 1971.

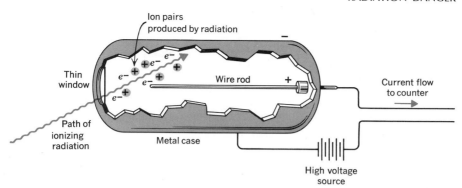

Figure 6-4
Operating principle of
a Geiger-Muller tube.
From *Introduction to
Chemistry* by T. R.
Dickson, John Wiley &
Sons, New York, 1971.

they act essentially like beta particles in that they produce more ion pairs. This produces a large number of electrons which quickly flow toward the center rod. The flow of electrons is equivalent to a small current passing through the tube. This flow of current is carried to an electronic device, called a counter, that is connected to the tube. The counter recognizes the current flow, which it registers as an ionizing event caused by a particle of radiation. Each time a particle enters the tube and causes the electron "avalanche," the counter will tally the event. In this way the Geiger-Müller tube and the counter serve as a means of measuring the activity of a radioactive substance. The counter is used to determine the number of disintegrations per second or the number of disintegrations per minute that are associated with a radioactive sample. This is one way in which the half-life of a radioactive nuclide can be determined. The **activity** of a sample expressed in disintegrations per second is directly related to the amount of the nuclide. So, by measuring the activity of a sample over a period of time and plotting the activity versus the time, it is possible to determine the half-life of the nuclide. Generally, a counter is a convenient way to determine the activity of any radioactive substance.

6-6 RADIATION DANGER

The ionizing effect of radiation is the factor which makes radiation dangerous to life. Exposure to too much high-intensity radiation can cause **radiation sickness** and death. Alpha particles cannot penetrate skin and are not dangerous externally. However, if some alpha emitter is eaten or inhaled, it can cause severe damage inside the body. Gamma radiation, x rays, and neutrons can penetrate the body and, thus, are the most harmful type of radiation. It is important to guard against undue exposure to them. Beta particles can penetrate the skin somewhat and may cause skin burns. Introduction of any source of radiation into the body can be dangerous. Ionizing radiation can cause the ionization of various compounds within the body, which will then not function correctly. It is thought that ionizing radiation interacts with water in the cells of the body to form unstable atomic aggregates called **free radicals.** These free radicals react with vital components of the cell such as the enzymes and the chromosomes, thereby preventing their normal function. This leads to the malfunction or destruction of the cell. Overexposure to radiation may cause nausea, vomiting, and weakness followed by a period (days or weeks) of feeling well and then a period characterized by weakness, weight loss, fever, diarrhea, internal bleeding, and loss of hair. Finally, the victim either slowly recovers or dies.

Another danger of ionizing radiation is that it can cause **genetic damage** to body cells. Sometimes this damage leads to cancer—in the early 1900s many scientists working with newly discovered radioactive materials died of cancer.

When genetic damage occurs in germ cells, cells that produce eggs or sperm, **mutations** of offspring can result. Many such mutations occurred in the children of Japanese citizens exposed to radiation from nuclear bombs dropped on Japan by the United States during World War II.

Humans are constantly being exposed to ionizing radiation from naturally occurring radioactive isotopes (**background radiation**), radioactive material which has entered the environment from nuclear power plants, and fallout from nuclear weapon testing. In addition, occasional doses of radiation are obtained from medical x rays and the use of radioactive materials in medicine. Much of this exposure is unavoidable but, since radiation can be deadly or cause cancer, it is important to keep close watch on the amounts of radioactive materials which enter the environment from man-made sources.

The amount of radiation absorbed by humans upon exposure to radiation can be expressed in terms of an absorbed dose unit called the **rem** (**roentgen equivalent man**). The technical definition of the rem is given in Table 6-4, but for our purposes we can consider the rem as a means of expressing radiation dosages to which we are exposed. Table 6-4 shows estimates of average yearly doses of radiation received by Americans (1970) according to the National Academy of Sciences. As the table shows, background radiation amounts to 102 millirems each year and the total average yearly dose is 182 millirems. The National Academy of Sciences Committee on the Biological Effects of Ionization Radiation recommends that the general population should not receive more than 170

Table 6-4 Estimated Average Dose Rate in the United States, 1970.

Roentgen equivalent man (rem) unit of radioactive dose in man. One rem is the quantity of any radiation that, when absorbed by the human body, causes an effect equivalent to the absorption of 1 roentgen. A roentgen is a unit of radioactive exposure. One roentgen (of gamma radiation or x rays) is the amount of radiation that will produce, in 1 cubic centimeter of air, ion pairs having a total of 2.1×10^9 units of electrical charge (electrostatic units). It is estimated that the minimum lethal dose for man (50 percent of those exposed would die within 30 days) is 500 rem.

Source	Average dose rate[a] (millirem per year)
Environmental	
Natural	102
Global fallout	4
Nuclear power	0.003
Subtotal	106
Medical	
Diagnostic	72
Radiopharmaceuticals	1
Subtotal	73
Occupational	0.8
Miscellaneous	2
Subtotal	3
Total	182

[a] 1 millirem = 10^{-3} rem

millirems of man-made radiation each year exclusive of background and medical radiation sources. The most likely source of additional radiation will be uranium mines, nuclear fuel processing plants, and nuclear power plants. Radioactive material introduced from these sources will have to be closely watched (See Section 6-12.) The committee further stated that the 170 millirem value is chosen as a balance between societal needs and genetic risk and that such a dose could lead to 6,000 extra deaths by cancer each year. They further state that improved x-ray equipment and the elimination of unnecessary x rays could greatly reduce the average yearly dose.

6-7 NUCLEAR TRANSMUTATIONS

A high-speed nuclear particle can, under certain conditions, collide with a nucleus to cause a nuclear reaction which produces a different nucleus. This process is called a **nuclear transmutation.** Some transmutations can occur naturally, but many are induced in the nuclear science laboratory. An example of a transmutation is

$$^{14}_{7}\text{N} + ^{4}_{2}\text{He} \longrightarrow ^{17}_{8}\text{O} + ^{1}_{1}\text{H}$$

In this reaction, the nuclear particle $^{4}_{2}\text{He}$ is called the **projectile,** and the $^{14}_{7}\text{N}$ is called the **target nuclei.** The transmutation process involves a collision of the target nuclei with the projectiles which results in new combinations of neutrons and protons corresponding to the products.

Nuclear scientists have used nuclear transmutations as a means of preparing artificial nuclides which are sometimes called **man-made isotopes.** Nearly all the naturally occurring nuclides have been used as targets, and a variety of nuclear particles such as protons ($^{1}_{1}\text{H}$), deuterons ($^{2}_{1}\text{H}$), neutrons ($^{1}_{0}n$), alpha particles ($^{4}_{2}\text{He}$), and electrons ($^{0}_{-1}e$) have been used as projectiles. Often, these projectiles must be given large kinetic energies before they are projected at the target nuclei. Particle-accelerating devices such as cyclotrons, linear accelerators, synchrotrons, and nuclear reactors are used to produce these high-energy projectiles. Through transmutations, nuclear scientists have been able to make about 1,000 different artificial nuclides. Some of the nuclides are useful in research and in medicine. For instance, carbon-14, made by the transmutation

$$^{14}_{7}\text{N} + ^{1}_{0}n \longrightarrow ^{14}_{6}\text{C} + ^{1}_{1}\text{H}$$

is a radioactive (α emitter) isotope of carbon. **Carbon-14** can be incorporated in organic compounds and serve as a **radioactive tag.** The radioactive carbon can then be traced through a series of chemical reactions. Using tagged compounds provides a convenient method of studying the way a series of reactions takes place. By use of carbon-14 tagged carbon dioxide, Melvin Calvin was able to accomplish a detailed study of photosynthesis, the process by which plants convert carbon dioxide and water into carbohydrates. The radioactive artificial nuclide **cobalt-60,** made by the transmutation

$$^{59}_{27}\text{Co} + ^{1}_{0}n \longrightarrow ^{60}_{27}\text{Co}$$

is used as a medical tool to destroy cancerous cells in the treatment of cancer.

Among the many interesting transmutations are those used to make the so-called **man-made elements.** Uranium is the element of highest atomic number found in nature. (Small amounts of plutonium have recently been found.) However, by carrying out certain transmutations, nuclear scientists have been able to make elements of higher atomic number than uranium. These are called the **transuranium elements.** The transmutations have been accomplished using special particle accelerators and targets of uranium and other previously synthesized transuranium elements. The transmutations by which the transuranium

Table 6-5
Nuclear Transmutations Used to Produce Transuranium Elements

Element	Atomic number	Reaction
Neptunium, Np	93	$^{238}_{92}U + {}^1_0n \longrightarrow {}^{239}_{93}Np + {}^0_{-1}e$
Plutonium, Pu	94	$^{238}_{92}U + {}^2_1H \longrightarrow {}^{238}_{93}Np + 2\,{}^1_0n$
		$^{238}_{93}Np \longrightarrow {}^{238}_{94}Pu + {}^0_{-1}e$
Americium, Am	95	$^{239}_{94}Pu + {}^1_0n \longrightarrow {}^{240}_{95}Am + {}^0_{-1}e$
Curium, Cm	96	$^{239}_{94}Pu + {}^4_2He \longrightarrow {}^{242}_{96}Cm + {}^1_0n$
Berkelium, Bk	97	$^{241}_{95}Am + {}^4_2He \longrightarrow {}^{243}_{97}Bk + 2\,{}^1_0n$
Californium, Cf	98	$^{242}_{96}Cm + {}^4_2He \longrightarrow {}^{245}_{98}Cf + {}^1_0n$
Einsteinium, Es	99	$^{238}_{92}U + 15\,{}^1_0n \longrightarrow {}^{253}_{99}Es + 7\,{}^0_{-1}e$
Fermium, Fm	100	$^{238}_{92}U + 17\,{}^1_0n \longrightarrow {}^{255}_{100}Fm + 8\,{}^0_{-1}e$
Mendelevium, Md	101	$^{253}_{99}Es + {}^4_2He \longrightarrow {}^{256}_{101}Md + {}^1_0n$
Nobelium, No	102	$^{246}_{96}Cm + {}^{12}_6C \longrightarrow {}^{254}_{102}No + 4\,{}^1_0n$
Lawrencium, Lr	103	$^{252}_{98}Cf + {}^{10}_5B \longrightarrow {}^{257}_{102}Lr + 5\,{}^1_0n$
Kurchatovium, Ku[a]	104	$^{242}_{94}Pu + {}^{22}_{10}Ne \longrightarrow {}^{260}_{104}Ku + 4\,{}^1_0n$

From *Introduction to Chemistry* by T. R. Dickson, John Wiley & Sons, New York, 1971.
[a] Not official.

elements have been made are given in Table 6-5. It is conceivable that more transuranium elements will be made in the future.

6-8 NUCLEAR FISSION

A very important type of nuclear reaction is **nuclear fission,** which was discovered by Hahn and Strassman in 1938. In fission, a neutron collides with and is captured by a heavy nucleus, causing the nucleus to become very unstable. The unstable nucleus then splits into two new nuclei plus a few neutrons. This fission is illustrated in Figure 6-5. The fission of uranium-235 can be represented as

$$^{235}_{92}U + {}^1_0n \longrightarrow \text{fission nuclei} + \text{neutrons} + \text{energy}$$

Various fission nuclei are produced, depending on how the original nucleus splits. A typical fission of $^{235}_{92}U$ is

$$^{235}_{92}U + {}^1_0n \longrightarrow {}^{90}_{38}Sr + {}^{144}_{54}Xe + 2\,{}^1_0n + \text{energy}$$

When fission occurs, the more stable product nuclei are formed from the less stable parent nuclei, which results in a large amount of energy being produced. This energy is the most important aspect of fission, since it represents **nuclear energy** that can be obtained by carrying out the fission process. The fission of 1 gram of uranium-235 could produce about 20 billion calories of energy. This means that 1 kilogram of uranium-235 can produce as much energy as the burning of about 2,600 tons of coal. The energy of fission is the basis for the production of atomic energy.

Figure 6-5
Fission of uranium-235. From *Introduction to Chemistry* by T. R. Dickson, John Wiley & Sons, New York, 1971.

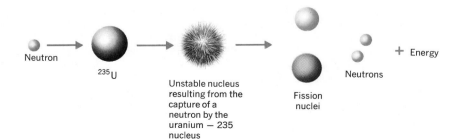

Neutron

^{235}U

Unstable nucleus resulting from the capture of a neutron by the uranium — 235 nucleus

Fission nuclei

Neutrons

+ Energy

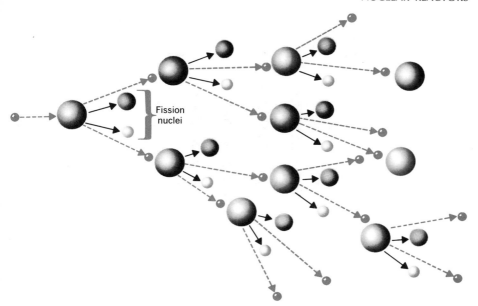

Figure 6-6
A nuclear chain reaction. The neutrons given off by the fission process can cause fission of other nuclei. Since each fission process produces about twice as many neutrons as are needed to cause fission, a nuclear chain reaction can result when a large amount of fissionable nuclei are present. From *Introduction to Chemistry* by T. R. Dickson, John Wiley & Sons, New York, 1971.

Some products of the fission process are neutrons, which are ejected during the fission process and are potentially capable of causing other nuclei to undergo further fission. Each time a fission occurs, more neutrons are produced and more fission can take place. In fact, each fission produces more neutrons than are consumed in the original fission process. Thus, if sufficient nuclei are present, it is possible to have an uncontrolled **chain (or sequential) reaction** of fission processes. This is illustrated in Figure 6-6. If insufficient nuclei are present in a sample of fissionable material (uranium-235 metal), then enough neutrons escape without causing further fission and the fission that does go on goes slowly enough to be under control. However, if enough fissionable material is present, a self-sustaining fission chain reaction can occur. The minimum amount of fissionable material in which a fission chain reaction can be self-sustaining is called the **critical mass** of the material. The critical mass of pure uranium-235 is less than 1 kilogram. A fission chain reaction can take place very quickly, and each fission process will release a large amount of energy. This accounts for the tremendous destructive power of **atomic bombs** (fission bombs).

6-9   NUCLEAR REACTORS

Nuclear chain reactions can be controlled, and the energy can be utilized as a source of power. A device in which a self-sustaining fission reaction can be carried out under controlled conditions is called a **nuclear reactor.** The key to using fission in a reactor is the control of the process so that a nuclear explosion can not occur. A typical nuclear reactor is illustrated in Figure 6-7. A reactor consists of five basic components. First, the **nuclear fuel** which contains sufficient amounts of fissionable material—typical nuclear fuel elements contain natural uranium in which the uranium-235 content has been increased from the natural level of about 0.7 percent to 3–4 percent. This uranium-235 enrichment of natural uranium is accomplished by special techniques specifically for use in nuclear reactors. There is no possibility of a critical mass of uranium-235 in a reactor and, thus, no chance of a nuclear explosion. In addition to uranium-235, the nuclides plutonium-239 and uranium-233 are also fissionable. Second, a **moderator** is used to slow down the fission neutrons. Graphite and heavy water

Figure 6-7
A simple nuclear reactor. From *Introduction to Chemistry* by T. R. Dickson, John Wiley & Sons, New York, 1971.

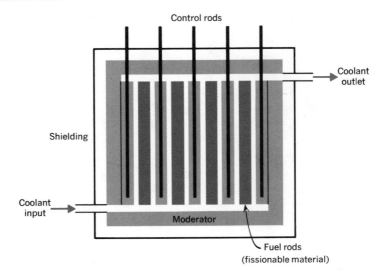

(water with hydrogen-2 isotope rather than hydrogen-1) are commonly used as moderators. Third, **control rods** made of cadmium or boron steel are used to control the fission. These rods are capable of absorbing neutrons. So, by moving these rods into the reactor, it is possible to carefully control the number of neutrons present. These control rods allow the correct number of neutrons to be available just to sustain continuous fission. Fourth, a **coolant** such as water or molten sodium metal is circulated about the reactor core to absorb the heat produced by fission. The heated coolant can be used to form steam which is used to generate electrical power. In fact, this is the purpose of an electrical power producing nuclear reactor. (See Section 6-11.) Fifth, some **shielding** material must be used to protect people from the highly radioactive core of the reactor. The shielding often consists of a thick concrete case around the reactor.

It is possible to design reactors which not only produce power but also are able to produce fissionable material. Reactors in which the fission process is used to produce more fissionable material are called **breeder reactors.** These reactors function by placing certain amounts of the naturally occurring isotopes uranium-238 or thorium-232 in specially designed reactors. These isotopes are ultimately transmuted to fissionable nuclides:

$$^{238}_{92}U + ^{1}_{0}n \text{ (from fission)} \longrightarrow ^{239}_{92}U \xrightarrow{\beta} ^{239}_{93}Np \xrightarrow{\beta} ^{239}_{94}Pu \text{ (fissionable)}$$

$$^{232}_{90}Th + ^{1}_{0}n \text{ (from fission)} \longrightarrow ^{233}_{90}Th \xrightarrow{\beta} ^{233}_{91}Pa \xrightarrow{\beta} ^{233}_{92}U \text{ (fissionable)}$$

The breeder reactor consumes uranium-235 as fuel but produces plutonium-239 (or uranium-233) during the operation of the reactor. The plutonium can be collected periodically and used in other reactors. A breeder reactor can produce more potential fuel than it consumes. Breeder reactors represent a vast potential source of energy, as discussed in Section 6-11.

6-10   FUSION

Fusion is another type of transmutation process which can produce energy. At high temperatures, it is possible for some lighter nuclei to fuse together to form heavier nuclei. The fusion process can produce large amounts of energy. In fact, the energy produced on the sun comes from fusion. Nuclear scientists found that the temperatures produced during fission (thermonuclear temperatures) were sufficient to cause certain fusion reactions involving various hydrogen nuclides and lithium nuclides:

$$^2_1H + ^3_1H \longrightarrow ^4_2He + ^1_0n + \text{energy}$$
$$^2_1H + ^2_1H \longrightarrow ^3_2He + ^1_0n + \text{energy}$$
$$^2_1H + ^2_1H \longrightarrow ^3_1H + ^1_1H + \text{energy}$$
$$^1_1H + ^7_3Li \longrightarrow ^4_2He + ^4_2He + \text{energy}$$

The fusion process can produce larger amounts of energy for a given amount of material than the fission process. Uncontrolled **thermonuclear** fusion releases a large amount of energy and accounts for the tremendous destructive power of **thermonuclear bombs** (H-bombs, hydrogen bombs, or fission–fusion bombs). No way has yet been found to carry out a controlled fusion process so that the energy released can be used. However, much research is being devoted to the study of **controlled fusion,** and some day fusion reactors may be developed to provide vast amounts of low-cost energy.

6-11 NUCLEAR ENERGY

A nuclear reactor in which uranium-235 undergoes fission can be designed to produce electrical power. Such a power reactor is illustrated in Figure 6-8. In a **power reactor** the energy of the fission is absorbed by the coolant circulating through the reactor. The hot coolant is used to heat water to generate steam. The steam turns a turbine which in turn drives the generator that produces electricity. The steam is condensed and cooled with a cooling water system. In some

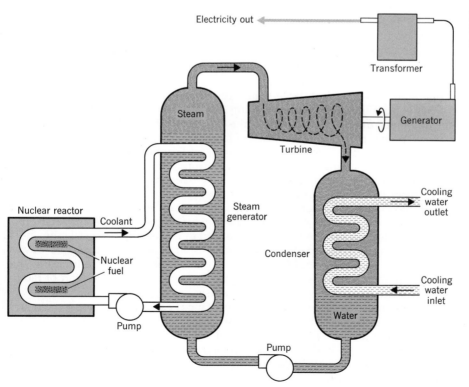

Figure 6-8
Nuclear power reactor.

power reactors, the coolant in the reactor is water which is converted directly to steam used in the turbine.

Nuclear power from fission is often predicted to be a main source of energy in the future. Currently, about 3 percent of the world's energy and less than 1 percent of the energy in the United States comes from nuclear sources. The International Atomic Energy Agency predicts that by the year 2000 about 60 percent of the energy will come from nuclear sources and by 2010 around 76 percent. It is further predicted that over 50 percent of the electricity used in the United States in 2000 will come from nuclear plants. However, these predictions are based on the important assumption that breeder reactors (see Section 6-9) will be functioning. More than 25 nuclear power plants are in operation in the United States, and around 117 are under construction or in the planning stage. The U.S. Atomic Energy Commission predicts that breeder reactors will be available in the 1980s. Currently, all the power reactors are designed to use uranium-235. Uranium-235 is the only common isotope that easily undergoes fission. Unfortunately, this isotope is quite rare in nature and comes only from high-quality uranium ores. Predictions of the amounts of such uranium ores in the world indicate that, if wide use of uranium-235 occurs in power reactors, the world's supply will diminish around the turn of the century. If this occurs, nuclear energy by fission will not be significant in the future. However, if breeder reactors are constructed that will produce fissionable uranium-233 and plutonium-239, the prospects for nuclear energy use will be increased. The isotopes uranium-238 and thorium-232 used in breeder reactors are very plentiful and would provide a vast source of nuclear fuel. In fact, it is predicted that such nuclear fuel would provide an amount of energy equal to 10–100 times the amount of energy produced by all the fossil fuels that ever existed. As was mentioned in Section 6-10, if controlled fusion reactors are ever developed, they could provide a vast source of energy.

6-12 PROS AND CONS OF NUCLEAR ENERGY

There are several advantages and disadvantages to the wide scale use of fission reactors for electrical power. Power reactors are capable of providing large amounts of energy and, thus, can replace conventional fossil fuel sources. At this time, power reactors seem to be the only feasible alternative to diminishing supplies of fossil fuels. A large-scale breeder reactor is not yet in operation. Consequently, the amounts of uranium-235 may limit nuclear fuels and make power plants uneconomical. Since the fission process is not a combustion process, power reactors do not give off large amounts of waste gases which can become air pollutants. However, some highly **radioactive gases** (such as krypton-85) are given off in small amounts during fission. These gases are not the same as normal air pollutants but are an especially dangerous new kind of pollutant with which we have to contend. In fact, the use of power reactors introduces several new kinds of pollution problems. Impurities in the reactor coolant can become radioactive when exposed to the reactor core. These are known as **low-level radioactive wastes;** they must be disposed of by sealing in special containers which are placed in semipermanent storage. Some low-level radioactive wastes leak into the secondary cooling waters and, when this water is returned to the source, the radioactivity is introduced into the environment. This kind of radioactive leakage has to be watched continuously.

Figure 6-9 illustrates how radioactive substances can enter the food chain and be transmitted to man. The inhalation of radioactive materials can induce lung cancer. Some radioactive isotopes are retained in the body because of chemical similarity to common body chemicals. Cesium-137, a beta emitter with a half-life of 30 years, is chemically similar to potassium and can become incorporated in all the cells of the body. Strontium-90, a beta emitter with a half-life of 28 years,

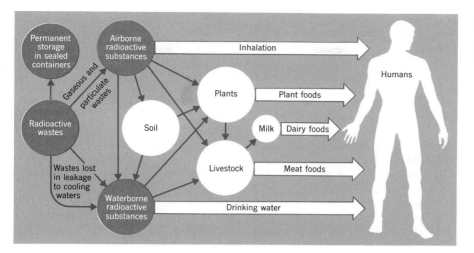

Figure 6-9
Transmission of radioactive substances to humans.

is chemically similar to calcium. Since calcium is a main constituent of bones and teeth, strontium-90 can accumulate in these areas of the body.

6-13 RADIOACTIVE WASTES

When fission occurs, the **fission products** accumulate in the reactor and have to be periodically removed. These fission products are extremely radioactive and hazardous. Such wastes are shipped in special containers to locations where they can be stored underground for safety reasons. Large amounts of such wastes are currently in storage, and if nuclear fuel usage increases, the amounts of these wastes will increase accordingly. If nuclear energy is used as predicted, it is expected that 27 billion curies of radioactive wastes will have accumulated by the year 2000. Transportation, handling, and storage problems will be significant. In fact, these wastes will have to be stored for thousands of years in safe areas. Some scientists have suggested the construction of large pyramid structures for storage. Archeologists of the future would be quite surprised if they opened these pyramids.

Another aspect of nuclear energy is the problem of **waste heat.** Currently, nuclear power plants are less efficient in energy conversion than fossil fuel plants. A conventional nuclear power plant produces 50 percent more waste heat than a comparable fossil fuel plant. However, more efficient nuclear plants are being developed, and breeder reactors should be as efficient as fossil fuel plants. Nevertheless, once nuclear energy begins to substantially replace fossil fuel energy, waste heat will be of greater concern.

6-14 REACTOR SAFETY

Another problem of nuclear reactors involves **reactor safety** and **emergency core cooling.** The core of a nuclear reactor in a power plant is maintained at a high temperature by the fission process. This temperature is kept at reasonable levels by the coolant. If the coolant flow stops, the temperature will rise. The fission process can be stopped or reduced by use of the reactor control rods (see Section 6-9). It is conceivable that the coolant flow may be cut off by accident and, before the control rods can be used, the temperature of the core could rise so that a melting of the inner portion of the reactor takes place. The probability of such an accident is low, but the results could be disastrous. Breeder reactors are more easily damaged by coolant effects than conventional reactors. In 1955 and 1966 experimental breeder reactors being tested in the United States suffered partial core melting and had to be abandoned. A **core melting** of a reactor could result in radioactive material from the fission products and fuel rods being released into the water and air environment around the reactor. Winds or waterways could spread this material about. The major problem with this type of pollution is that it is extremely dangerous to life and cannot be seen, tasted, or smelled. It takes special instruments to detect radiation. Furthermore, some of the radioactive isotopes have long half-lives, which means that a contaminated area would have to be evacuated and would not be usable for living or agriculture for a period of time. Plutonium-239 used and produced in breeder reactors has a half-life of about 24,000 years. A nuclear accident involving plutonium could contaminate an area for an indefinite length of time. The chance of such a disaster means that the safety factors in nuclear reactors have to be carefully developed and continuously reviewed. However, the effects of earthquakes, human error of plant operators, and even the chance of sabotage are unpredictable factors. Overall, nuclear fission is potentially the most directly hazardous source of energy.

6-15 PLUTONIUM

A final concern over the large-scale use of nuclear energy involves **plutonium.** A large breeder reactor requires 1,000 kilograms of plutonium as fuel. Breeder reactors produce more fuel than they consume, which means that if such reactors are operating as expected, they will produce 80,000 kilograms of plutonium per year by the year 2000. At the current price of $10,000 per kilogram or even a lower price, plutonium sources would be attractive to criminal elements of society and a black market might develop. Moreover, the critical mass of plutonium-239 is about 5 kilograms. It is conceivable that a crude atomic bomb could be constructed from this amount of plutonium.

The desirable aspects and possible necessity contrasted with the many disadvantages of nuclear power comprise a significant environmental dilemma. We will have to make many decisions on nuclear power use in the near future.

BIBLIOGRAPHY   BOOKS

Alexander, P. *Atomic Radiations and Life.* Baltimore: Penguin, 1965.
Collins, G. C. *Radioactive Wastes: Their Treatment and Disposal.* New York: Wiley, 1961. 239 pp.
Curtis, R., and E. Hogan. *Perils of the Peaceful Atom.* New York: Doubleday, 1969. 274 pp.
Glassner, A. *Introduction to Nuclear Science.* New York: Litton Educational Publishing, Van Nostrand Reinhold, 1961.
Scientific American, ed. *Atomic Power.* New York: Simon and Schuster, 1955.

ARTICLES AND PAMPHLETS

Choppin, G. R. "Nuclear Fission," *Chemistry* **40** (7), 25 (1967).
Forbes, I. A., et al. "Cooling Water." *Environment* **14** (1), (1972).
Ford, D. F., and H. W. Kendall. "Nuclear Safety." *Environment* **14** (7) (1972).
Gilinsky, V. "Bombs and Electricity." *Environment* **14** (7) (1972).
Gough, W. C., and B. J. Eastlund. "The Prospects of Fusion Power." *Scientific American* 50 (Feb. 1971).
Johnsen, R. H. "Radiation Chemistry." *Chemistry* **40** (7), 31–36 (1967).

"The Nuclear Industry and Air Pollution." *Environmental Science and Technology* **4**, 392–395 (1970).

"Nuclear Power: the Social Conflict." *Environmental Science and Technology* **5**, 404–410 (1970).

U.S. Atomic Energy Commission. *Atoms, Nature and Man-Made Radioactivity in the Environment,* by N. O. Hines. Oak Ridge, Tenn.: U.S. AEC, 1969. 57 pp.

U.S. Atomic Energy Commission. *Nuclear Power and the Environment.* Oak Ridge, Tenn.: U.S. AEC, 1969. 30 pp.

U.S. Atomic Energy Commission. *Radioactive Wastes,* by C. H. Fox. Oak Ridge, Tenn.: U.S. AEC, 1965. 46 pp.

Wood, L., and J. Nuckolls. "Fusion Power." *Environment* **14** (May 1972).

Woodwell, G. M. "The Ecological Effects of Radiation." *Scientific American* **208**, 40–49 (June 1963).

QUESTIONS AND PROBLEMS

1. Describe the components of an atomic nucleus.

2. What is the meaning of the atomic number and mass number of a nuclide?

3. How many protons and neutrons are in each of the following nuclides?

    $^{2}_{1}H$ $^{12}_{6}C$ $^{16}_{8}O$ $^{238}_{92}U$

4. What are isotopes?

5. The three natural isotopes of oxygen (atomic number 8) have nuclei which contain eight, nine, and ten neutrons, respectively. Give the symbols for each of these oxygen isotopes.

6. What is radioactivity?

7. Describe each of the following:

    (*a*)   alpha decay   (*c*)   gamma emission
    (*b*)   beta decay

8. What is meant by the term half-life?

9. Why is ionizing radiation dangerous?

10. Describe how a Geiger-Müller tube functions.

11. What is radiation sickness?

12. What is a nuclear transmutation?

13. What are the transuranium elements?

14. Describe nuclear fission.

15. What is meant by the critical mass of fissionable material?

16. Describe and list the basic components of a nuclear reactor.

17. What is a breeder reactor?

18. Describe nuclear fusion.

19. Describe a nuclear–electric power plant.

20. Why are nuclear power plants being considered as a major source of electricity in the near future?

21. Why will large-scale use of nuclear power plants depend upon the development of breeder reactors?

22. List some of the advantages and disadvantages of nuclear power plants.

7-1 INTRODUCTION

When you feel the wind blowing in your face or smell the fragrance of perfume you are experiencing matter in the **gaseous** state. **Gases** have interested and stimulated the imaginations of scientists for centuries. The fascination of this state is that we can experience it without seeing it. In fact, investigations of gases were fundamental to the development of the atomic theory. As will be discussed in this chapter, the constituent particles (molecules or atoms) of gases are darting about at high speeds, and it is this motion which contributes to the interesting properties of gases.

As we look around us, the three states of matter are very apparent. Solid objects are firm and occupy definite regions of space. Liquids are more mobile. Liquids flow and have forms which depend upon the shape of a container, a bottle, or a river bottom. Without a container, liquids evaporate or trickle away. The gaseous form of matter is the most elusive form.

The wind blowing against us or the firmness of a filled balloon suggest the existence of matter in the form of gases. To trap a gas we must keep it in a closed container. If we open the container, the gas quickly escapes into the immediate atmosphere.

In previous chapters we have established the idea that matter is made up of atoms or combined atoms in the form of ions or molecules. A solid substance consists of a vast collection of atoms, molecules, or ions. A nail is made up of numerous iron atoms. Ice consists of an aggregation of water molecules. Table salt is made up of visible crystals, each of which is an aggregation of sodium ions and chloride ions. Pure liquids are collections of molecules. Liquid water consists of a vast collection of water molecules. Likewise, common gases are collections of molecules.

The description of matter as collections of tiny particles allows us to imagine the submicroscopic make-up of matter but does not provide any explanation of the three possible states of matter. Matter displays various properties and behavior, depending upon whether it is in the solid state, liquid state, or gaseous state. In fact, it was the observation and description of these properties and behavior that provided evidence of the atomic theory and the particulate nature of matter. However, in addition to providing evidence for the particulate nature of matter, these observations established the view that the particles of matter are in motion. That is, these particles are not frozen in static positions in space but rather are capable of moving about in space. To have a satisfactory description of matter, the dynamics of the particles must be included. Such a model or description of gases has been established. Before we describe this model let us consider the behavior and properties of gases.

The gaseous state is a very common state of matter. We are quite familiar with gases, especially the mixture of gases known as air. Substances in the gaseous state are mobile and compressible. A gas will expand to occupy any container into which it is placed. Consider the behavior of a gas sample in a balloon. The balloon serves as a container for the gas. If the balloon is opened, the gas will quickly diffuse and dissipate into the surrounding air. If a tiny hole is made in the balloon, the gas will slowly leak out. More gas can be added to the balloon with an air pump, a compressor, or by just blowing by the mouth. Different gases can mix with one another in any proportions to form gaseous mixtures. When we add more gas to the balloon, it will increase in volume until it pops. If we tie off the balloon, we will have a sample of a gas having a certain volume. If we warm the balloon, the volume increases, and if we cool the balloon, the volume decreases. If we squeeze the balloon (increase the pressure), the volume decreases; the gas is compressed. When we stop squeezing (decrease the pressure and release our grip), the volume increases. Many important elements and compounds exist in the gaseous state under normal conditions of room temperature and atmospheric pressure. Furthermore, many substances can be converted to the gaseous state merely by heating them.

If we capture a gas in a container and keep the container closed, we can maintain the sample of the gas indefinitely. To describe such a gas sample we can state the **volume** of the container and the **temperature.** Moreover, the sample is said to exert a specific pressure on the container. When a pressure-measuring device is attached to the container, the device will register a specific pressure. We shall find out shortly why gases exert pressures, but first let us consider what a pressure is. The pressure exerted on an object by a gas is the force exerted on a unit area of the object. That is, **pressure** is defined as force per unit area and can be expressed in terms of dynes per square centimeter or pounds per square inch. To illustrate the idea of pressure, place the tip of a pen or pencil on a magazine or notebook and push down gently. Note the slight indentation in the paper resulting from your push (force) exerted over the area of the tip. Now lift the pen or pencil above the magazine or notebook and jab it into the paper with more force. The greater pressure (force per unit area) will result in a larger indentation in the paper. As we shall see, the force exerted on objects by a gas results from constituent particles of the gas "pushing" on the objects. When a gas is cooled to a specific temperature and compressed by application of a specific pressure, it will condense into a liquid. This is called the **liquefaction** of the gas. The particular combination of temperature and pressure needed to liquefy a gas is different for the various substances which exist as gases.

7-2   GAS LAWS

The systematic study of the behavior of gases has been of interest to scientists for centuries. Many scientists have contributed to our understanding of gases, however, the names of a few are prominent, since they stated certain relations concerning the behavior of gases.

In 1662 Robert Boyle, an English scientist, experimented with the compressibility of air. Boyle, using an apparatus similar to that shown in Figure 7-1, observed the relation between the volume and the pressure of a sample of gas. He found that if a sample of a gas is kept at a constant temperature, then the gas can be compressed by increasing the pressure on the gas. He also noted that

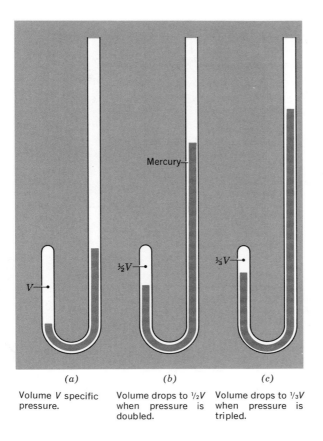

Figure 7-1
Apparatus illustrating
Boyle's Law.

Mercury

½V

⅓V

V

(a)   (b)   (c)

Volume V specific pressure.

Volume drops to ½V when pressure is doubled.

Volume drops to ⅓V when pressure is tripled.

when the pressure is released, the volume of the gas sample increases. As illustrated in Figure 7-1, Boyle observed that if the pressure on the gas is doubled, the gas will compress to one-half the original volume, and if the pressure is tripled, the volume will drop to one-third the original volume. Likewise, he found that if the pressure is cut down by one-half, the gas will expand to double the volume. To explain the compressibility of air, Boyle suggested that air is made up of particles having weight and a spring-like nature. He further suggested that pressure of gases could result from the motion of these particles.

Today, after numerous conformations of Boyle's work and additional experimentation, a generalization concerning the relation between the volume and the pressure of a gas has become known as **Boyle's Law,** which can be stated as follows:

> The volume of a sample of a gas kept at a constant temperature
> is inversely proportional to the pressure.

As illustrated in Figure 7-2, this means that if the pressure of such a gas sample increases, the volume will decrease proportionately, and if the pressure decreases, the volume will increase. If two quantities are inversely proportional, when one increases the other decreases. Consider the filled balloon. When the balloon is squeezed (pressure increased) the volume decreases. When the balloon is released (pressure decreased) the volume increases. A very widely used method to show the relation between two quantities is to construct a graph showing how the two quantities vary with respect to one another. Such a graph illustrating Boyle's Law is shown in Figure 7-2.

In 1787, Jacques Charles first studied the relation between the volume and the

Figure 7-2
Boyle's Law. From
*Introduction to
Chemistry* by T. R.
Dickson, John Wiley &
Sons, New York, 1971.

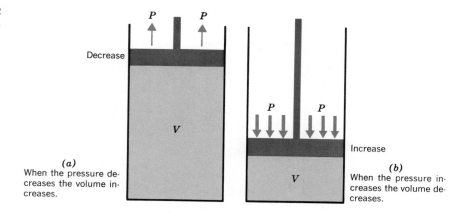

(a)
When the pressure decreases the volume increases.

(b)
When the pressure increases the volume decreases.

(c)
A plot of the pressure, P and the corresponding volume, V of a sample of an ideal gas.

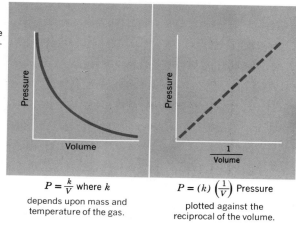

$P = \frac{k}{V}$ where k depends upon mass and temperature of the gas.

$P = (k)\left(\frac{1}{V}\right)$ Pressure plotted against the reciprocal of the volume.

temperature of a sample of a gas at a constant pressure. He used an apparatus similar to that shown in Figure 7-3. By keeping the pressure constant, the volume is affected only by the temperature change. Charles observed that when the temperature of such a gas sample is increased, the volume increases, and when the temperature decreases the volume decreases. Specifically, he noted a direct linear proportionality between the volume and the temperature. Today, this proportionality has become known as **Charles' Law** and can be stated as follows (see Figure 7-4):

> The volume of a sample of a gas at a constant pressure is directly proportional to the temperature.

A direct proportion exists between two quantities when if one is increased, the other increases, and if one is decreased, the other decreases. Charles' Law can be remembered if we consider that a filled balloon will increase in volume when warmed and decrease in volume when cooled. A graphic illustration of Charles' Law is shown in Figure 7-4.

An additional relation involving gases is that between the pressure and the temperature of a sample of a gas of constant volume. In such a case, it is observed that if the temperature of the gas sample is increased, the pressure increases, and if the temperature is decreased, the pressure decreases. This rela-

Figure 7-3
Apparatus illustrating
Charles' Law.

(a)   (b)   (c)

Gas sample occupies vol-
ume V_1, at temperature T_1,
and constant pressure P.

Gas sample occupies larger
volume V_2, at higher temperature T_2, and constant
pressure P.

Gas sample occupies still
larger volume V_3, at still
higher temperature T_3, and
constant pressure P.

Figure 7-4
Charles' Law. From
*Introduction to
Chemistry* by T. R.
Dickson, John Wiley &
Sons, New York, 1971.

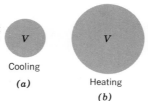

When a sample of a gas at constant
pressure is cooled the volume decreases.

Cooling

(a)

When a sample of a gas at constant pressure is heated the volume increases.

Heating

(b)

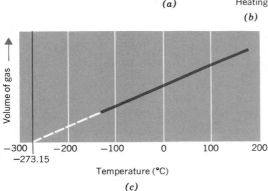

A plot of the volume of a sample of an
ideal gas versus the temperature.
(Note the point at which the volume of
the gas would become zero is the temperature of absolute zero.)

(c)

tion can be stated as follows:

> The pressure of a sample of a gas of fixed volume is directly
> proportional to the temperature.

This relation is illustrated in Figure 7-5. As a practical example of this relation,
consider what happens when a gas-fillled aerosol can is heated in a fire. The volume of the can is fixed. Thus, as the temperature increases, the pressure of the
gas increases. When the temperature is high enough, the pressure of the gas is
so great that the can ruptures and an explosion results. This is the reason why
such cans should never be put into an open fire. Incidentally, numerous people
have been injured or killed by exploding aerosol cans. Some consumer groups
claim that aerosol cans should be equipped with inexpensive pressure-relief
valves to prevent accidents.

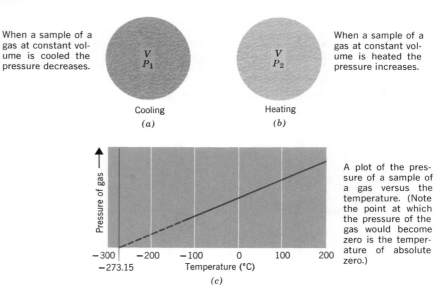

When a sample of a
gas at constant vol-
ume is cooled the
pressure decreases.

V
P_1

V
P_2

When a sample of a
gas at constant vol-
ume is heated the
pressure increases.

Cooling
(a)

Heating
(b)

Pressure of gas

−300   −200   −100   0   100   200

−273.15

Temperature (°C)

(c)

A plot of the pres-
sure of a sample of
a gas versus the
temperature. (Note
the point at which
the pressure of the
gas would become
zero is the temper-
ature of absolute
zero.)

7-3   KINETIC MOLECULAR THEORY OF GASES

We can observe that gases occupy any volume into which they are placed, are
easily compressed, exert pressures, and have measurable temperatures. These
observations and the various gas laws can be explained by establishing a general
theory of the behavior of gases. The model that is used to explain the dynamic
behavior of gases is called the **Kinetic Molecular Theory (KMT).** This theory is
based on the idea that gases are composed of particles (atoms or molecules)
which are in a state of continuous motion. The Kinetic Molecular Theory can be
expressed in terms of the following basic postulates which are illustrated in Figure 7-6:

1. Gases consist of particles (molecules) which are so small and the average
   distance between them is so great that the actual volume occupied by
   the particles is negligible compared to the empty space between them.
2. There are no attractive or repulsive forces between the particles comprising a gas, and they can be considered to behave like very small
   masses.
3. The particles are in rapid, random, continuous motion and are constantly
   colliding with one another and with any object in their environment, such
   as the walls of the container. As a result of this motion, the particles
   possess **kinetic energy,** KE ($KE = \frac{1}{2}mv^2$, where m is the mass of a particle
   and v is the velocity).
4. The collisions result in no net loss in the total kinetic energy of the particles and are said to be perfectly elastic.

These postulates provide a good model of a gas. We can now view a gas as a
collection of molecules which are rapidly moving about, constantly colliding
with one another and with any object in the vicinity of the particles. Even though
you do not feel it, you are constantly being bombarded by the molecules of the
air, as are all the objects within the atmosphere. Gases are quite dynamic, and
this will dictate the behavior of gases. A gas will occupy any volume in which it is
placed, because the particles are in rapid motion and will travel to all parts of
the container even if other gas particles are present in the container. As an example of the rapid motion of gas molecules, the average velocity of a hydrogen

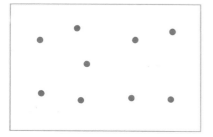

Gases consist of molecules (or atoms) which are so small and so far apart that the volume occupied by the particles is negligible compared to the empty space between them.

There are no appreciable attractive forces between the particles and they behave as tiny masses.

Figure 7-6
The postulates of the Kinetic Molecular Theory. From *Introduction to Chemistry* by T. R. Dickson, John Wiley & Sons, New York, 1971.

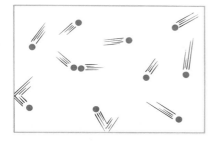

The particles are in rapid, random straightline motion and are constantly colliding with one another and with the walls of the container. As a result of this motion the particles possess kinetic energy ($KE = \frac{1}{2}mv$ where m is the mass and v is the speed.)

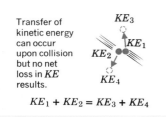

Transfer of kinetic energy can occur upon collision but no net loss in KE results.

$$KE_1 + KE_2 = KE_3 + KE_4$$

The collisions result in no net loss in kinetic energy of the particles.

molecule at room temperature is about 2×10^5 centimeters per second or 3,600 miles per hour. A sample of a gas can be easily compressed, because the sample consists mainly of empty space. The pressure of a gas results from the collisions of gas particles with the objects with which the gas is in contact. The constant collisions of the particles in a sample of a gas with the walls of a container will result in a certain force per unit area (or pressure) being exerted on the surface of the container. The pressure will be directly related to the rate at which the collisions are occurring and the kinetic energies of the particles. When a thermometer is placed in a sample of gas, the gas particles exchange kinetic energy with the thermometer, causing it to register a specific temperature. If the gas sample is heated, increasing the average kinetic energy of the particles, the temperature increases. If the sample is cooled, decreasing the average kinetic energy of the particles, the temperature decreases. In other words, the measurable temperature of a gas is a direct result of the kinetic energy of the particles. In fact, the average kinetic energy of the particles is directly proportional to the temperature.

When a gas is heated, the speeds of the molecules are increased, and the pressure must increase if the volume of the sample is constant. That is, the heating increases the number of collisions of gas molecules with the container and, thus, the pressure increases. If a sample of a gas is heated in a flexible container, the increase in the number of collisions causes the volume of the container to increase.

7-4 MEASUREMENT OF GAS PRESSURE

The volume of a sample of a gas depends upon the temperature and pressure of the sample. Volumes of gas samples are usually expressed in terms of liters or cubic feet. The measurement and expression of the pressure of a gas is often related to the pressure of the air in the atmosphere.

The earth is surrounded by a layer of air contained about the earth by gravitational forces. This air layer is more dense near the lower surfaces of the earth and becomes more diffuse at distances removed from the earth's surface. The molecules and atoms that are present in the atmosphere exert pressures on all objects exposed to the atmosphere. This pressure is called air pressure or atmospheric pressure, and it arises from the continuous bombardment of objects by molecules and atoms in the air. Atmospheric pressure is often measured with a device known as a barometer, invented by E. Torricelli in the seventeenth century. A barometer can be constructed by filling a glass tube, which is sealed on one end, with liquid mercury. This tube is then inverted and supported in a container of mercury which is open to the atmosphere. See Figure 7-7. The pressure exerted on the surface of the mercury in the container by the atmosphere will support a column of mercury in the tube. The height of the mercury column is directly proportional to the atmospheric pressure. In fact, as the atmospheric pressure changes as a result of temperature and weather changes, the height of the mercury column will fluctuate accordingly. The height of the mercury column supported by any gas is proportional to the pressure of the gas. Consequently, gas pressures are often expressed in terms of the number of **millimeters** (or inches) **of mercury** (Hg) the gas will support. A special unit used to express pressure is the **torr.** A torr (named after Torricelli) is defined as follows:

$$1 \text{ torr} = 1 \text{ millimeter Hg}$$

A gas pressure can be expressed in terms of millimeters of mercury or number of torr. Atmospheric pressures given on weather reports are usually stated in inches of mercury (1 inch = 25.4 millimeters).

Atmospheric pressures generally decrease with altitude as the air becomes less dense. Atmospheric pressure around sea level is in the range of 760 torr. Another unit of pressure called the **atmosphere** (atm) is defined as follows:

$$1 \text{ atmosphere} = 760 \text{ torr}$$

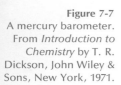

Figure 7-7
A mercury barometer. From *Introduction to Chemistry* by T. R. Dickson, John Wiley & Sons, New York, 1971.

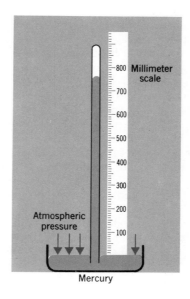

Note that 1 atmosphere is the typical pressure of the atmosphere at sea level. This unit is often used to express high pressures used in industrial and experimental work. For instance, synthetic diamonds can be made by subjecting graphite (a common form of carbon) to high temperatures (2,000°C) and pressures around 70,000 atmospheres.

Pressures of contained gas samples are often measured by attaching special gas pressure gauges to the container. Usually, these gauges are calibrated so that pressures can be read in units of pounds per square inch (psi) or dynes per square centimeter (dyne/cm²). These units express pressures in terms of force per unit area which, as you will recall, is how a pressure is defined. The relation between these units and the atmosphere unit is

$$1 \text{ atm} = 14.7 \text{ psi} = 1.0 \times 10^6 \text{ dyne/cm}^2$$

As this shows, typical atmospheric pressure at sea level is 14.7 pounds per square inch. Imagine a 1 square inch column of air extending from the surface of the earth at sea level to the upper regions of the atmosphere. The air in this column will have a weight of 14.7 pounds. Comparison of the various pressure units is given in Table 7-1.

Unit	Abbreviation
Millimeter of mercury	mm Hg
Inch of mercury	in. Hg
Torr	torr
Atmosphere	atm
Pounds per square inch	psi
Dynes per square centimeter	dynes/cm²

Table 7-1
Various Pressure Units

Equivalents

1 mm Hg = 1 torr
25.4 mm Hg = 1 in. Hg
1 atm = 760 torr
1 atm = 14.7 psi
1 atm = 29.9 in. Hg
$1 \text{ atm} = 1.01 \times 10^6 \text{ dyne/cm}^2$
$1 \text{ atm} = 1,033 \text{ cm H}_2\text{O} = 34 \text{ ft H}_2\text{O}$

7-5 IMPORTANT GASES

Many common molecular substances exist as gases. A few of the most important gases are discussed below.

OXYGEN Oxygen gas, O_2, makes up about 21 percent by volume of the atmosphere. Oxygen liquefies to a pale blue liquid at −183°C, and the liquid solidifies at −218°C. Pure oxygen is used in the manufacture of steel, in large rocket engines, and in sewage treatment. Oxygen in the air reacts with metals to corrode or rust them, and, of course, oxygen is utilized by plants and animals in respiration.

Mixtures of pure oxygen and helium are used medically for patients suffering from anoxia (lack of sufficient oxygen in body tissue). In atmospheres of pure oxygen or those containing high percentages of oxygen, a high risk of fire exists. In 1967, three United States astronauts perished in a fire in a space capsule containing pure oxygen. Now space capsules are filled with 60 percent oxygen and 40 percent nitrogen during preparation for lift-off.

Deep sea divers normally use helium–oxygen mixtures for breathing, and scuba divers use compressed air. One problem of diving using air is that of nitrogen narcosis. The high pressures under water cause some nitrogen to dissolve in the blood. Too much nitrogen can produce an anesthetic effect which has been described as much like alcohol intoxication. Nitrogen narcosis can be very dangerous to divers, and becomes more significant at greater depths, limiting air diving to around 300 feet.

Another problem associated with deep sea and scuba diving is decompression sickness, or the bends. Diving at depths below about 35 feet for long time periods can result in significant amounts of nitrogen (or helium if oxygen–helium mixtures are used) becoming dissolved in the blood through the lungs. More becomes dissolved at greater depths. If the diver quickly surfaces, some of the gases dissolved in the blood or body tissue will form bubbles which cause the decompression sickness. Symptoms include intense pain, dizziness, weakness, numbness, and convulsions. Decompression sickness can cause permanent injury or death. To prevent this sickness, a diver has to make decompression stops at various depths when resurfacing from a dive.

In the presence of an electrical spark, ultraviolet radiation, or certain catalysts, oxygen can be converted to ozone gas, O_3. Ozone is a poisonous, irritating gas that is found in Los Angeles-type smog and in the higher regions of the atmosphere (the ozonosphere).

NITROGEN Nitrogen gas, N_2, makes up about 78 percent by volume of the atmosphere. Nitrogen is a colorless, odorless, and tasteless gas that condenses to a liquid at −196°C, and the liquid solidifies at −210°C. Nitrogen forms several gaseous oxides (N_2O, NO, N_2O_3, NO_2, and N_2O_5). Nitrogen oxide (nitric oxide), NO, is a colorless gas formed when oxygen and nitrogen combine at high temperatures. Nitrogen dioxide, NO_2, is a highly poisonous red-brown gas with an irritating odor. It is formed when nitrogen oxide gas combines with oxygen gas. Both nitrogen oxide and nitrogen dioxide are involved in certain kinds of air pollution.

AMMONIA Ammonia, NH_3, is a colorless gas having a strong odor. You have probably smelled ammonia-containing household cleaners. Ammonia condenses to a liquid at −33°C and solidifies at −78°C. Some refrigeration systems use ammonia. Ammonia is very soluble in water and forms water solutions called aqua ammonia which are sometimes used as cleaning solutions. Ammonia is manufactured by allowing hydrogen gas and nitrogen gas to react in the presence of a catalyst. Ammonia is used in the manufacture of fertilizers, medicines, explosives, and synthetic fibers and plastics. It is used in dyeing, tanning, papermaking, and electroplating processes.

CARBON DIOXIDE Carbon dioxide, CO_2, is a colorless, odorless gas found in the atmosphere (about 0.03 percent by volume). It is produced when organic materials undergo combustion in air and is the product of plant and animal respiration. Solid carbon dioxide is known as "dry ice." Carbon dioxide is used in fire extinguishers, in carbonated beverages, and in the manufacture of some chemicals (sodium carbonate, sodium bicarbonate, and organic chemicals). Carbon dioxide is used by plants in the process of photosynthesis.

CARBON MONOXIDE Carbon monoxide, CO, is a colorless, odorless, highly poisonous gas. It is often formed when incomplete combustion occurs. Carbon monoxide combines with hemoglobin in the blood and interferes with respiration, causing death due to lack of oxygen.

METHANE Methane, CH_4, is a colorless, odorless gas which is the main component of natural gas. Methane is formed in the decomposition of vegetable matter and is found in large quantities in natural gas deposits and in some petroleum deposits. Interestingly, methane is an odorless gas of low toxicity. An especially odiferous gas, methyl mercaptan, is added in minute quantities (part per billion amounts) to natural gas, which is used as fuel. This is done so that we will be aware of any gas leaks that could result in explosions. In addition to its use as a fuel, methane is used to produce carbon black ($CH_4 + O_2 \rightarrow C + 2H_2O$). Carbon black is a powdered form of carbon used in the manufacture of inks, paints, and automobile tires. Tires typically contain 1 part carbon black to every 2 parts rubber. This gives tires the normal black color, but more importantly the carbon black reinforces rubber and increases the wear-resistance. Without the carbon black, a tire would wear out after 500 miles of use. Methane is also used industrially to prepare hydrogen gas ($2 CH_4 + O_2 \rightarrow 2 CO + 4 H_2$).

HYDROGEN Hydrogen, H_2, is a very low-density gas which liquefies at $-253°C$ and solidifies at $-259°C$. Hydrogen is almost always found in the combined state, but a very small amount of hydrogen gas is found in the atmosphere (0.00005 percent). Hydrogen combines explosively with oxygen when a mixture of the gases is ignited. Hydrogen gas is used industrially in the manufacture of ammonia gas, and in the hydrogenation of vegetable oils used for foods and for making soaps. It is also used in the manufacture of hydrochloric acid, methyl alcohol, and other chemicals.

CHLORINE Chlorine, Cl_2, is a green-yellow gas with an irritating odor. It liquefies at $-35°C$ and solidifies at $-101°C$. Chlorine is a highly poisonous gas. Its use as a poisonous gas in World War I resulted in the 1925 Geneva Protocol prohibiting the use of chemical and biological weapons in warfare. Chlorine is used to kill bacteria in water to make it safe for drinking. It is also used in the manufacture of paper, dyes, textiles, medicines, insecticides, paints, plastics, rubbers, chloroform, carbon tetrachloride, bleaching solutions, and bromine.

HYDROGEN CHLORIDE Hydrogen chloride, HCl, is a colorless gas having a sharp odor. It liquefies at $-85°C$ and solidifies at $-115°C$. This gas is manufactured by combining hydrogen gas and chlorine gas ($H_2 + Cl_2 \rightarrow 2$ HCl) and it is formed in large amounts as a by-product in the manufacture of some organic chemicals. Hydrogen chloride is mixed with water to form hydrochloric acid, which is also known as muriatic acid. Hydrochloric acid is used in the manufacture of metal chlorides, dyes, glues, glucose, and various other chemicals. It is also used in electroplating processes and steel manufacturing.

HYDROGEN SULFIDE Hydrogen sulfide, H_2S, is a colorless, highly poisonous gas having a terrible rotten egg-like odor. It liquefies at $-61°C$ and solidifies at $-86°C$. Hydrogen sulfide is found in some petroleum deposits and is a product of the decomposition of proteins. It is produced in large amounts by anaerobic decay of organic matter in coastal mudflats and boggy areas (see Section 11-5).

SULFUR DIOXIDE Sulfur dioxide, SO_2, is a colorless, poisonous gas with a choking odor. This gas is often produced upon the striking of a match. It condenses to a liquid at $-10°C$ and solidifies at $-73°C$. It is manufactured by burning sulfur or pyrites (FeS_2) in air. Sulfur dioxide is used as a bleach for paper, textiles, and flour and sometimes in small amounts as a food preservative in dried fruits. It is also used in the manufacture of paper. The most important use of sulfur dioxide is to produce sulfur trioxide, SO_3, which is used in the manufacture of sulfuric acid. Sulfur dioxide is produced when sulfur-containing fuels are burned and is involved in air pollution.

INERT GASES   The important inert gases include helium, He; neon, Ne; argon, Ar; krypton, Kr; and xenon, Xe. All these chemically unreactive gases occur in slight amounts in the atmosphere. It was not until 1962 that it was found that some of the inert gases can form chemical compounds with other elements. Thus, the term inert is not completely accurate. Most of the helium used in the United States is obtained from helium gas deposits found in Texas, Oklahoma, and Kansas. Helium is used as an inert gas for arc welding and in several industrial processes requiring an inert atmosphere. It is used as a lifting gas in lighter-than-aircraft and as a gaseous coolant in certain nuclear reactors. Neon is obtained from liquified air, and its main use is in neon signs. It is also used in certain electronic tubes and television tubes. Liquid neon is used as a refrigerant in special cooling operations. Argon is also obtained from liquefied air, and its main use is in electric lightbulbs, fluorescent bulbs, and other electronic tubes. It is also used as an inert gas in arc welding and certain industrial processes. Krypton and xenon are obtained from liquid air and are both used in special lamps and electronic tubes.

HALOALKANES   A few of the variety of gases called the haloalkanes are commercially useful. These gases or easily vaporized liquids are generally nontoxic and chemically inert. Freon-12 (dichlorodifluoromethane) is used as a refrigerant in refrigerators and air conditioners. It is widely used as a propellant in aerosol spray cans. Freon-114 (1,1,2,2-tetrafluoroethane) is used as a refrigerant in household refrigerators. Fluothane or halothane (1-bromo-1-chloro-2,2,2-trifluoroethane) is a very common inhalation anesthetic used in surgery.

7-6   MIXTURES, AEROSOLS, AND PARTICULATES

Most of the gases with which we are familiar occur as mixtures—intermingled molecules of various gases. Gases easily form mixtures by diffusing into one another. Since there is much empty space between molecules of a gas, other gas particles can be present without squeezing one another out. Air and natural gas are both mixtures of gases. Of course, it is possible to have a sample of pure gas. Bottled samples of pure gases are available. However, gases that occur in the environment are normally in the form of mixtures.

One obvious characteristic of gases is that they are clear. "Clear" means that we can see through a sample of a gas. Most common gases are colorless and only a few have characteristic colors. But, even if a gas has a color, a sample of such a gas will still be clear. For instance, the gas nitrogen dioxide has a red-brown color. A sample of nitrogen dioxide in a glass container has the red-brown color, but you can look directly through the gas sample.

Of course, we have all seen cloudy or opaque gas samples. Think of the clouds in the sky, a hazy horizon, or a smoke-filled room. Such a cloudy or hazy appearance results from the presence of small pieces of liquids and/or solids intermixed with the gases. The term "small pieces" does not refer to individual molecules or ions but to clumps or droplets containing numerous molecules or ions. Often, these clumps and droplets are large enough to be visible to the eye, others could be seen with a microscope, while some are submicroscopic. Such particles of solids and liquids in the air are noticeable because they reflect and scatter some light passing through the air. This scattering and reflection of light produces the cloudy and hazy effect.

A liquid dispersed as tiny droplets in a gas is called an **aerosol.** An aerosol spray can is designed to disperse a liquid in the air to form an aerosol. Aerosols can take the appearance of mists, fogs, and clouds. Tiny pieces of solids dispersed in a gas are called **particulates.** Particulates occur as smokes or dusts. Dispersed particles of solids and liquids in air play a significant part in air pollution (see Chapter 8) and are often referred to collectively as **particulate pollutants.**

The sizes of particulates vary over a wide range. Some are so large that they quickly settle out from the air as a result of gravitational attraction. Smaller particulates settle out more slowly, since they are buoyed up by wind currents. When particulates have diameters below 0.000001 meter they are beyond the view of the naked eye. The distance 0.000001 (10^{-6}) meter is called a **micron,** and particulates with diameters less than 1 micron are called **submicron** particulates. Submicron particulates are so small that they settle out from air at an extremely slow rate, if at all. These particulates are small enough so that the constant collisions of the air molecules keep them dispersed in the air.

7-7 PARTS PER MILLION AND MICROGRAMS PER CUBIC METER

When dealing with mixtures of gases it is often desirable to state the composition of the mixture in terms of the relative amount of each component that is present in the mixture. One expression of the composition of a mixture of gases is **percentage by volume.** This, of course, expresses the number of parts per hundred parts, where "part" means "volume" (i.e., number of volumes of one constituent in 100 volumes of the whole mixture). Suppose we had 100 liters of air at some temperature and pressure. If the oxygen is removed from the air, 79 liters remain. This means that there are 21 liters of oxygen in 100 liters of air, which is 21 volumes in 100 volumes. Thus, we can say that air is 21 percent by volume oxygen.

To express small concentrations of components of a gas mixture, **parts per million parts of volume** can be used. One part per million by volume would correspond to 1 milliliter in 1,000 liters. For instance, since there is 1 milliliter of krypton gas in each 1,000 liters of air, we say that the concentration of krypton in the atmosphere is 1 part per million (ppm). The relation between percentage by volume and parts per million by volume is the same as that between percentage by mass and parts per million by mass. That is, a percentage can be converted to parts per million by multiplying by 10,000. For instance, the percentage by volume of carbon dioxide in air is about 0.03. Thus, the parts per million amount of carbon dioxide in air is 300 ppm ($0.03 \times 10,000$).

Another method used to express the concentrations of minor components in air samples is to state the mass of the component in a specific volume of air. The mass is normally expressed in micrograms, μg (0.000001 gram = 1 microgram), and the volume in cubic meters, m^3, of air. For instance, the atmosphere contains about 1,400 $\mu g/m^3$ of methane, CH_4. The **microgram per cubic meter** term is often used when stating the concentrations of certain kinds of pollutants in air.

BOOKS   **BIBLIOGRAPHY**

Simpson, C. H. *Chemicals from the Atmosphere.* Garden City, N.Y.: Doubleday, 1969.

ARTICLES AND PAMPHLETS

Broecker, W. S. "Man's Oxygen Reserves." *Science* 1537 (June 26, 1970).
Farber, E. "Oxygen: The Element with Two Faces." *Chemistry* 17 (May 1966).
Hall, M. B. "Robert Boyle." *Scientific American* **217,** (Aug. 1967).

1. Give a statement of Boyle's Law.

2. Give a statement of Charles' Law.

3. A gas sample is maintained in a flexible container at a constant temperature. How will the volume of the gas change when the pressure exerted on the gas is tripled?

4. A gas sample is maintained in a flexible container at a constant pressure. How will the volume of the gas change when the temperature is

   (a) tripled?
   (b) decreased to one-half the original temperature?

5. Why is it dangerous to heat an aerosol can in an open fire?

6. Describe the Kinetic Molecular Theory of gases.

7. Why will a gas occupy any volume into which it is placed?

8. Why are gases easily compressed?

9. How do gases exert pressures on objects?

10. How do you suppose it is possible to smell an odor at a distance from the source of the odor?

11. Describe how a barometer functions.

12. Suppose that the atmospheric pressure is 30.0 inches of mercury.

    (a) What is this pressure in torr?
    (b) What is this pressure in atmospheres?

13. Why would you expect atmospheric pressures to be lower in a mountainous region than in a coastal region?

14. List some uses of oxygen gas.

15. Describe two oxides of nitrogen.

16. What are some uses of ammonia?

17. What is "dry ice"?

18. Does methane gas have an odor? How are natural gas leaks detected?

19. List some uses of hydrogen gas.

20. What are some uses of chlorine gas?

21. What gas is associated with a rotten egg-like odor?

22. Describe two oxides of sulfur.

23. What are two uses of haloalkanes?

24. What is an aerosol?

25. What is a particulate?

26. What is meant by a submicron particulate?

27. Concentrations of gases are sometimes expressed in terms of percent by volume. What does this mean?

28. Look up Robert Boyle in an encyclopedia or history of science book. What is he noted for?

AIR POLLUTION

8

8-1 INTRODUCTION

Throughout history, man has been plagued by the smokes and fumes of burning fuels. As large cities developed and the industrial revolution began, air pollution due to the burning of coal was common to cities like London. In 1661, during the reign of Charles II of England, John Evelyn, a scientist, wrote an essay entitled "Fumifugium; or the Inconvenience of Aer and Skome of London Dissipated; together with Some Remedies Humbly Proposed." Air pollution caused by the burning of coal was common to London for centuries following the rise of industrialization. Serious air pollution episodes were recorded in 1880, 1882, 1873, 1891, 1892, 1952, 1959, and 1962. The worst episode, however, was the "killer fog" in December of 1952. For five days the city was blanketed by a dense smokey fog so thick that the sun was obscured and traffic was stopped. It is estimated that around 4,000 deaths resulted from the episode. After a similar episode in 1962, the English acted to cut down air pollution. Today, the burning of coal is banned in London and, in terms of air pollution, it is one of the cleanest cities of its size in the world.

The United States has also had significant air pollution episodes. In 1948, the city of Donora, Pa. was covered by a layer of smog for several days. Nearly one-half of the population became ill and at least 20 died. In 1963, an air pollution episode involving a portion of the eastern United States was blamed for 400 deaths in New York City.

The dramatic episodes are newsworthy, but many of our cities are plagued by low levels of air pollution during numerous days each year. According to the U.S. Environmental Protection Agency, more than 300 American cities suffer from severe air pollution periodically. We are not alone. Industrial cities around the world including Mexico City, Rio de Janeiro, Paris, Ruhr, Milan, Ankara, New Delhi, Melbourne, Tokyo, and Moscow report smog problems. Apollo astronauts reported seeing yellowish concentrations of smog 100 miles below on the surface of the earth.

Air pollution is known to affect the health of humans and plants. Furthermore, smoggy days have psychological effects on humans. Depression, forgetfulness, and irritability are known to increase on smoggy days. In this chapter we consider the nature and sources of air pollution. The effects of various pollutants on humans and plants are discussed. Finally, various methods of air pollution control are covered.

8-2 THE U.S. ENVIRONMENTAL PROTECTION AGENCY AND THE FEDERAL CLEAN AIR ACT

The **U.S. Environmental Protection Agency** (**EPA**) is a federal agency which includes various subagencies involved in environmental problems such as air pollution, water pollution, solid wastes, noise pollution, pesticide control, and radiation health. Among numerous activities, the EPA is charged with the implementation of the 1972 Water Pollution Control Amendments to the Water Pollution Control Act of 1965 (see Section 11-2) and the 1970 Clean Air Amendments to the Federal Clean Air Act of 1963. As far as the Clean Air Amendments are concerned, the activities of the EPA include the following:

1. Establish **national air quality standards** for various air pollutants. (This has been done and the standards are mentioned in various sections of this chapter.)
2. Specify specific air quality regions in the country.
3. Solicit air pollution control plans from various regions and states, and accept, reject, or alter these plans.
4. Establish regulations for air pollutant emissions from any new stationary sources.
5. Establish air pollutant emission standards for new automobiles which must be achieved by 1975 or 1976.
6. Propose regulations of aircraft air pollutant emissions.
7. Report to Congress on noise pollution.
8. Oversee the achievement of national air quality as specified by the air quality standards by 1976.

The accomplishment of some of these activities by the EPA has generated much controversy. Several automobile manufacturer are challenging the EPA enforcement of the law in court. Another problem of the implementation of the law is best illustrated by the situation in the Los Angeles area, a chronic air pollution region in which a direct conflict with certain national air quality standards exists. In fact, within this region, smoggy days are common and several smog alerts in which school children are not allowed to play outdoors occur each year. The controversy in the Los Angeles region revolves around the fact that no general air pollution control plan has been implemented to improve the air quality. For years, strict control of certain air pollution sources and some control of automobile emissions has been applied in this region. However, the Clean Air Act requires the implementation of a general plan to improve air quality. In the absence of an effective state plan, the EPA is required to devise a plan. In 1973, the EPA suggested a potential plan which was quite dramatic. The fact is that about 70 percent of air pollutants in the Los Angeles region come from automobiles. The use of all possible emission controls on automobiles would cut the pollution by only 25 percent. The EPA stated that the air quality could be improved to acceptable levels by rationing gasoline by up to 80 percent during the smoggy season. Gasoline rationing could be accomplished by use of ration coupons or by selling only a limited amount of gasoline on a first-come, first-served basis.

Needless to say, the EPA suggestion of gas rationing has stimulated a discussion of the problem. The Los Angeles region situation is a real test of the Clean Air Act. The question is, do the people of the region want to sacrifice to obtain air quality as dictated by the law or would they prefer a less stringent approach and air of poorer quality?

8-3 THE ATMOSPHERE

The air of the atmosphere is a mixture of gases which apparently has evolved to the present composition over a period of billions of years. In fact, it is thought that the composition of the atmosphere has not changed for the last 50 million years. However, it now appears that the activities of man are altering the entire atmosphere to a certain extent. Furthermore, it is obvious that the endeavors of an industrial society alter specific regions of the atmosphere, resulting in periodic air pollution episodes.

The vast majority of the estimated 5.5×10^{21} grams of air which make up the atmosphere is located within less than 100 kilometers of the surface of the earth. Atmospheric air moves over the earth by a wind system that changes strength and direction daily. Wind activity occurs in the troposphere (up to about 10 kilometers), and a given air parcel can circle the earth in a matter of days. Of course, air parcels move about somewhat randomly, resulting in a complete intermixing of the gases of the global atmosphere.

Air is a mixture of ten to twenty different gases. The major components of dry air in the troposphere in terms of percentage by volume are the following:

Component	Percentage by volume	Component	Percentage by volume
Nitrogen, N_2	78.08	Neon, Ne	0.00182
Oxygen, O_2	20.95	Helium, He	0.000524
Argon, Ar	0.934	Krypton, Kr	0.000114
Carbon dioxide, CO_2	0.0314		

Note that the air is 99.03 percent nitrogen and oxygen, and 99.96 percent nitrogen, oxygen, and argon.

The minor components of the air are listed in Table 8-1. Most of these minor components appear to be present as a result of natural biological processes and volcanic activity. Air also contains water vapor, H_2O. However, the concentration of water vapor in the atmosphere is variable and ranges from very small concentrations in desert regions to large concentrations in tropical areas. Typically, the concentration of water vapor in air is 1–3 percent by volume. Within a local region, the water content varies daily, depending on temperature and weather conditions. The water vapor content of the atmosphere within a local region is often expressed in terms of **relative humidity.**

Air completely saturated with water vapor at a particular temperature is said to have 100 percent relative humidity. The percent relative humidity expresses the amount of water vapor in the air as a percentage of the amount of water in the air under completely saturated conditions. For example, if the relative humidity

Component	Percentage by volume	Parts per million[a]
Nitrous oxide, N_2O	0.000025	0.25
Hydrogen, H_2	0.00005	0.5
Methane, CH_4	0.00015	1.5
Nitrogen dioxide, NO_2	0.0000001	0.001
Ammonia, NH_3	0.000001	0.01
Ozone, O_3	0.000002	0.02
Sulfur dioxide, SO_2	0.00000002	0.0002
Carbon monoxide, CO	0.00001	0.1

Table 8-1
Trace Gases in Normal Air

[a] The concentrations of some of these gases vary and some concentrations given are approximate.

is 60 percent, this means the air has 60 percent of the water vapor it could hold under saturated conditions.

8-4   AIR SHEDS

Weather changes and winds cause air masses to move from one place to another. However, on land the movement of air masses is dictated to a certain extent by the presence of mountains, hills, valleys, and bodies of water. In other words, the movement of air in the lower altitudes of the troposphere is governed by the topography of the landscape. In fact, as a result of the topography of a region and weather conditions, air masses can be temporarily trapped or isolated within the region. Such a region is called an **air shed** or **air basin.** Temporary immobility of the air in an air shed may last for only a few hours or for several days.

A glance at a topographical map of the United States indicates that major air sheds exist west of the Rocky Mountains, in the great plains area of the midwest, and on the east coast. However, smaller regional air sheds exist all over the country. A topographical map of the state of California reveals the presence of a huge air shed including the Sacramento and San Joaquin Valleys and smaller air sheds such as the Los Angeles basin bounded by mountains to the north and east and the ocean on the west and south.

The existence of these air sheds has an important effect upon the occurrence of air pollution episodes. Air sheds in various regions of the country are currently being studied in an attempt to discover the boundaries of the sheds, how air flows occur within the sheds, and what weather conditions are responsible for the immobility of the air masses within the sheds.

8-5   PRIMARY AIR POLLUTANTS

The activities of an industrial society produce waste gases. Many industrial processes generate by-product gases which are not useful. The automobile produces exhaust gases; most manufacturing processes and the burning of trash produce gases and smoke. When these gaseous products are mixed with the atmosphere, they can become semipermanent components. Just because the products are released into the air does not mean that they are gone. The fact is, they can produce serious air pollution. It is estimated that, in the United States, over 200 million tons of pollutants are released into the atmosphere each year. Air pollution problems arise when these pollutants accumulate in specific geographical areas. The gases that are produced by an industrial society and released into the atmosphere are called **primary** air pollutants. There are five major kinds of gaseous primary pollutants. These pollutants, listed in Table 8-2, are **carbon monoxide, sulfur oxides, nitrogen oxides, hydrocarbons,** and **particulates.** It is worth noting that each of these pollutants is produced in significant amounts by natural biological, volcanic, and geological sources. Table 8-3 compares approximate annual production of natural and man-made amounts of some of these pollutants. In the following sections, each of the primary pollutants will be considered, followed by a discussion of air pollution sources.

Table 8-2 Primary Air Pollutants (as Classified by the U.S. Environmental Protection Agency)			
Carbon monoxide CO		Nitrogen oxides NO_x	
			nitric oxide, NO
			nitrogen dioxide, NO_2
Sulfur oxides	SO_x	Hydrocarbons	HC
	sulfur dioxide, SO_2		
	sulfur trioxide, SO_3		
Particulates			

Gas	Natural production (tons per year)	Man-made production (tons per year)
Carbon monoxide	3.5×10^9	3×10^8
Nitrogen oxides	1.4×10^9	1.5×10^7
Sulfur oxides (and hydrogen sulfide)	1.42×10^8	7.3×10^7

Table 8-3
Annual Natural and Man-Made Production of Some Gases

8-6 CARBON MONOXIDE

Carbon monoxide is an unwanted product of the combustion of fossil fuels. Carbon monoxide is formed by incomplete combustion of carbon or carbon-containing compounds:

$$2\,C + O_2 \longrightarrow 2\,CO$$

The formation of carbon monoxide is common in the internal combustion engines of motor vehicles. Consequently, carbon monoxide accumulates in urban areas near freeways and busy streets and varies in concentration as the amount of traffic varies. Figure 8-1 lists the daily variation of carbon monoxide levels on a New York City street.

Carbon monoxide at the levels found in urban air does not appear to have an effect on plants, but it is poisonous to humans. Carbon monoxide interferes with the transport of oxygen in the bloodstream. **Hemoglobin** (symbolically represented as Hb) is the compound in blood which is involved in oxygen transport. Hemoglobin combines with oxygen in the lungs to form **oxyhemoglobin** (HbO_2) which flows through the blood to carry oxygen to the cells of the body. Carbon monoxide entering the lungs from polluted air or cigarette smoke can combine with hemoglobin to form **carboxyhemoglobin** (HbCO). The carbon monoxide binds strongly with the hemoglobin, preventing the normal transport of oxygen.

The national ambient air quality standards for carbon monoxide as established by the U.S. Environmental Protection Agency (EPA) call for a maximum 8 hour concentration of 10 milligrams per cubic meter, or 9 parts per million, not to be exceeded once a year, and a maximum 1 hour concentration of 40 milligrams per cubic meter, or 35 parts per million, not to be exceeded once a year. The basic purpose of these standards is to protect public health. Tests of air in specific urban areas are carried out periodically. Current and past tests indicate that the air of several cities does not meet the EPA standards. The intent of the Clean Air Act of 1970 is to have the states devise air pollution control plans (subject to EPA approval) which will allow the achievement of the national air quality standards.

The level of carbon monoxide absorbed in the blood depends upon the carbon monoxide content of the air. After breathing air containing carbon monoxide over a period of time, the carbon monoxide level in the blood reaches a specific level as shown in Figure 8-2. The blood levels will decrease, however, when the carbon monoxide air level decreases. Occasionally, traffic policemen in Tokyo have to breath pure oxygen to remove the accumulated carbon monoxide in their blood. Incidentally, since cigarette smoke contains carbon monoxide, a smoker may have from two to five times as much carbon monoxide in the blood as a nonsmoker.

The health effects of various carbon monoxide blood levels are shown in Table 8-4. Low levels of carbon monoxide can decrease a person's ability to perceive light, or reduce the ability to estimate passages of time. Slightly higher levels reduce the supply of oxygen to the cardiac muscles and may cause heart

Figure 8-1
Variation of carbon
monoxide levels on a
New York City street.
From "Carbon
Monoxide:
Measurement and
Monitoring in Urban
Air," by Philip Wolf,
*Environmental Science
and Technology* **5** (3)
(Mar. 1971).

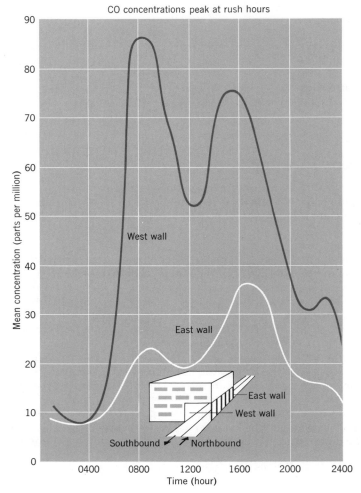

Peaks: CO concentrations, taken at three different levels, 75 feet inside the FDR Drive at 54th Street in New York City, show the differences in carbon monoxide buildup in completely enclosed areas versus partially open areas

Figure 8-2
Toxicity of carbon
monoxide. From
"Carbon Monoxide:
Measurement and
Monitoring in Urban
Air," by Philip Wolf,
*Environmental Science
and Technology* **5** (3)
(Mar. 1971).

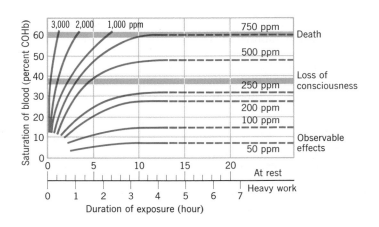

COHb level (percent)	Demonstrated effects
Less than 1.0	No apparent effect
1.0–2.0	Some evidence of effect on behavioral performance
2.0–5.0	Central nervous system effects; impairment of time interval discrimination, visual acuity, brightness discrimination, and certain other psychomotor functions
Greater than 5.0	Cardiac and pulmonary functional changes
10.0–80.0	Headaches, fatigue, drowsiness, coma, respiratory failure, death

Table 8-4
Health Effects of Carbon Monoxide—as COHb levels or Duration of Exposure Increase, Health Effects Become More Serious

From "Carbon Monoxide: Measurement and Monitoring in Urban Air," by Philip Wolf, *Environmental Science and Technology* **5** (3) (Mar. 1971).
Original source: National Academy of Science and National Academy of Engineering.

attacks in vulnerable persons. Higher concentrations of blood carbon monoxide can cause headaches, fatigue, drowsiness, coma, and death. Research on the health effects of low levels of carbon monoxide is currently being carried on. Recent studies at the Long Beach, Cal., Veterans Administration Hospital have shown that heart patients are more likely to suffer heart pain after a 90 minute freeway ride as compared to patients who have not traveled the freeways.

8-7 SULFUR DIOXIDE

Sulfur dioxide, SO_2, is a primary pollutant produced from the combustion of sulfur-containing compounds. It is formed when sulfur-bearing coal and oil are burned:

$$S \text{ (fuels)} + O_2 \longrightarrow SO_2$$

Sulfur dioxide is also produced in the refining of certain ores which are sulfides. For example, one of the first steps in the refining of lead ore (PbS) is the roasting process:

$$2\,PbS + 3\,O_2 \longrightarrow 2\,PbO + 2\,SO_2$$

Sulfur dioxide is the most common sulfur oxide involved in air pollution. However, sulfur trioxide, SO_3, is emitted to a certain extent by industrial process and is also formed in the atmosphere in small amounts by a reaction between sulfur dioxide and oxygen:

$$2\,SO_2 + O_2 \longrightarrow 2\,SO_3$$

This reaction is catalyzed by certain particulates in the air. Sulfur dioxide and sulfur trioxide are sometimes referred to collectively as the sulfur oxides, SO_x. When water vapor and water aerosols (fogs and mists) are present in air, sulfur trioxide reacts to form sulfuric acid:

$$SO_3 + H_2O \longrightarrow H_2SO_4$$

Sulfuric acid, H_2SO_4, is a very dangerous secondary air pollutant. A **secondary** air pollutant is one which is produced by chemical reactions of primary pollutants with one another or other components of the atmosphere. See Section 8-19 for additional discussion of secondary pollutants.

The EPA national ambient air quality standards for sulfur dioxide are an annual daily average of 80 micrograms per cubic meter, or 0.03 part per million, and a 24 hour maximum concentration of 365 micrograms per cubic meter, or 0.14 part

Table 8-5
Health Effects of
Sulfur Dioxide

Effects on humans	Concentration (parts per million)
Causes bronchoconstriction	1–6
Lowest concentration for detectable odor	3–5
Throat irritation begins	8–12
Eye irritation and coughing begin	20
Maximum concentration for short (30 minute) exposure	50–100
Can be deadly even for short exposure	400–500

Adapted from "Physiological Effects of Sulfur Dioxide Gas," in R. E. Kirk and D. F. Othmer, eds., *Encyclopedia of Chemical Technology*, 2nd ed., vol. 19, John Wiley & Sons, New York, 1969, p. 417.

per million, not to be exceeded more than once a year. Sulfur dioxide content of the air is quite variable in different regions of the country, but it is most prevalent in eastern states where large amounts of coal are burned.

Sulfur oxides in air can inhibit plant growth and be lethal to certain plants. Leaf areas die and dry out when plants are exposed to sulfur dioxide over long periods at sublethal levels. Plants may be affected when sulfur dioxide levels exceed 0.3–0.5 part per million, but various plant species differ in their response to sulfur dioxide poisoning.

Noticeable effects of sulfur dioxide on humans occur at higher levels than those causing plant damage. The physiological effects of various levels of sulfur dioxide on man are listed in Table 8-5. It is thought that long-term exposure to concentrations above 1 part per million are required before any effects are apparent. The long-term effects on humans of low levels of sulfur dioxide are not known. There is no definite evidence that sulfur dioxide causes respiratory ailments, but there is a definite correlation between the incidence of sulfur oxides in the atmosphere and the death rate of persons suffering from chronic cardiovascular and respiratory diseases. In fact, it is the sick and old suffering from these diseases who are affected the most by sulfur oxide air pollution episodes. In the "killer" fogs of London the death rate among the sick and old was higher than the general population, and the most frequent causes of death were bronchitis, bronchopneumonia, and heart disease.

8-8 NITROGEN OXIDES

The oxides of nitrogen, nitric oxide (NO) and nitrogen dioxide (NO_2), are primary air pollutants. These oxides are referred to collectively as **nitrogen oxides,** NO_x. Nitric oxide is a colorless, odorless gas, while nitrogen dioxide is a red-brown gas with a strong, choking, chlorine-like odor. Nitric oxide is formed by reaction of nitrogen and oxygen in the air:

$$N_2 + O_2 \longrightarrow 2\,NO$$

This reaction occurs at high temperatures during the combustion of fossil fuels. Nitrogen dioxide is formed by the reaction of nitric oxide with oxygen in the air.

$$2\,NO + O_2 \longrightarrow 2\,NO_2$$

An average annual level of 100 micrograms per cubic meter, or 0.05 part per million, is the EPA national air quality standard for nitrogen oxides.* As is the case with other pollutants, the concentration of nitrogen oxides in the air can vary daily and differ widely in various regions of the country. Figure 8-3 shows the

* The air quality standard for nitrogen oxides is being revised.

Figure 8-3
Regions in the United States with periodically high nitrogen oxide levels.

regions of the United States which have chronically high nitrogen oxide levels (1972).

The effects of nitrogen oxides on plants is not known with certainty. A certain amount of plant damage occurs with nitrogen dioxide levels higher than those found in air. However, some secondary pollutants formed from nitrogen oxides are major plant-pathogenic air pollutants (see Section 8-19).

The physiological effects of various nitrogen dioxide levels on humans are listed in Table 8-6. There is little information available concerning the effects on humans of nitrogen oxide levels of less than 1 part per million. Nitrogen dioxide affects the lungs and is toxic. Currently, the relation between public health and long-term exposure to low levels of nitrogen oxides is being researched.

8-9 HYDROCARBONS

The hydrocarbons (represented by the symbol HC) are classed as primary pollutants. Recall that the hydrocarbons (see Chapter 5) are organic compounds comprised of various combinations of carbon and hydrogen. Four common hydrocarbons, methane, ethane, propane, and butane, are gases, and most of the other hydrocarbons are liquids. Incidentally, methane, because of its natural abundance in air and its chemical behavior, is not included in the category of hydrocarbon air pollutants. Many of the liquid hydrocarbons are highly **volatile** and can be changed to gases by **evaporation** at normal atmospheric temperatures. Hydrocarbons enter the air by evaporation of fossil fuel products such as

Effects on humans	Concentration (parts per million)
Lowest concentration for detectable odor	1–3
Nose, throat, and eye irritation begin	13
Causes lung congestion and disorder	25
Can be deadly even for short exposure	100–1,000

Table 8-6
Health Effects of Nitrogen Dioxide

gasoline. Maybe you have noticed the gasoline fumes entering the atmosphere from the gas tank of a car as it is being filled with gasoline. Unburned hydrocarbons also escape during the combustion of gasoline, oil, coal, and wood.

The EPA national air quality standard for hydrocarbons is 160 micrograms per cubic meter, or 0.24 part per million, maximum for 3 hour concentration (6–9 AM) not to be exceeded more than once a year. Common hydrocarbons are toxic to plants and animals at relatively high concentrations (500 parts per million and above), but, at the levels at which they exist in air, there is no evidence that they have a significant effect on humans. Unsaturated and aromatic hydrocarbons have greater toxicities than other hydrocarbons, but no effects are apparent below about 25 parts per million concentration.

8-10 PARTICULATES

In addition to gaseous pollutants, polluted air may contain suspended and dispersed particles of solids and liquids. This particulate material is classed as a primary pollutant. These particulates account for the hazy, cloudy appearance of polluted air. The composition of particulates is quite variable and complicated. Some take the form of true liquid aerosols and appear as mists or fogs. Some take the form of solid smokes, dusts, and fly ash. On the other hand, particulates may be a conglomeration of solids and liquids. Particulates are found to contain a variety of substances such as aluminum, calcium, iron, lead, magnesium, and sodium in various states of combination with ions such as nitrate ion, sulfate ion, and chloride ion. Some are found to contain organic compounds, sulfuric acid, and nitric acid. Most of the lead in particulate pollution comes from lead alkyl fuel additives. Some particulates absorb gaseous pollutants from polluted air.

The EPA national air quality standards for particulates call for a 75 micrograms per cubic meter yearly average and a 260 micrograms per cubic meter 24 hour maximum not to be exceeded more than once a year.

Particulates are great health hazards to humans. Breathing particulate polluted air can result in particulates being deposited in the lungs. Actually, particulates larger than about 5.0 microns (see Section 7-7 for a discussion of particulate sizes) are too large to enter the lungs and are trapped in the nose and throat. Particulates from about 1.0 to 5.0 microns in size can be deposited in portions of the lungs. The submicron particulates are the most dangerous, since they can be deposited in tiny alveoli (air sacs) in the lungs. When particulates are deposited in the lungs, they may irritate the tissue or they may contain toxic material which affect the respiratory process. Levels of particulate pollution have been directly correlated to respiratory disorders. Particulates can carry acids, toxic gases, and radioactive materials to the lungs. Also of great public health concern are particulates containing toxic metals, beryllium and asbestos. In fact, asbestos, beryllium, and mercury are classified by the EPA as hazardous air pollutants, and emission standards for them are pending.

8-11 AIR POLLUTION SOURCES

Each year around 1 ton of air pollutants are emitted into the atmosphere for each person in the United States. This adds up to over 200 million tons of pollutants annually. These pollutants come mainly from transportation, electrical power plants, industrial processes, and solid waste incineration. Table 8-7 lists the sources of pollutants and the percentage each source contributes to the total amount of pollution. Of course, the effect each source has on air pollution problems depends upon many factors. For instance, in California, transportation emissions provide most problems, while in parts of the eastern United States,

Source	Percent by weight of grand annual total (over 200 million tons)					
	Carbon monoxide	Sulfur oxides	Nitrogen oxides	Hydro-carbons	Partic-ulates	Total
Transportation	29.8	0.4	3.8	7.75	0.56	42.3
Fuel combustion (power plants, space heating, etc.)	0.89	11.4	4.67	0.3	4.2	21.4
Industrial processes	4.5	3.4	0.09	2.1	3.5	13.6
Solid waste disposal	3.6	0.05	0.3	0.75	0.51	5.2
Miscellaneous (forest and agricultural fires, etc.)	7.89	0.3	0.79	4.0	4.5	17.5
Total	46.7	15.5	9.6	14.9	13.3	100.0

Table 8-7
Sources of Air Pollutants in the United States

Data from U.S. Department of Health, Education and Welfare, 1968.

power plant emissions are more significant. Some experts feel that air pollutants should be considered with respect to their effects on humans, plants, and materials rather than a weight percentage basis. Lyndon Babcock, Jr., has developed a weighted listing of air pollution sources which considers air pollution standards and other factors. According to this listing, shown in Table 8-8, particulates are the pollutants of greatest concern, and electrical power plants and other fuel-burning stationary sources are the most significant pollution sources. It is interesting to compare Tables 8-7 and 8-8. Let us take a closer look at the sources of each of the primary pollutants.

8-12 CARBON MONOXIDE SOURCES

As Table 8-9 shows, the greatest single source of carbon monoxide by weight is transportation. Over 90 percent of this carbon monoxide comes from auto-

Source	Percentage of grand annual total (over 200 million tons) on weighted basis using air quality standards and other factors					
	Carbon monoxide	Sulfur oxides	Nitrogen oxides	Hydro-carbons	Partic-ulates	Total
Transportation	3.0	1.0	6.9	4.6	0.9	16.4
Fuel combustion (power plants, industrial space heating, etc.)	0.1	19.8	3.7	1.1	10.2	34.9
Industrial processes	0.3	6.1	0.1	0.1	20.4	27.0
Solid waste disposal	0.2	0.1	0.3	0.2	2.0	2.8
Miscellaneous (forest and agricultural fires, etc.)	0.5	0.2	1.3	0.8	16.1	18.9
Total	4.1	27.2	12.3	6.8	49.6	100.0

Table 8-8
Sources of Air Pollutants in the United States on a Weighted Basis

Adapted from Lyndon Babcock, Jr., and Niren Nagada, University of Illinois at Chicago Circle, *Chemical and Engineering News* 33 (Jan. 10, 1972).

Table 8-9
Carbon Monoxide
Sources

Source	Percent of total annual CO emissions	
Transportation	63.8	
Motor vehicles (gasoline)		59.0
Motor vehicles (diesel)		0.2
Airplanes		2.4
Nonhighway use of motor fuels		1.8
Marine vessels		0.3
Trains		0.1
Fuel combustion (stationary sources—power plants, industrial space heating, etc.)	1.9	
Wood		1.0
Coal		0.8
Fuel oil		0.1
Industrial processes	9.6	
Iron and steel refining and manufacturing		6.0
Petroleum refineries		2.2
Kraft pulp and paper mills		0.8
Others		0.6
Solid waste disposal	7.8	
Conical burners		3.6
Open-pit burning		3.4
Incinerators		0.8
Miscellaneous	16.9	
Agricultural burning		8.3
Forest fires		7.2
Coal waste burning		1.2
Structural fires		0.2
Total	100.0	

Based on U.S. Department of Health, Education and Welfare data, 1968.

mobiles, and the remainder comes from such vehicles as airplanes and diesel trucks and trains. A small amount of carbon monoxide comes from electrical power plants, while a larger amount is produced in industrial endeavors such as steel manufacturing, petroleum refining, and paper mills. The burning of solid wastes accounts for still more carbon monoxide. The miscellaneous sources of carbon monoxide are forest fires, domestic fires, coal waste burning, and agricultural burning.

Gases which enter the atmosphere as pollutants can become widely dispersed and diluted in the air. Air pollution problems arise when pollutants become trapped in confined air masses. In any case, most primary air pollutants reside in the air for a certain period but are removed by various processes. This removal involves the chemical reaction or biological use of the pollutant. The ultimate region in which the pollutant ends up is called a **sink.** Carbon monoxide, for example, appears to be removed from the air by some microorganism in the soil. Since this microorganism resides in the soil, carbon monoxide is said to have a **soil sink.** The existence of a sink for a pollutant is important, since, in the absence of a sink, the pollutant would accumulate in the atmosphere.

8-13 NITROGEN OXIDE SOURCES

A large amount of nitric oxide is emitted into the atmosphere by certain kinds of bacteria. This is natural emission which cannot be controlled or stopped. As

Source	Percent of total annual NO_x emissions	
Transportation	39.3	
Motor vehicles (gasoline)		32.0
Motor vehicles (diesel)		2.9
Trains		1.9
Nonhighway use of motor fuels		1.5
Marine vessels		1.0
Fuel combustion (stationary sources— power plants, industrial space heating, etc.)	48.5	
Natural gas		23.3
Coal		19.4
Fuel oil		4.8
Wood		1.0
Industrial processes (nitric acid plants, etc.)	1.0	
Solid waste disposal	2.9	
Miscellaneous	8.3	
Forest fires		5.8
Agricultural burning		1.5
Coal waste burning		1.0
Total	100.0	

Table 8-10
Nitrogen Oxide Sources

Based on U.S. Department of Health, Education and Welfare data, 1968.

Table 8-10 shows, most of the man-made nitrogen oxides come from electrical power plants in which the high temperature of the burning of fossil fuels induces the oxide formation. In a like manner, the high temperature of the internal combustion engine results in transportation vehicles being the second major source of nitrogen oxides. Industrial processes and refuse disposal account for only a small fraction of the nitrogen oxides. Miscellaneous sources are forest fires, domestic fires, coal waste burning, and agricultural burning. Nitrogen oxides are involved in the formation of secondary air pollutants, which tends to remove a small portion from the atmosphere. Most of the oxides are ultimately converted to nitric acid, HNO_3, and nitrates, NO_3^-. In this form they end up deposited on the land or in the ocean by rainfall or the settling of particulates. Thus, the soil and the ocean are the ultimate sink for the nitrogen oxides once they have been converted to nitrates.

8-14 SULFUR OXIDE SOURCES

The nationwide emissions of sulfur oxides are listed in Table 8-11. As can be seen, the greatest source of man-made oxides of sulfur comes from the combustion of coal and oil in electrical generating plants. Coal and oil both contain small percentages of sulfur-containing compounds. Thus, the sulfur oxides are an unwanted by-product of fossil fuel use. Industrial processes account for most of the remaining emissions. The industrial processes that are the major contributors are the roasting of sulfide ores, petroleum refining, sulfuric acid manufacturing, and the manufacturing of coke from coal. Transportation sources and miscellaneous sources account for only a small fraction of the sulfur oxides in the atmosphere. Sulfur oxides are removed from air by conversion to sulfuric acids and sulfates. In this form they end up deposited on the land or ocean by

Table 8-11
Sulfur Oxide Sources

Source	Percent of total annual SO$_x$ emissions	
Transportation	2.4	
Motor vehicles (gasoline)		0.6
Motor vehicles (diesel)		0.3
Marine vessels		0.9
Nonhighway use of motor fuel		0.3
Trains		0.3
Fuel combustion (stationary sources—power plants, industrial space heating, etc.)	73.5	
Coal		60.5
Fuel oil		13.0
Industrial processes	22.0	
Metal smelting		11.8
Petroleum refineries		6.3
Sulfuric acid plants		1.8
Coal coking		1.8
Others		0.3
Solid waste disposal	0.3	
Miscellaneous (waste coal burning)	1.8	
Total	100.0	

Based on U.S. Department of Health, Education and Welfare data, 1968.

rainfall or the settling of particulates. As was the case with nitrogen oxides, the ultimate sink for the sulfur oxides is the soil or the ocean.

8-15   HYDROCARBON SOURCES

As might be expected, the greatest single source of hydrocarbons is transportation. Table 8-12 lists the sources of hydrocarbons. Small amounts of hydrocarbons are emitted from electrical power plants and refuse disposal. A larger amount comes from industrial processes. Undoubtedly the greatest portion of this is produced by the petroleum industry. Miscellaneous sources account for over one-quarter of hydrocarbon emission. These sources include forest fires, domestic fires, coal waste burning, agricultural burning, solvent evaporation, and gasoline evaporation from gas stations. The automobile is the greatest single source of hydrocarbons. As shown in Figure 8-4, an automobile emits hydrocarbons by evaporative loss and as unburned hydrocarbons in the exhaust.

8-16   PARTICULATE SOURCES

Particulates are produced along with gaseous air pollutants by the various activities of an industrial society. Table 8-13 lists the sources of particulates. The burning of fossil fuels and especially coal in electrical power plants accounts for a large portion of the particulates. A similar portion of the particulates is produced by miscellaneous burning such as forest fires. Transportation sources contribute a small amount of particulates. The remainder of the particulates come from industrial sources. The greatest industrial contributors are iron and steel manufacturing, cement production, rock and sand mining and marketing, grain storage and marketing, and pulp and paper manufacturing. Typical particulate concentrations and sizes are given in Table 8-14.

Source	Percent of total annual HC emissions	
Transportation	51.9	
Motor vehicles (gasoline)		47.5
Motor vehicles (diesel)		1.3
Nonhighway use of motor fuels		1.0
Airplanes		0.9
Trains		0.9
Marine vessels		0.3
Fuel combustion (stationary sources—power plants, industrial space heating, etc.)	2.2	
Wood		1.3
Coal		0.6
Fuel oil		0.3
Industrial processes (evaporation during manufacture, storage, and transfer of hydrocarbon-containing products)	14.4	
Solid waste disposal	5.0	
Miscellaneous	26.5	
Organic solvent evaporation		9.7
Forest fires		6.9
Agricultural burning		5.3
Gasoline transfer and marketing		3.7
Coal waste burning		0.6
Structural fires		0.3
Total	100.0	

Table 8-12
Hydrocarbon Sources

Based on U.S. Department of Health, Education and Welfare data, 1968.

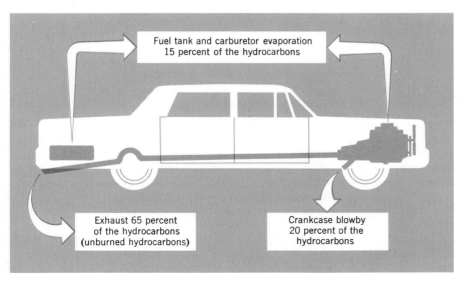

Figure 8-4
Sources of hydrocarbons in an uncontrolled automobile. Before emission controls, automobiles emitted an average of 11 grams of hydrocarbons per mile of driving.

Table 8-13
Particulate Sources

Source	Percent of total annual particulate emission	
Transportation	4.3	
Motor vehicles (gasoline)		1.8
Motor vehicles (diesel)		1.0
Trains		0.7
Marine vessels		0.4
Nonhighway use of motor fuels		0.4
Fuel combustion (stationary sources—power plants, industrial space heating, etc.)	31.4	
Coal		29.0
Fuel oil		1.0
Natural gas		0.7
Wood		0.7
Industrial processes	26.5	
Iron and steel refining and manufacturing		7.3
Other metals refining and manufacturing		0.4
Cement		3.1
Rock, sand, and gravel		3.1
Grain transfer and storage		2.8
Pulp and paper		2.5
Asphalt processing		1.9
Lime		1.6
Flour and feed milling		1.1
Phosphate rock		0.7
Coal cleaning		0.6
Other mineral processing		0.6
Petroleum refineries		0.4
Other chemical industries		0.3
Others		0.1
Solid waste disposal	3.9	
Miscellaneous	33.9	
Forest fires		23.7
Agricultural burning		8.4
Coal waste burning		1.4
Structural fires		0.4
Total	100.0	

Based on U.S. Department of Health, Education and Welfare data, 1968.

8-17   AIR POLLUTION PHENOMENA AND SMOG

If the 600,000 tons of air pollutants produced in the United States each day were completely mixed with the entire air mass above the United States, air pollutants would be present at very low concentrations. Of course, this does not occur. The accumulation of air pollutants within a restricted geological area can bring about an **air pollution episode.** The existence of specific air sheds in various geological regions was discussed in Section 8-4. As a result of weather fluctuations and winds, air masses can be swept horizontally from an air shed to another region of the atmosphere. When such **horizontal** movement of air masses is unrestricted, air pollutants are quickly dispersed. However, topological factors of valleys, hills, and mountains can restrict the movement of air masses, especially if winds are minimal. Whenever an air mass within an air shed is not replenished by horizontal air movement, it is called a **stagnant air mass.** Stagna-

Table 8-14
Typical Particulate
Concentrations and
Sizes

Location	Concentrations (micrograms per cubic meter)
Rural–nonurban regions	10
Suburban–near urban regions	60
Urban regions	60–220

Processes	Particle size (microns)
Grinding, spraying, wind, erosion, crushing of materials by vehicular traffic, certain industrial processes	10 and above
Industrial dusts, soil dusts, agricultural dusts, some combustion processes	1–10
Combustion of fuels and materials, secondary particulates formed in photochemical smog	submicron to 1

Based on U.S. Department of Health, Education and Welfare data, 1961–1965.

tion of air masses can occur when specific unchanging weather patterns exist. A stagnant air mass often remains only for a few hours, but such stagnation can last for weeks. Air pollutants can accumulate within a stagnant air mass, resulting in an air pollution episode. The London "killer" fogs of 1952 and 1962 were air pollution episodes of this type. The air pollutants (mostly sulfur oxides) accumulated in a moist, cold, immobile air mass shrouding the city of London for weeks.

8-18 TEMPERATURE INVERSIONS

Another means by which air mass movement can occur is by the rising of warm air to higher regions of the atmosphere. When air near the surface of the earth is heated, it becomes less dense and rises vertically; it is replaced by cooler air from the higher regions of the atmosphere. This **vertical** movement of the surface air can disperse air pollutants into higher regions of the atmosphere. Normally, the temperature of the air decreases with altitude, as illustrated in Figure 8-5. The air mass near the earth's surface is warmer. Sometimes a cooler mass of air moves in at a low altitude and underlies the warm air mass. This results in a layer of cooler air under a layer of warm air. As shown in Figure 8-5, the layer of warm air becomes trapped between the cool air mass and the cooler air of higher altitudes. This phenomenon is called a **temperature inversion.** When an inversion occurs, the temperature of the air decreases with altitude up to the warm air layer. At this level, the temperature begins to increase with altitude until the overlying cooler air is reached. Beyond this point the temperature decreases with altitude as usual. A temperature inversion can occur in a sheltered air shed and result in an **immobile air mass** in which pollutants can accumulate. Often, temperature inversions will break up at night, but a new inversion may form the next day. Sometimes an inversion may persist for days. Temperature inversions often occur during cloudless sunny days which, as discussed below, compounds the air pollution episode.

Once a temperature inversion has formed, primary air pollutants can become trapped and accumulate in localized areas. If the inversion occurs during a warm, cloudless day, the primary air pollutants in the presence of sunlight can form secondary air pollutants. The sunlight-induced chemical reactions which produce secondary pollutants are called **photochemical reactions.** The inversion layer acts as a large reaction vessel in which photochemical reactions and

Figure 8-5
Normal and inversion
conditions in an air
basin.

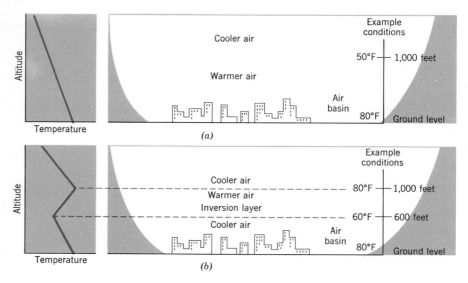

(a)

(b)

subsequent reactions produce a variety of secondary pollutants known collectively as **smog.** Such **photochemical smog** is characteristic of air pollution episodes in various regions of the country and especially regions of California. Photochemical smog is a complex mixture of chemical compounds produced from the air pollutants originating mainly from the automobile. However, other sources contribute to a certain extent.

8-19 PHOTOCHEMICAL OXIDANTS

The **secondary** pollutants of greatest concern are called **photochemical oxidants.** Let us consider some of the photochemical oxidants and the chemistry of their formation. Nitrogen dioxide, a primary pollutant, in the presence of sunlight is involved in a photochemical cycle illustrated in Figure 8-6a. As can be seen in the figure, ultraviolet radiation from sunlight causes the nitrogen dioxide to split into nitric oxide, NO, and oxygen atoms, O. The oxygen atoms react with oxygen molecules, O_2, in the atmosphere to produce ozone, O_3. **Ozone** is a secondary pollutant. The cycle is completed when ozone reacts with nitric oxide to give nitrogen dioxide and molecular oxygen. Since the basic nitrogen dioxide cycle uses as much ozone as it produces, it cannot produce the levels of ozone observed in photochemical smog. Such a buildup in ozone arises when hydrocarbons disrupt the cycle by entering into chemical reactions which allow nitric oxide to be reconverted to nitrogen dioxide without using up the ozone. The cycle is unbalanced, and nitric oxide is reconverted to nitrogen dioxide at a faster rate than nitrogen dioxide breaks up. The net result is an accumulation of ozone. Furthermore, as shown in Figure 8-6b, the introduction of hydrocarbons in the nitrogen dioxide cycle produces many other secondary pollutants comprising smog.

Smog contains over 50 different components. The chemistry involved in the formation of many of these components is complex and is incompletely known. It is known that hydrocarbons react with atomic oxygen, ozone, and molecular oxygen to form species known as **radicals.** These radicals react with other hydrocarbons, oxygen species, and nitrogen oxides to produce the components of smog. One notable secondary pollutant produced in such reactions is **peroxyacetylnitrate,** the simplest of a group of compounds, the peroxyacylnitrates, referred to collectively as **PAN.**

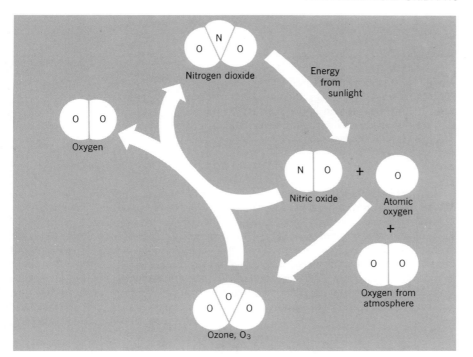

Figure 8-6a
The nitrogen dioxide–ozone cycle. Nitrogen dioxide decomposes in the presence of sunlight to form atomic oxygen and nitric oxide. The atomic oxygen combines with oxygen to form ozone. To complete the cycle the nitric oxide and ozone combine to form nitrogen dioxide and oxygen.

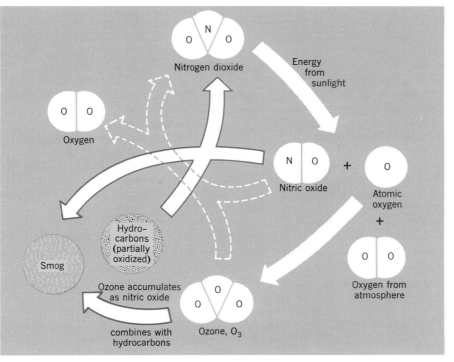

Figure 8-6b
The nitrogen dioxide–ozone cycle in the presence of hydrocarbons. Hydrocarbons in air combine with nitric oxide as one step in smog formation. Removal of the nitric oxide from the nitrogen dioxide–ozone cycle allows ozone to accumulate and participate in smog formation.

Table 8-15
Health Effects of
Ozone

Effects on humans[a]	Concentration (parts per million)
Lowest concentration for detectable odor	0.02–0.05
Nose and throat irritation begin	0.3
Causes fatigue, headache, and loss of coordination	1–3
Causes lung congestion and disorder	9
Can be deadly even for short exposure	15

[a] Quite variable on various individuals.

The mixture of substances involved in photochemical smog include nitrogen oxides, carbon monoxide, hydrocarbons, ozone, PAN, aldehydes, and a variety of other organic compounds. Particulates are also associated with photochemical smog and account for the hazy or cloudy appearance. It appears that primary particulates may be involved in photochemical smog formation and others, called **secondary particulates,** may be produced as the products of smog formation accumulate. The basic ingredients for photochemical smog formation include a temperature inversion, sunlight, nitrogen oxides, hydrocarbons, and possibly carbon monoxide, particulate catalysts, water vapor, and oxygen. Research into the nature, formation, and long-term effects of photochemical smog is now underway.

The EPA national air quality standards for photochemical oxidants, the most notable of which are ozone and PAN, have been established. The standards call for 160 micrograms per cubic meter, or 0.08 part per million, maximum 1 hour concentrations not to be exceeded more than once a year.

Ozone is a gas with a sharp odor. High levels of ozone are present in photochemical smog and, thus, the measurement of ozone levels gives an indication of the extent of photochemical activity. Ozone can cause nose and throat irritation, loss of muscular coordination, and tiredness. Table 8-15 lists the known effects of various ozone concentrations on humans. The typical ozone levels found in polluted air (below 0.1 part per million) are just below the levels which cause irritation. Unfortunately, the long-term effect of breathing low levels of ozone is not known. Many plants are very sensitive to ozone and can be injured or killed by low levels. Concentrations as low as 0.05 part per million are known to cause injury to sensitive plant species. Certain crops of leafy vegetables cannot be grown in areas of California. In Southern California, numerous ozone-sensitive pine trees are being killed. Ozone also attacks fabrics and rubber. Many rubber products develop cracks and deteriorate upon exposure to ozone. Antioxidants are added to tires to retard ozone attack.

PAN is a tear-producing (lacrimator) compound and is thought to be responsible for much of the eye irritation experienced from smog. PAN and related compounds have not been shown to be significant health hazards at the levels found in smog. However, research concerning the effects of such compounds has only recently begun. It is known that many plants, such as citrus, leafy vegetables, and coniferous trees, are very sensitive to PAN and related compounds. Exposure of plants to PAN concentrations of 0.02–0.05 part per million is known to injure plants.

8-20 KINDS AND COSTS OF SMOG

There are generally two distinct types of smog. These are compared in Table 8-16. The first is **London-type** smog, which is formed when pollutants (especially particulates and sulfur oxides) accumulate in a cool, moist, stagnant air mass

Table 8-16
Major Types of Smog

Characteristics	London type	Los Angeles type (photochemical smog)
When peak intensities occur	Early in the morning; wintertime	Around noontime
Ambient temperature	30–40°F	75–90°F
Humidity	Humid and foggy	Low humidity
Thermal inversion	Close to ground	Overhead (varies)
Causes irritation principally to	Bronchia and lungs (also eyes)	Eyes (but also bronchia and lungs)
Chief irritants	Soot and other particulates; sulfur oxides	Ozone; PAN; aldehydes; oxides of nitrogen; also sulfur dioxide and particulates (carbon monoxide, which has no odor, is also present)
Major source of irritants	Coal fires	Vehicular traffic
Other effects	Severe haze	Haze; severe damage to crops, pines, ornamentals; cracking of rubber

From *Elements of General and Biological Chemistry*, 3rd ed., by John Holum, John Wiley & Sons, New York, 1972.

characteristic of wintertime. The sulfur dioxide content of the air is directly related to the formation of this kind of smog. Consequently, the sulfur dioxide concentration in many parts of the country are monitored, especially during the winter months.

The second kind of smog is the **photochemical** or **Los Angeles type.** This smog forms on warm, sunny days when pollutants accumulate in a temperature inversion. Carbon monoxide and ozone concentrations are monitored as a measure of photochemical smog. In Los Angeles County a first-stage **smog alert** is called (children are not to enter into strenuous outdoor activity) when the ozone level is 0.35 part per million. A second-stage alert will be called if the ozone level reaches 1 part per million. To date, this level has never been reached. A carbon monoxide alert is called in Los Angeles County when the carbon monoxide concentration reaches 50 parts per million. There is clear evidence that smog is a health hazard and can have a significant effect on agriculture. In fact, it is estimated by the U.S. Department of Agriculture that smog caused agricultural losses amounting to over $500 million annually. Further, Lester B. Lave of Carnegie-Mellon Institute estimates from an extensive study that a 50 percent reduction in air pollution levels in major urban areas could annually save over $2 billion in medical and related costs. Finally, it is estimated that $1.5 billion in damage to buildings and bridges occurs each year from air pollution.

8-21 AIR POLLUTION CONTROL

As long as the major source of energy is the combustion of fossil fuels, air pollution will be a potential problem. To control the emission of air pollutants, each source must be considered. Among the general approaches to **air pollution control** are (*1*) fuel substitution, (*2*) changes in process to minimize emission, (*3*) removal of pollutant from emissions, (*4*) replacement of process with a less polluting alternative, (5) relocation of stationary source, (6) changes in mode of transportation, and (7) changes in land use procedures.

8-22 CONTROL OF AUTOMOBILE EMISSIONS

Gasoline-powered motor vehicles are the largest single source of air pollutants. Under the Clean Air Act Amendments of 1970, the EPA has established current and future **automobile emission standards.** Table 8-17 lists these emission standards. Note that they are given in terms of grams of pollutant per mile of driving. Figure 8-7 indicates the sources of pollutants from the **internal combustion engine.** Attempts are being made to control these sources. (See Figure 8-7.) The crankcase vent, which allowed unburned hydrocarbons blown by the piston rings to be emitted into the atmosphere, was the first source to be controlled. **Positive crankcase ventilation** (PCV) systems which recycle the hydrocarbons are now installed on all new cars. Evaporative losses can be cut down by directing gasoline vapors into a storage area rather than venting them to the atmosphere. The vapors are collected in activated charcoal canisters or are directed to the crankcase.

Emissions from the exhaust system of the internal combustion engine are the most significant. Under the impetus of federal emission standards, automobile manufacturers are attempting the control of these emissions in various ways. Some control is accomplished by redesign and modification of the carburation system, combustion chambers, ignition system, spark timing, and other engine components. The removal of carbon monoxide and hydrocarbons from the exhaust can be accomplished by an **exhaust manifold thermal reactor** and/or a **catalytic converter.** The manifold thermal reactor takes the place of the manifold and provides a high-temperature chamber in which air can be mixed with the exhaust to convert the carbon monoxide and hydrocarbons to carbon dioxide and water. The catalytic converter is a device in which air is mixed with exhaust gases in the presence of a catalyst which catalyzes the conversion of the gases to carbon dioxide and water. The presence of lead in the gasoline can interfere with the operation of the catalytic converter. Furthermore, some catalysts being tested do not last more than 5,000 miles. Research and development of catalytic mufflers is being carried out, and it is likely that 1975 or 1976 cars will be equipped with these devices.

The most difficult pollutants to control in the internal combustion engine are the nitrogen oxides. High combustion temperatures and high air-to-fuel ratios tend to produce more nitrogen oxides. The recycling of some exhaust gases which decreases the peak combustion temperatures can cut down nitrogen

	Hydrocarbons (grams per mile)	Carbon monoxide (grams per mile)	Nitrogen oxides (grams per mile)
Before emission controls	11	80	4
1970 standards (some emission controls)	2.2	23	—
1976 standards (additional controls are needed to meet these standards)	0.5	11.0	0.9

Table 8-17
Federal Emission
Standards for
Automobiles

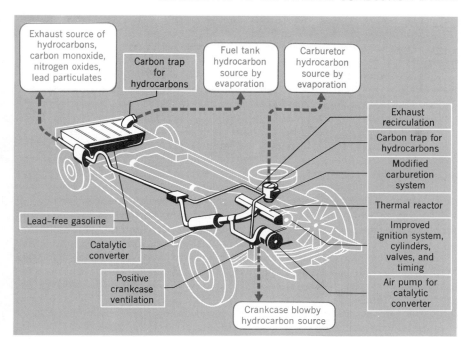

Figure 8-7
Pollutant sources and possible control methods in automobiles.

oxide emissions. Catalytic conversion of nitrogen oxides to nitrogen or ammonia is being investigated.

One of the major problems of all of the various modifications of the internal combustion engine is that of fuel economy. The number of miles per gallon of gasoline decreases when many of the emission control techniques are used on an automobile. In any case, it appears that modified internal combustion engines will be in use for many years to come.

The use of fuels other than gasoline could cut down emissions. Such fuels as ammonia, ethyl alcohol, liquefied petroleum gases (LPG), liquid hydrogen, and natural gas have been suggested. Of these, natural gas is the most feasible alternate fuel. Some fleet vehicles have been converted to burn natural gas. Unfortunately, natural gas is one of our most limited fossil fuels, and supplies are quickly being depleted. Synthetic natural gas will be more expensive, but could still be used. Liquid hydrogen has long-term possibilities as a substitute fuel.

8-23 ALTERNATIVES TO THE INTERNAL COMBUSTION ENGINE

It has been suggested that the internal combustion engine could be replaced by an alternative power device.

Gas turbines are being investigated for use in automobiles. They have few moving parts, have no cooling system, and can burn a variety of fuels. Indications are that the emission levels from turbines are low. Turbines are now being used in heavy trucks for diesel engine replacement. The major problems for use in automobiles appear to be high cost, size, and noise considerations, but they are promising.

Steam engines are another possibility being researched. Steam engines are external combustion engines using vapor from water or some other fluid to drive pistons or to spin turbines. They use a variety of fuels and appear to have good emission characteristics. Steam engines are being tested in buses, but it appears that it may be a few years before they can be considered as satisfactory substitutes for internal combustion engines.

The **Wankel** or **rotary engine** is an internal combustion engine of radically different design than the standard internal combustion engine. Of course, these engines are being used in some cars today. The rotary engine produces high levels of unburned hydrocarbons, moderate levels of carbon monoxide, and low levels of nitrogen oxides. With catalytic or manifold reactors the hydrocarbon and carbon monoxide levels could be further reduced. The rotary engine has a good potential for replacing the standard internal combustion engines.

Battery-operated **electric cars** are also mentioned as possible alternatives to standard cars. Currently, it appears that such cars could be useful for short-trip commuting and local driving. However, it does not appear that these vehicles will be mass-produced within the next few decades.

Other alternatives include the **Stirling cycle engine,** which involves the heating and cooling of a trapped gas; **fuel cell**-powered cars; and **hybrid** cars involving **electric motors** or **flywheels** in combination with standard engines. These alternatives are just being researched and developed and are long-term possibilities.

The remaining alternative to automobile emissions is the development of **rapid transit** systems in urban areas. Rapid transit systems could provide for more efficient transportation of large numbers of people, diminishing the need for excessive automobile use. In terms of energy efficiency or the number of gallons of fuel per 100 passenger miles, trains and buses are more than twice as efficient as automobiles.

8-24   INDUSTRIAL EMISSION CONTROLS

Now that we have considered emission control in automobiles, let us consider the problem in the **stationary sources** of industry, electrical power plants, and refuse burning. Control of each type of pollutant is possible. However, as mentioned previously, it may be necessary in some cases to change a pollution-producing process entirely or relocate the source.

Emissions of hydrocarbons can be controlled by use of an afterburner, by condensation, or by scrubbing. In an **afterburner,** waste hydrocarbons are burned as they leave the stack. Some hydrocarbons will condense to liquids at lower temperatures. Thus, the cooling of waste gases allows for the collection of the hydrocarbons by **condensation.** Sometimes it is possible to collect hydrocarbons by passing waste gases through a liquid which will dissolve the hydrocarbons. This is called **scrubbing.**

Sulfur oxide emissions can be controlled by use of low-sulfur fuels or the removal of oxides from the waste gases. Natural gas and certain kinds of petroleum have low sulfur content. It is possible, although expensive, to lower the sulfur content (desulfurize) of some fuels. Since most sulfur oxides come from electrical power plants, the use of nuclear plants would cut down such emissions. It is possible to remove much of the sulfur oxides from waste gases by chemical means. The sulfur dioxide can be catalytically converted to sulfuric acid, which can be removed:

$$2 \, SO_2 + O_2 \longrightarrow 2 \, SO_3$$
$$SO_3 + H_2O \longrightarrow H_2SO_4 \text{ (sulfuric acid)}$$

Other methods of removing sulfur dioxide from waste gases are being investigated. One of the most promising of these methods uses a limestone slurry to scrub the waste gases. This removes the sulfur dioxide by conversion to calcium sulfate:

$$2 \, CaCO_3 + 2 \, SO_2 + O_2 \longrightarrow 2 \, CaSO_4 + 2 \, CO_2$$

Figure 8-8
Common methods of
particulate removal
from waste gases.

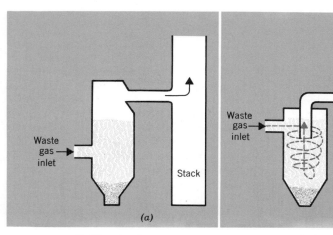

(a)

Settling chamber

(b)

Cyclone chamber

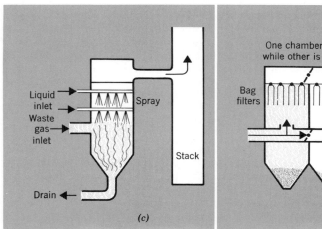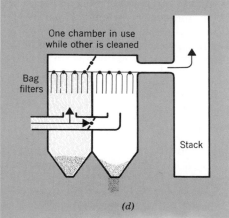

(c)

Scrubber

(d)

Filter system

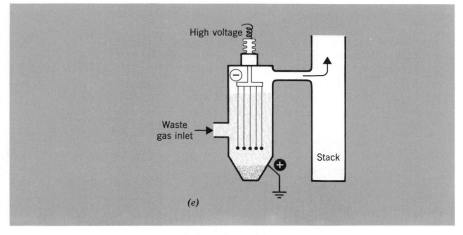

(e)

Electrostatic precipitator

This does produce a problem of the disposal of large amounts of the solid calcium sulfate.

Nitrogen oxides are difficult to control in high-temperature burning processes. One approach is to decrease the temperature of the combustion by recirculating exhaust gases to the burning chamber or by injecting water into the burning chamber. Another approach is to pass the waste gases through an absorbing medium designed to remove the nitrogen oxides. Catalytic conversion of nitrogen oxides to nitrogen gas is currently being investigated.

Several methods are available for the removal of particulates from waste gases. Larger particulates usually are removed with ease, while smaller particulates are more difficult. The various methods are illustrated in Figure 8-8. In **settling chambers,** waste gases are passed into a large chamber where particles settle out by gravity. In the **cyclone chamber,** the waste gases are swirled so that smaller particles collide and stick to the sides of the chamber. In the **scrubbers,** the waste gases are brought into contact with a liquid such as water which tends to remove the solid or liquid particles. In a **filtering device,** the waste gases are passed through a solid, porous medium which traps the particles. The **electrostatic precipitators** function by passing the waste gases through a series of electrically charged plates. The particulates acquire electrical charges and are attracted to the plates, where they can be collected.

The removal of submicron particulates is the most difficult problem of particulate control. A smokestack emitting large amounts of black or grey smoke can often be cleaned up using one of the techniques discussed above. However, the more dangerous submicron particles will still be emitted and usually cannot be seen in the stack emissions.

BIBLIOGRAPHY

BOOKS

American Chemical Society. *Cleaning Our Environment: The Chemical Basis for Action.* section 1. Washington, D.C.: ACS Special Issues Sales, 1969.
Carr, D. E. *The Breath of Life.* New York: Berkeley Publishing, 1970.
Esposito, J. C. *Vanishing Air.* New York: Grossman, 1970.
National Tuberculosis and Respiratory Disease Association. *Air Pollution Primer.* New York: National Tuberculosis and Respiratory Disease Association, 1969. 104 pp.

ARTICLES AND PAMPHLETS

"Air Pollution and Future Climates." *Chemistry* 3 (Jan. 1971).
Brodine, V. "Point of Damage." *Environment* **14** (4) (1972).
"Chemistry and the Atmosphere." *Chemical and Engineering News.* Reprint.
The Effects of Air Pollution. Public Health Service Publication No. 1556. Washington, D.C.: U.S. Government Printing Office, 1966.
Haagen-Smit, A. J. "The Control of Air Pollution." *Scientific American* 25 (Jan. 1964).
Hall, S. K. "Sulfur Compounds in the Atmosphere." *Chemistry* (Mar. 1972).
Lave, L. B., and E. P. Seskin. "Air Pollution and Human Health." *Science* **169** (3947), 723–733 (1970).
McDermott, W. "Air Pollution and Public Health." *Scientific American* **205,** 49–57 (1961).
Medeiros, R. W. "Smog Formation Simplified." *Chemistry* (Jan. 1972).
"New Blueprint Emerges for Air Pollution Controls." *Environmental Science and Technology* (Feb. 1971).

The Sources of Air Pollution and Their Control. Public Health Service Publication No. 1548. Washington, D.C.: U.S. Government Printing Office, 1969.

U.S. Department of Commerce. *Automotive Fuels and Air Pollution.* Washington, D.C.: U.S. Government Printing Office, 1971.

Wolf, P. C. "Carbon Monoxide—Measurement and Monitoring in Urban Air." *Environmental Science and Technology* **5,** 212–218 (1971).

QUESTIONS AND PROBLEMS

1. Which six gases comprise over 99.9 percent of the atmosphere?

2. Which two gases comprise over 99 percent of the atmosphere?

3. Describe an air basin or air shed.

4. List the five primary air pollutants.

5. What kind of chemical reaction accounts for man-made carbon monoxide?

6. Why is carbon monoxide toxic to humans?

7. Why does a cigarette smoker often have 2–5 times as much carbon monoxide in the blood than a nonsmoker?

8. What kind of chemical reactions give rise to man-made sulfur dioxide?

9. What is a secondary air pollutant? Give an example.

10. What chemical reactions and what reaction conditions give rise to man-made nitrogen oxides?

11. Describe particulate air pollutants.

12. What are the four main sources of air pollutants in the United States (not including miscellaneous sources)?

13. In the United States, what is the major source of each of the following:
    - (a) carbon monoxide
    - (b) nitrogen oxides
    - (c) sulfur oxides
    - (d) hydrocarbons
    - (e) particulates

14. What is a stagnant air mass?

15. Describe a temperature inversion.

16. How does the formation of photochemical smog take place?

17. What are photochemical oxidants? Give two examples.

18. What are the two general types of smog and what climate conditions are conducive to the formation of each type?

19. List some emission control devices which are used or may be used on internal combustion engines in automobiles.

20. List some of the possible replacements for the internal combustion engine.

21. What are some methods for the removal of particulates from waste gases?

22. According to Table 8-7, what is the major air pollutant and the major source of air pollution? What is the major air pollutant and major source according to Table 8-8? Why do the two tables differ?

23. Suppose some factory had a smokestack that emitted a thick, black smoke. After the installation of particulate control devices, the emission from the stack was no longer black. What does this indicate?

WATER AND SOLUTIONS

9

9-1 WATER

Water, the ubiquitous compound that covers nearly three-quarters of the surface of the earth, is thought to have existed in the celestial material from which the earth might have been formed billions of years ago. Water, through erosion and glacial action, has played a significant part in forming the physical features of the earth. Photographs from Apollo spacecraft have emphasized the beauty and turbulence of the ever-changing cloud cover shrouding the globe. Water is the only common substance that exists in the environment in all of the three states of matter—solid, liquid, and gas.

The waters of the primeval oceans are thought to be the birthplace of primitive life forms. All forms of life are dependent upon water and, to varying degrees, are composed of water. Ancient civilizations developed in regions of abundant water. The Egyptian civilization of the Nile and Mesopotamian civilization of the Tigris and Euphrates river valleys are familiar examples. Thousands of years ago similar civilizations developed around Indus and Ganga rivers of India and the Yellow and Yangtze rivers of China. Of course, today these two areas support about 40 percent of the entire population of the world.

9-2 WATER IN HUMANS

The human body averages about 65 percent by mass water. The percentage of water is quite variable among the population. Men generally contain a higher percentage of water than women. Various parts of the body contain differing amounts of water. Blood is 83 percent water, muscle contains 76 percent water, and bone contains only 22 percent water. Water is incorporated in all cells of the body and in the extracellular body fluids. Some 63 percent of the body water is intracellular, 23 percent is in the interstitial fluids between cells, 6.3 percent is in the blood, and 7.7 percent is found in body cavities such as the intestines and eyeballs.

A typical person loses about 2.5 liters of body water each day. About 1.5 liters is excreted in the urine, 0.5 liter is lost in perspiration, and 0.5 liter is lost in the breath. To replenish this water it is necessary to drink about 1.25 liters of water or water-containing liquids; the water content of food supplies 1 liter; and metabolism of food produces about 250 milliliters.

9-3 WATER AND SOCIETY

In addition to being vital to life, water is necessary for agriculture, industry, and transportation. Ancient and present-day civilization developed extensive canals and aqueduct systems to transport water for irrigation and city use. Some canals were designed for transportation purposes. The Grand Canal of China once spanned 1,000 miles across China. Today, only a portion is still in use for transportation. The Romans developed an intricate aqueduct system to supply water to their cities. Some of these aqueducts are still in use.

Water has served as a source of energy and a medium of energy transfer which has allowed the development of industrial societies. The waterwheel and steam engine were fundamental to the Industrial Revolution. Today, hydroelectric power, steam engines, and steam turbines are basic to our industrial society.

In this chapter we will consider the liquid state of matter and the nature and properties of water. This will be followed by a discussion of the nature and properties of aqueous solutions.

9-4 THE LIQUID STATE

When a gas is cooled, the average kinetic energy of the particles decreases. At high pressures, the gaseous particles move closer together and the attractive forces become important. These attractive forces tend to pull the molecules to one another. Under the proper conditions of temperature and pressure, a gas can be liquefied; converted from the gaseous to the liquid state. **Liquefaction** occurs when the attractive forces between molecules overcome the forces of kinetic motion, which causes the molecules to aggregate into the **liquid state.** The liquefaction process is illustrated in Figure 9-1. Depending upon the chemical nature of the substance, its liquid state may consist of molecules or atoms or oppositely charged ions which are randomly packed in a relatively close manner. However, the liquid state is dynamic, and the particles move about randomly,

Figure 9-1
The liquefaction of a gas. From *Introduction to Chemistry* by T. R. Dickson, John Wiley & Sons, New York, 1971.

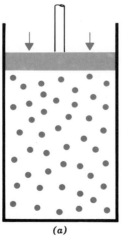

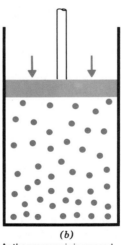

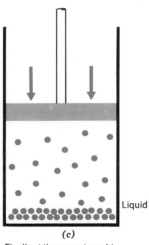

Liquid

(a)

In the gaseous state the molecules are rapidly moving and are far apart.

(b)

As the pressure is increased and the temperature decreased the molecules slow down and are forced closer together.

(c)

Finally at the correct combination of temperature and pressure most of the molecules aggregate into a collection of molecules characteristic of the liquid state; liquefaction has taken place.

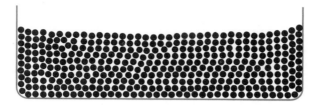

with short distances between collisions with other molecules and the container. Figure 9-2 gives an illustration of the liquid state.

The simple view of the liquid state as a collection of particles which are constantly moving about and mutually interacting can be used to describe and rationalize some of the properties of liquids. As a result of the freedom of movement and the attractive forces which exist between particles, all liquids can be poured. The attractive forces allow the aggregation of molecules to be transferred as a body. However, these same attractive forces cause a certain internal resistance to flow, which is called **viscosity.** Some liquids, such as oils, can be quite viscous, while others have very low viscosities and flow readily. The viscosity of a liquid depends on the nature of its particles.

The kinetic energies of the molecules in a liquid tend to oppose the attractive forces. Some molecules which possess sufficient kinetic energy and are on or near the surface of the liquid can break away from the attractive forces of the other molecules and escape into the vapor state. This is called **evaporation.** Since the higher-energy molecules are those which can evaporate, the average kinetic energy of the molecules in the liquid decreases and thus the temperature of the liquid phase is lowered by the evaporation process. This is why you often feel cooler when water evaporates from your skin.

When an open container of a liquid is exposed to the atmosphere, the molecules of the air exert a pressure on the surface of the liquid. This pressure inhibits the evaporation of the liquid. Evaporation still occurs under these conditions, but does not occur as rapidly as it would if fewer air molecules were present. Let us consider what happens when we heat a sample of a liquid in a container which is open to the atmosphere. As the liquid is heated, the temperature will increase. The increase in temperature corresponds to an increase in the kinetic energy of the particles in the liquid. As the heating is continued, eventually a point is reached at which further heating does not increase the temperature of the liquid and bubbles of vapor are rapidly formed throughout the entire liquid volume. These bubbles rise to the surface and burst as the vapor escapes. When this point of rapid evaporation is reached, the liquid is said to be **boiling.** Boiling will not take place until the pressure exerted by the vaporized particles of the liquid is equal to the prevailing atmospheric pressure to which the liquid is exposed. The vapor particles formed by the liquid exert a pressure on the air molecules and prevent them from restraining the evaporation process. Of course, the boiling point of a liquid depends upon the atmospheric pressure to which it is exposed. As you may have noticed when you cook in the mountains, the lower atmospheric pressures cause water to boil at a lower temperature. On the other hand, higher pressures result in higher boiling points. The normal **boiling point** of a liquid is the temperature at which the liquid boils when exposed to a pressure of 1 atmosphere.

9-5 THE NATURE OF WATER

Pure water is a colorless, odorless, tasteless liquid which boils at 100°C and freezes at 0°C. Chemically, water exists as molecules consisting of two hydrogen atoms joined to an oxygen atom by covalent bonds. These molecules can be

represented as

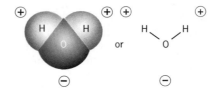

One distinct characteristic of a water molecule is the angular shape. The hydrogens protrude from one side of the oxygen at an angle of 105° to each other.

The covalent bonds in water involve the sharing of electrons between the hydrogen and oxygen. The oxygen attracts the shared electrons more strongly than the hydrogen. This results in the electrons being pulled toward the oxygen. Since electrons are negatively charged, this produces a slight negative charge around the oxygen and a positive charge around the hydrogens (due to the positive protons in the hydrogen nuclei). This situation in a water molecule can be pictured as

This separation of charges makes water molecules **polar.** That is, the molecules have a positive end and a negative end, much like a magnet has a north pole and a south pole. Molecules of many other substances are polar, but water molecules are highly polar. The polarity of water molecules accounts for some of the unique properties of water.

The polar nature of water molecules results in strong intermolecular attraction. The hydrogen ends of the water molecules, with partial positive charges, are attracted to the oxygen region of neighboring molecules, a region of partial negative charge. Perhaps you have noticed an analogous attraction between two bar magnets. When the two like poles of the magnets are brought together, the magnets repel, but when opposite poles are brought together, the magnets attract to form a loose attachment. Water molecules interact in a somewhat similar manner except that the attractive and repulsive forces result from partial electrical charges and not magnetic forces. The attraction between water molecules is strong enough to result in a loose aggregation or clumping of the molecules. In fact, this force of attraction between the hydrogen of one molecule and the oxygen of another is strong enough to be considered to be a type of chemical bond called the **hydrogen bond.** The hydrogen bond is also common to collections of other kinds of molecules in which hydrogen is bonded to such elements as oxygen, nitrogen, and fluorine.

Liquid water can be viewed as a loose aggregate of water molecules which are hydrogen bonded to one another:

These hydrogen bonds are continuously being formed and broken in a collection of water molecules.

9-6 THE PROPERTIES OF WATER

Hydrogen bonding causes the boiling point and freezing point of water to be much higher than expected. The boiling and freezing points of most chemically similar compounds follow a pattern of increasing as the molecular weights of the compounds increase. As shown in Figure 9-3, if water fit this pattern when compared to chemically related compounds, it should boil at about $-90°C$ and freeze at $-100°C$. The clumping together of water molecules causes the shift in the boiling and freezing points.

Hydrogen bonding also accounts for the fact that ice floats on water. Most substances expand when heated and contract when cooled. Water follows this behavior except that at 4°C it no longer contracts but begins to expand. As it is cooled further and freezes into ice it suddenly expands by 10 percent to form a solid which is less dense than the water from which it was formed. Let us consider how hydrogen bonding explains this behavior. Hot water contains numerous clumps of hydrogen bonded molecules. These clumps break and reform in fractions of seconds. As the water is cooled, the kinetic energies of the molecules decrease and larger hydrogen bonded aggregations are formed. The molecules come closer together and the liquid contracts in volume. At 4°C the kinetic energies of the water molecules are low enough so that hydrogen bonding becomes the dominant force which dictates the arrangements of the molecules in the liquid. The molecules begin to arrange themselves in a pattern in which the maximum amount of hydrogen bonding can occur. As freezing occurs, the water molecules form an open cage-like network of hydrogen bonded water molecules, characteristic of ice. This structure allows for the maximum amount of hydrogen bonding in which each water molecule is bonded to others. The structure of ice is such that the water molecules are further apart than they are in liquid water and, thus, the ice is less dense than the water. As ice melts to a liquid, some of the hydrogen bonds break, the open cage-like structure is disrupted, and the water molecules move closer to one another.

Incidentally, this uncommon behavior of water is of great importance to us. If ice were more dense than water, it would sink to the bottom of lakes and rivers.

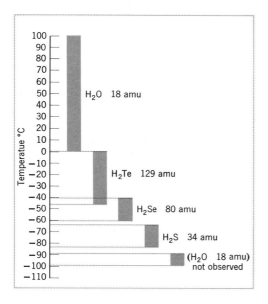

Figure 9-3

The boiling and freezing behavior of water. The temperature range between the freezing and boiling points of water deviates from the expected behavior of water compared to chemically similar compounds. The deviation is attributed to hydrogen bonding effects which allow the water molecules to clump together and behave as more massive molecules Ammonia, NH_3, and hydrogen fluoride, HF, also display this anomolous behavior when compared to chemically similar compounds. Hydrogen bonding accounts for the deviant properties of these molecules.

Ice would build up from the bottom and ultimately freeze over the lakes, rivers, and oceans of the world.

Some other properties of water such as its ability to dissolve many substances, its surface tension, and capillary action can be explained on the basis of the polar nature of water molecules. The dissolving process is discussed in Section 9-8 and explanations of the other two properties are illustrated in Figure 9-4.

9-7 SOLUTIONS

A **solution** consists of two or more components which are so intimately mixed that the particles making up the components are essentially intermingled on an atomic, molecular, or ionic basis. The types of particles involved depend on the natures of the components. As a result of the great degree of intermingling of the particles of the components of a solution and the forces of interaction which exist between them, the components of a solution cannot be separated by filtration. In fact, solution components can only be separated by more involved separation methods. For instance, if we want to get the salt out of sea water, we have to heat the solution to boil away the water.

Solutions are very important in chemistry and industry, and are used in everyday life. Many food products and medicines are solutions, as are many household chemicals such as cleaning fluids and rubbing alcohol. The gasoline we use in our car is a solution, and the water we drink from the tap is a solution. Special solutions are often prepared for specific uses, while many occur naturally.

Figure 9-4
Some properties of water.

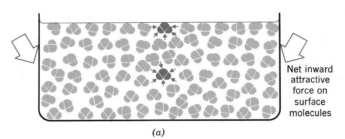

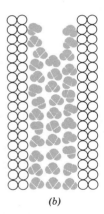

Net inward attractive force on surface molecules

(a)

(b)

Surface tension—A molecule in the body of the liquid is attracted equally by the molecules around it. The molecules at the surface of the liquid are attracted only by the molecules below them and beside them. As a result, unbalanced attractive forces exist on the surface molecules. The unbalanced attraction results in a net inward pull tending to draw the surface molecules into the body of the liquid. Consequently, the liquid surface is under a certain strain or tension called the surface tension.

Capillary action—Water wets glass as a result of an attraction of water molecules to the ions in the glass. In a narrow diameter tube (capillary tube) the water molecules wet the tube, and surface molecules attract other water molecules a certain distance up the tube. This ascension of water up a narrow tube is called capillary action. In a very narrow tube water may ascend many centimeters.

A solution composed of two substances is called a **binary** solution. A ternary solution has three components, and it is possible to have a solution consisting of four or more substances. However, not all substances mix with one another to form solutions. Usually, solutions are prepared by dissolving one substance in another. When a solution is prepared using two substances of different phases, the substance that is of the same phase as the resulting solution is called the **solvent** and the substance that has been dissolved in the solvent is called the **solute.** For example, if we prepare a liquid solution by dissolving sugar in water, the water is the solvent and the sugar is the solute. If the two substances are of the same phase, the solvent is usually considered to be the component present in the greater amount, but in such solutions the distinction is not important. In a solution made up of water and alcohol, the water is considered to be the solvent if it is present in the greater amount. It is possible to have liquid-phase solutions, solid-phase solutions, and gaseous-phase solutions. The most commonly encountered type is the liquid-phase solution. Since water is one of the most useful solvents, the most common solutions are aqueous solutions. All solutions involving water as the solvent are called **aqueous** solutions, while liquid solutions not involving water are called **nonaqueous** solutions.

9-8 THE DISSOLVING PROCESS AND SOLUBILITY

The dissolving of a substance in water can be viewed as an interaction between the solute particles and the polar water molecules. As shown in Figure 9-5, when a solute is added to water the water molecules penetrate between the solute particles, break them apart, and surround them as they enter the solution. If the force of interaction between the water molecules and the solute particles is great enough to overcome the forces that hold the solute particles together, the substance will dissolve. Once in solution, the solute particles become intimately mixed with the water molecules.

Numerous ionic and molecular compounds will dissolve in water, but there

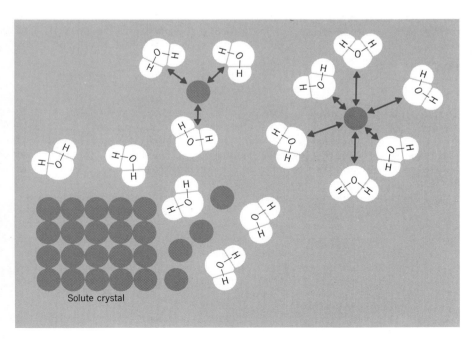

Solute crystal

Figure 9-5
The dissolving of a solute in water. Water dissolves a solute by attracting the solute particles from the crystalline solute. The nature of the interaction between the polar water molecules and the solute particles depends upon whether the particles are positive ions, negative ions (ionic solids), or molecules.

are many compounds which will not dissolve. However, given enough time, water will dissolve almost any inorganic compound to varying degrees. Rocks, minerals, and metals will dissolve in differing amounts in water. As a result of the dissolving ability of water, pure water is seldom found in nature. Natural waters are complex solutions containing a variety of dissolved materials.

The amount of a substance that can be dissolved in a given amount of solvent at a specific temperature is called the **solubility** of the solute at that temperature. Each substance has a characteristic solubility in a given solvent. Sometimes solubilities are expressed in terms of the number of grams of solute that can be dissolved in 100 grams of solvent at a specific temperature. For example, the solubility of sodium chloride in water at 0°C is 35.7 grams per 100 grams of water.

A **saturated** solution contains the maximum amount of solute that can be dissolved in the given amount of solvent to form a stable solution at that temperature. When a solution contains less solute than could be contained in solution at saturated conditions, it is referred to as an **unsaturated** solution. The solubilities of many solid substances increase with an increase in temperature. Incidentally, it is interesting to note that gases behave in the opposite manner; that is, the solubilities of gases decrease with an increase in temperature. It is possible to prepare solutions of solid substances that contain fairly large amounts of solute by elevating the temperature of the solution so that more solute will dissolve. Usually, when such solutions are cooled down, the excess solute will recrystallize into the solid phase. Sometimes, however, when such a solution is cooled, a solution is formed that contains more solute than a saturated solution would contain at the same temperature. These solutions that contain more solute than would be expected are called **supersaturated** solutions. Supersaturated solutions are quite unstable and can be converted easily to saturated solutions when the excess solute recrystallizes. The crystallization process which involves the formation of the solid solute from the solution requires the proper arrangement of the solute particles into a form which produces the crystal. If the solution is free of tiny solid particles and the solution container has a smooth surface, such as a glass or metal surface, no crystal growth can be initiated. Thus, upon cooling, the solution will contain more solute than it would under normal saturated solution conditions. Such a supersaturated solution can be converted to a saturated solution by initiating the crystallization process. This initiation is accomplished by adding a small seed crystal to the solution or by agitating the solution.

9-9 COLLIGATIVE PROPERTIES OF SOLUTIONS

Certain properties of solutions, called **colligative properties,** are related more to the presence of solute particles dispersed among the solvent particles than to the identities of the particles. The term colligative refers to the collective effect of the solute particles. These properties are generally the same no matter what solute is involved, and they depend on the concentration and not the nature of the solute particles.

One important colligative property is the effect of a nonvolatile solute on the freezing point and boiling point of the solvent. When we dissolve a nonvolatile solute in a solvent, the resulting solution will boil at a higher temperature than the pure solvent and will freeze at a lower temperature than the pure solvent. These phenomena are called **boiling point elevation** and **freezing point depression.** (See Figure 9-6.) We take advantage of the freezing point depression of water by a solute when we put antifreeze in the radiator of our car. The antifreeze is a solute which causes the freezing point of the water in the radiator to be lowered so that it does not freeze in cold weather.

Another colligative property of aqueous solutions that is very important in

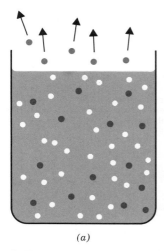
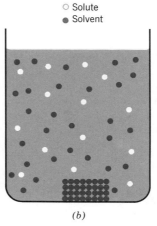

○ Solute
● Solvent

Figure 9-6
Boiling point elevation
and freezing point
depression

(a)

(b)

Boiling point elevation—The presence of solute molecules interferes with the ability of the solvent molecules to evaporate. Consequently, a higher temperature is needed to make the solution boil. The increase in the boiling point of a solvent caused by the addition of a nonvolatile solute is called boiling point elevation.

Freezing point depression—The presence of solute particles interferes with the ability of the solvent molecules to freeze to ice crystals. Consequently, a lower temperature is needed to reach the freezing point. The decrease in the freezing point of a solvent caused by the addition of a solute is called freezing point depression. Freezing point depression is used in the making of homemade ice cream. Salt is added to an ice–water mixture to lower the freezing temperature enough to freeze the ice cream mixture.

chemical and biological processes is **osmosis.** Osmosis is a phenomenon involving solutions separated by a membrane. The membrane acts as a barrier between the solutions and has the property of allowing certain types of molecules to pass through, while preventing the passage of other species in solution. This membrane is called a **semipermeable membrane,** since it is only permeable to selected species. Semipermeable membranes that allow the passage of the solvent but not the solute are called **osmotic membranes.** When a solution is separated by an osmotic membrane from a sample of pure solvent (or a like solution of lower concentration), the solvent molecules will spontaneously penetrate the membrane from both directions, but not at equal rates. There is a net migration of solvent molecules from the pure solvent side of the membrane to the solution side. This phenomenon is called osmosis and is illustrated in Figure 9-7. The net result of osmosis is the transfer of solvent across the membrane to the side having the lower concentration of solvent. The mechanism of osmosis is not completely understood, but it appears to occur between aqueous solutions separated by an osmotic membrane no matter what nonvolatile solute is involved. Osmosis can be stopped if a certain opposing pressure or force per unit area is exerted on the solution side of the membrane, as illustrated in Figure 9-7. This means that osmosis results in a pressure being exerted on the solution of higher concentration. This pressure is called the **osmotic pressure.** The osmotic pressure can be quite large and sometimes involves pressures of hundreds of atmospheres. Osmotic pressure is one of the factors involved in the uptake of water by trees and plants. Osmotic pressures of body fluids are important to the distribution of body water. Pure water has an osmotic pressure of zero. Any dis-

Figure 9-7
Osmosis and osmotic
pressure.

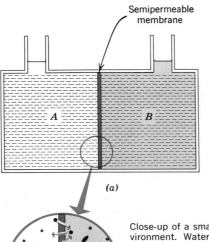

Osmosis: Enlarged view of two fluid compartments separated by a semipermeable membrane.

(a)

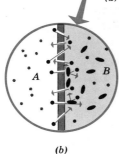

(b)

Close-up of a small section of membrane and its immediate environment. Water molecules are represented by dots; sugar molecules in compartment *B* are represented by shaded ovals. In the "sieve" theory of osmosis, the membrane is said to have pores large enough to permit passage of water molecules but small enough to stop solute molecules (or ions). As drawn, of every five molecules of water that get from *A* to *B*, only three return. Two others are shown colliding with sugar molecules. The result is a net flow of water from *A* to *B*, and level *B* rises. Parts *a* and *b* from *Elements of General and Biological Chemistry*, 3rd ed., by John Holum, John Wiley & Sons, 1972, 138.

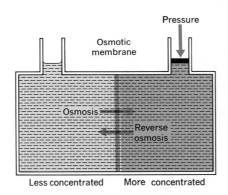

Less concentrated   More concentrated

(c)

Osmotic pressure: In the osmosis process, solvent molecules pass from the less concentrated solution to the more concentrated solution. The process of osmosis can be reversed by applying pressure to the more concentrated side. The amount of pressure needed just to counteract the osmosis process is called the osmotic pressure. Pressures in excess of the osmotic pressure can cause reverse osmosis.

solved nonvolatile solute will increase the osmotic pressure. Two solutions which have the same osmotic pressure are called **isotonic** solutions. Intravenous solutions (IV solutions) used in hospitals to carry nutrients into the blood are formulated to be isotonic with the blood so that they do not upset the water balance in the blood cells.

9-10 COLLOIDS AND SUSPENSIONS

The diameters of most ions, atoms, and molecules in solution range from about 0.5 angstrom to about 3 angstroms (an angstrom is 10^{-8} centimeter). In a solution these particles are intermingled with the solvent particles. All liquid solutions are clear. Keep in mind that just because a solution is clear, this does not mean that it cannot be colored. Occasionally, a solution is so deeply colored it is difficult to see through it. Nevertheless, it is still a solution. When some

Figure 9-8
The Tyndall effect.
The photograph is a
top view of three
beakers through
which a laser beam is
being passed. The
beakers on the right
and left contain a
colloidal dispersion of
starch in water and
the beaker in the
middle contains a
pure solution. Note
that as a result of the
Tyndall light-scattering
effect the laser beam
can be seen passing
through the colloidal
dispersions and
cannot be seen
passing through the
pure solution.
(Photograph by David
Crouch.) From
*Introduction to
Chemistry* by T. R.
Dickson, John Wiley &
Sons, New York, 1971.

substances are mixed with water, a mixture is formed which is not a true solution nor is it simply a crude suspension with components that will more or less quickly separate like a fine sand stirred in water. In such a case, when the substance is mixed with water, the particles of the substance tend to be attracted to one another, and they cluster together to form relatively large aggregates. These clusters, called **colloidal particles,** may contain hundreds or thousands of molecular-sized particles and may range in size from 10 to 1,000 angstroms. A mixture in which one substance is dispersed in another so that the particles of the first substance form colloidal particles is called a **colloidal dispersion.** Colloidal dispersions occur naturally, but in many cases are manufactured industrially. For example, milk, cheese, metal alloys, and some paints are types of colloidal dispersions. Colloidal dispersions may appear to the eye to be solutions, but they are not true solutions.

How can we tell whether a liquid mixture is a solution or a colloidal dispersion? A beam of light cast on a solution passes through it, and we say that the solution is transparent. However, when we cast a beam of light on a liquid colloidal dispersion, the colloidal particles are just the right size to cause reflection or scattering of the light. Air contains colloidal particles which reflect and scatter light. Thus, a searchlight beam appears as a column of light. This phenomenon is called the **Tyndall effect** and is illustrated in Figure 9-8.

When a substance is dispersed in another substance in a form in which the dispersed particles are larger than colloidal particles (larger than 1,000 angstroms), the dispersion is called a **suspension.** This means that the particles (large clumps of molecules or ions) are suspended in the other substance. Unlike colloidal particles, the suspended particles give a cloudy appearance to the mixture and will settle out if we let the mixture stand.

9-11 DIALYSIS

An important property of colloidal dispersions involves a phenomenon called **dialysis.** Certain membranes that are permeable, not just to solvent molecules but also to other small molecules and ions, are not permeable to colloidal particles. If we place a colloidal dispersion in such a membrane, as shown in Figure 9-9, the ordinary size particles (not colloidal) can pass through the membrane, but the colloidal particles will be retained. This phenomenon is called dialysis. A membrane in which dialysis occurs is called a **dialyzing membrane.** Dialyzing

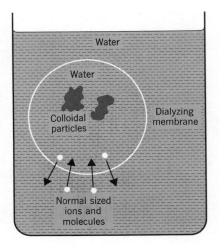

membranes are present in animals and plants, and dialysis is a very important biological process. In fact, the membranes of body cells are dialyzing membranes. These membranes provide for the transfer of water, normal sized molecules, and ions to and from body cells, while colloidal particles and very large molecules are kept within cells or not allowed to enter the cells.

9-12 METHODS OF DESCRIBING SOLUTION CONCENTRATIONS

An aqueous solution can be described in terms of the kinds of solutes present. This is a qualitative description and states nothing about how much solute is present in the solution. Of course, the amount of solute in a solution depends upon the amount of solution considered. The ocean as a large salt solution contains an amount of dissolved salt totaling about 4×10^{22} grams, while a cup of sea water contains about 8 grams of dissolved salt. Obviously, just stating the amount of solute in a solution is not a useful description.

The quantitative make-up of a solution is best described by stating the concentration of the solute. **Concentration** is an expression of the amount of solute contained in a unit amount of solution. For instance, the concentration of dissolved salt in sea water can be expressed as 30 grams of sodium chloride per liter of sea water. The concentration of a solution is independent of the amount of solution. The concentration of a large volume of a given solution is the same as the concentration of a small volume. The concentration of dissolved salt in a cup of sea water is the same as the concentration of salt in a bucket of sea water.

It is important to distinguish between amount and concentration. The amount of dissolved gold in the oceans of the world is great enough to provide each person in the world with around 2 kilograms of gold. However, gold is present in sea water at such a low concentration that it is not economically feasible to extract it.

Solution concentrations provide a basis of comparison. Sea water contains 30 grams of dissolved salt per liter and typical drinking water contains less than 0.4 gram of dissolved salt per liter. The sea water is more concentrated in salt than the drinking water. Concentrations are also needed to state tolerable levels of chemical pollutants and other substances in water. Two common methods of expressing concentration are parts per million and molarity, which are discussed in the following sections.

9-13 PARTS PER MILLION

There are about 1.4×10^{21} grams of dissolved aluminum in the oceans of the world. This is indeed a large amount of aluminum, but its concentration is so low we find it convenient to refer to the concentration in terms of parts per million. Solutes found in very low concentrations in solutions are sometimes expressed in terms of the **number of milligrams of solute per kilogram** of solution, or the **number of milligrams of solute per liter** of solution. The densities of dilute aqueous solutions are approximately 1 gram per milliliter, so a kilogram of solution and a liter are approximately the same. A milligram is one-millionth part of a kilogram, thus, this expression of concentration corresponds to **parts per million concentration.** The parts per million (ppm) concentration of a species is found by determining the number of milligrams of the species in a known number of liters or kilograms of the solution. For instance, if a 0.50 liter sample of a water solution is found to contain 2.2 milligrams of fluoride ion, the parts per million of fluoride is

$$\text{ppm } F^- = \left(\frac{2.2 \text{ mg } F^-}{0.50 \ \ell} \right) = \left(\frac{4.4 \text{ mg } F^-}{1 \ \ell} \right) = 4.4 \text{ ppm } F^-$$

9-14 THE CHEMICAL MOLE

Before considering the concentration term called molarity we need to know about the chemical mole. It is sometimes desirable to know the number of chemical particles in a sample of a substance. It is not possible to work with individual molecules, atoms, or ions. A typical amount of a substance will contain around one septillion (10^{24}) chemical particles. Chemists have developed a method of relating the mass of a sample of a substance to the number of chemical particles in the sample. This method is based on a term called the mole, which is defined as follows:

> **Mole** — The amount of a substance in grams that contains as many chemical particles as there are carbon atoms in exactly 12 grams of carbon-12. The chemical particles must be specified and may be atoms, molecules, or ions, or any specified groups of such entities.

As an explanation of this definition consider that the atomic weights and formula weights express the relative masses of the chemical particles of elements and compounds in terms of atomic mass units. The mole allows for the use of relative masses on a gram basis. This can be illustrated as follows:

$$\left(\frac{12 \text{ amu}}{1 \text{ atom carbon-12}} \right) \qquad \left(\frac{16 \text{ amu}}{1 \text{ atom O}} \right)$$

$$\left(\frac{12 \text{ grams}}{\text{certain number of carbon-12 atoms}} \right) \qquad \left(\frac{16 \text{ grams}}{\text{certain number of O-atoms}} \right) \qquad \begin{array}{l} \text{The number of carbon atoms} \\ \text{is the same as the number of} \\ \text{oxygen atoms} \end{array}$$

Twelve grams of carbon-12 will contain a specific number of carbon-12 atoms. Since the relative mass of carbon-12 to oxygen is 12 : 16, then 16 grams of oxygen will contain the same number of atoms as 12 grams of carbon-12. This number of atoms is a mole of atoms. The amount of any element or compound corresponding to a mole will be the number of grams numerically the same as atomic weight or formula weight. This is the basic reason for the definition of the mole. That is, an amount of an element in grams numerically the same as the atomic weight will contain a mole of atoms of that element. Likewise, the amount of a compound in grams numerically equal to the formula weight will contain a mole of that compound.

The mole is somewhat like our use of dozen to refer to a specific number of objects. When we order a dozen eggs or a dozen donuts this refers to different amounts in terms of weight but the same number of each. Further, it is not really important that a dozen is 12. It could just as well correspond to 11 or 13 objects. Chemists use the mole in a similar manner. That is, they are not concerned with the actual number of chemical particles in a mole but rather with the fact that a mole is a specific number. For instance, when we have a mole of iron and a mole of sodium we know that the number of atoms in each is the same.

The actual number of particles in a mole has been experimentally determined to be **6.02 × 10²³**. It is interesting that this is a very large number (called **Avogadro's number**), but chemists are seldom concerned with the actual number of particles in a sample. The number of particles can be expressed in terms of moles. For example, a sample of iron may contain 2.3 moles. This is sufficient information to express the number of atoms present in the sample. Again, this is somewhat like the use of dozen. We usually can say that we have 2 dozen eggs rather than specifying 24 eggs.

The usefulness of the mole is that it allows the deduction of the number of chemical particles in a sample of a substance just by determining the mass of the sample. For each element or compound we can express the **number of grams per mole** or **molar mass** by using the numerical value of the atomic or formula weight. For example, iron has 55.8 grams per mole or

$$\left(\frac{55.8 \text{ g}}{1 \text{ mole Fe}}\right)$$

and sodium chloride has 58.5 grams per mole [23.0(Na) + 35.5(Cl) = 58.5(NaCl)] or

$$\left(\frac{58.5 \text{ g}}{1 \text{ mole NaCl}}\right)$$

To illustrate how the molar mass can be used to deduce the number of moles in a sample of a substance consider an analogous example in which we are working with screws. Suppose we want to count out a large number of screws for some purpose. Obviously, it is not convenient to count them out individually. The problem can be approached by determining the mass of a dozen screws. Suppose that the mass of a dozen screws is 20 grams:

$$\left(\frac{20 \text{ g}}{1 \text{ dozen screws}}\right)$$

We can use this factor to determine the number of screws in a sample of screws of known mass. Let us deduce the number of dozen screws in 1,000 grams of screws. This is done by dividing the mass by the number of grams per dozen:

$$\left(\frac{1,000 \text{ g}}{20 \text{ g per dozen screws}}\right) = 50 \text{ dozen screws}$$

By use of the mass per dozen screws we are able to determine the number of dozen screws in any sample of screws as long as we know the mass of the sample.

Since the mass of a mole of a substance will always be given by the formula weight in grams, it is possible to determine the number of moles of a substance in any known mass of the substance. For example, suppose we had a 700 gram sample of sodium chloride (approximately the mass of sodium chloride in a typical carton of salt). The formula weight of NaCl is 58.5. We can determine the number of moles of NaCl in the sample by dividing the sample mass by the formula weight in grams:

$$\left(\frac{700 \text{ g}}{58.5 \text{ g per mole NaCl}}\right) = 12.0 \text{ moles NaCl}$$

Thus, the sample contains 12.0 moles of sodium chloride. The same procedure can be used to determine the number of moles of any substance in a sample of the substance. The sample mass is divided by the formula weight in grams (grams per mole):

$$\left(\frac{\text{sample mass}}{\text{formula weight in grams}}\right)$$

9-15 MOLARITY

Molarity is a chemical term defined to provide the expression of solution concentrations in terms of the number of solute particles in a liter of a solution. The definition of molarity uses the chemical mole as an expression of the number of chemical particles of solute. Molarity (symbolically represented by M) is defined as follows:

> **Molarity** (M)—The number of moles of a solute per liter of solution.

This definition is devised so that solution concentrations can be expressed in terms of the number of solute particles (moles) that will be present in each liter of solution. Let us consider an example of a solution containing 117 grams of sodium chloride dissolved in enough water to make 2.0 liters of solution. To express the molarity of such a solution we determine the number of moles of solute involved and then divide by the solution volume in liters:

$$M = \left(\frac{\text{number of moles of solute}}{\text{volume of solution in liters}}\right)$$

To determine the molarity of the sodium chloride solution we first determine the number of moles of sodium chloride by dividing the 117 grams by the formula weight in grams:

$$\left(\frac{117 \text{ g}}{58.5 \text{ g per mole NaCl}}\right) = 2.0 \text{ moles NaCl}$$

Now we divide the number of moles by the volume of the solution:

$$M = \left(\frac{2.0 \text{ moles NaCl}}{2.0 \text{ liters}}\right) = \frac{1.0 \text{ mole NaCl}}{\text{liter}}$$

Such a concentration is often expressed as $1.0M$ NaCl and means that the solution contains a concentration of sodium chloride so that each liter of solution will contain 1 mole of sodium chloride.

Chemists use molarities quite often when dealing with solutions. In a chemistry laboratory you often see solutions with labels which express the concentration of a solute in terms of molarity.

BOOKS

Davis, K. S., and J. A. Day. *Water: The Mirror of Science*. Garden City, N.Y.: Doubleday, 1961.

Leopold, L. B., and W. B. Langheim. *A Primer on Water*. Washington, D.C.: U.S. Government Printing Office, 1960.

Overman, M. *Water*. Garden City, N.Y.: Doubleday, 1968.

ARTICLES AND PAMPHLETS

Buswell, A. M., and W. H. Rodebush. "Water." *Scientific American* (Apr. 1956).

QUESTIONS AND PROBLEMS

1. What percentage of the human body is water?

2. Describe the liquefaction of a gas.

3. Describe the evaporation of a liquid.

4. What is boiling? What is the normal boiling point of a liquid?

5. Describe the molecular structure of water.

6. What is a hydrogen bond?

7. Why does ice float on water?

8. What is a solution?

9. Define the following terms:

   (a) solvent   (d) unsaturated solution
   (b) solute   (e) supersaturated solution
   (c) saturated solution

10. Describe osmosis and an osmotic membrane.

11. What is a colloidal particle? What is a colloidal dispersion?

12. Describe dialysis.

13. If a 0.25 liter sample of water is found to contain 7.5 milligrams of nitrate ion, what is the parts per million nitrate ion in the water?

14. If a 2.0 liter sample of water is found to contain 500 milligrams of sulfate ion, what is the parts per million sulfate ion in the water?

15. What is a mole?

16. Give a definition for molarity.

17. Suppose a solution contains 84 grams of sodium hydrogen carbonate, $NaHCO_3$, in 0.50 liter of solution. What is the molarity of the solution?

18. A solution contains 60 grams of acetic acid, $HC_2H_3O_2$, in 2.0 liters of solution. What is the molarity of the solution?

WATER
IN THE
ENVIRONMENT

10

10-1 INTRODUCTION

The **natural waters** of the hydrosphere are normally solutions of varying complexity. This results from natural water's intimate contact with the chemicals of the lithosphere, atmosphere, and biosphere. Rain water contains dissolved gases of the atmosphere and, on occasion, dissolved air pollutants. River and lake waters contain dissolved minerals, atmospheric gases, and a variety of chemicals contributed by humans. Ocean water is a complex solution consisting of a variety of chemicals. Even drinking water contains some dissolved chemicals. When you look at a glass of water imagine that it might contain a variety of dissolved gases, dissolved molecules, and positive and negative ions.

   The chemicals in natural waters contribute important properties to the waters. Some of these chemicals are vital to aquatic plants and animals. On the other hand, some of these chemicals interfere with the intended use of the water and, thus, are considered to be **water pollutants.** This chapter deals with the types of substances that may be found in the various kinds of water found in the natural environment.

10-2 DISTRIBUTION OF WATER IN
THE ECOSPHERE

The waters of the hydrosphere are continuously being transferred from the oceans to the land areas and back to the oceans in cyclic process called the **water cycle.** Since relatively little new water is generated by volcanic activity, the same water has been passing through the water cycle for ages. The same water molecule found in a drop of your saliva may have spent millions of years in the primeval oceans and thousands of years in a polar ice cap. It may have been carried through the atmosphere as vapor and deposited in the Nile River during the reign of a great Pharoah. The same molecule may have drifted in the Mediterranean Sea and through evaporation and condensation entered a Roman aqueduct where it was consumed by Caesar, becoming part of his body water. It may have left his decaying body and spent centuries in the local water cycles of medieval Europe. In 1492, this molecule may have traveled with Columbus to be urinated in the New World. It may have become part of the underground water supply of the American colonies and been carried west by the pioneers. The

Table 10-1
Distribution of Water
in the Hydrosphere

	Percentage of 1.5 billion cubic kilometers
Oceans	97.2
Polar icecaps and glaciers	2.15
Subsurface water (soil moisture and ground water)	0.63
Surface water (fresh-water lakes, rivers, and streams, and saline lakes and landlocked seas)	0.019
Atmospheric water	0.001

molecule may have circled the globe countless times or been caught in local water cycles, finally entering your body yesterday.

Before discussing the details of the water cycle, let us consider the distribution of the earth's water. It is estimated that the total volume of the water of the hydrosphere is 1.5 billion cubic kilometers. As shown in Table 10-1, over 97 percent of this water is in the oceans which cover nearly three-quarters of the surface of the earth. Slightly more than 2 percent is in the form of ice and snow in the polar icecaps and glaciers. The remainder is made up of the surface and subsurface waters of the lithosphere and the water vapor and clouds of the atmosphere. Only about 0.001 percent of the earth's water is in the atmosphere.

We are familiar with ocean water and surface waters in lakes, rivers, and streams, but most of the water in the lithosphere is **subsurface ground** water. When water falls on soil it wets the soil. When the soil is saturated, the excess water may run off as streams, and some percolates down through the soil particles and porous subsurface rocks until it reaches some impermeable rock. At this point it accumulates, forming a saturated subsurface zone. The top of this zone is called the **water table,** as shown in Figure 10-1. Ground water can flow horizontally when it accumulates in porous geological formations called **aquifers.** Aquifers may be a few meters to hundreds of meters thick and may underlie an area of a few hundred square meters or many square kilometers. In fact, within

Figure 10-1
Ground water, the
water table, and
aquifers.

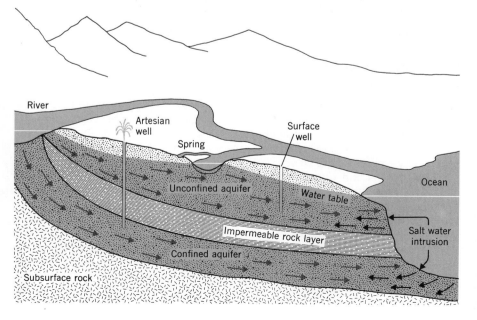

the aquifer ground water can flow for hundreds of miles to fill spring-fed lakes or to empty into the ocean. However, ground water movement is quite slow and is often measured in meters per year as compared to river flows of meters per second or minute. Some relatively confined aquifers contain water which was deposited 10,000–30,000 years ago.

When the land surface dips below the water table, the water may become surface water as a spring or accumulate as a lake as illustrated in Figure 10-1. An **unconfined aquifer** may be tapped by use of a surface well; a shaft sunk into the ground to allow access to the subsurface water table. A **confined aquifer** is one in which the vertical movement of the water is prevented by an impermeable rock layer. (See Figure 10-1.) Often, water in a confined aquifer is under pressure, and when a well is sunk into the aquifer the pressure will drive the water upward as an **artesian well.**

Surface water is overflow from aquifers or water which has not reached an aquifer. It is estimated that 97 percent of the fresh water in the United States is underground. About one-fifth of the fresh water used in the United States is ground water. Rural areas and much of the Southwest are almost entirely dependent upon ground water. In regions of high annual rainfall, aquifers are recharged by the rains. In regions of low rainfall, such as the Western United States, aquifers are charged by the melting snow packs of mountains or contain ancient water thousands of years old. In some regions, aquifers are artificially recharged by spreading water in porous regions or by reverse pumping through wells. In the Los Angeles area, large amounts of Colorado River water, drainage from irrigation, and effluent from sewage-disposal plants are used to replenish the aquifer. In coastal regions of the Los Angeles area and Long Island, N.Y., aquifers are being recharged to protect against **intrusion** of salt water from the ocean into the aquifer.

10-3 THE WATER CYCLE

The water cycle is illustrated in Figure 10-2. The oceans of the world are exposed to large amounts of solar radiation. About one-half of the solar radiation absorbed by the sea results in the evaporation of the ocean water. Solar radiation is the source of energy for the water cycle and the weather fluctuations and

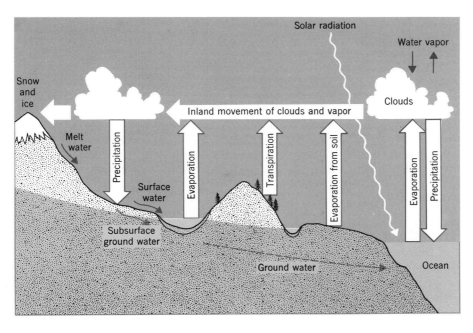

Figure 10-2
The water cycle.

turbulance which is associated with the cycle. Water vapor formed by evaporation from the oceans, lakes, rivers, or soil by solar radiation or transpired from the leaves of plants may travel long distances in the atmosphere. **Transpiration** is the phenomenon by which plants release excess water to the atmosphere through tiny leaf pores called stomata, which also act as exchange pores for the oxygen and carbon dioxide involved in photosynthesis.

Weather fluctuations are accompanied by increasing and decreasing concentrations of atmospheric water vapor. Some of the water vapor condenses into clouds of water droplets or ice crystals. Interestingly, the formation of clouds occurs when condensation occurs on tiny particulates of dust, smoke, and sea salt in the air. **Precipitation** depends upon these particulates, which serve as nuclei upon which the cloud particles grow. When the particles are large enough, they fall to the earth as rain, hail, or snow.

Much of the precipitation occurs over the ocean and does not reach the land surface. However, a portion of the evaporated waters of the oceans is deposited as fresh water on the land. The water that reaches the lithosphere may almost immediately reevaporate or accumulate in lakes, streams, or rivers as surface water. However, much of the precipitation percolates through the soil and becomes ground water. A portion of the precipitation is captured in the snow packs of mountains and in the polar icecaps and glaciers. Some of the water passes through the life cycle of animals and plants, serving as a source of hydrogen in photosynthesis and as a component of living cells.

Ultimately, much of the fresh water of the lithosphere flows back into the oceans to complete the water cycle. (See Figure 10-2.) Of course, before the water reaches the oceans it may be used numerous times by humans for agricultural irrigation, industrial needs, domestic purposes, and to carry away our wastes and sewage.

10-4 ELECTRICAL CONDUCTIVITY OF WATER SOLUTIONS

A rule of electrical safety is to never use an electrical appliance when you are wet or are standing in water. The purpose of the rule is to avoid an electrical shock caused by electricity flowing from the appliance through your body to the ground. You might wonder how water can conduct electricity. Actually, it is the presence of dissolved ions in water which make water an electrical conductor.

Some substances dissolve in water to form ions. Other substances dissolve so that their molecules become intimately mixed with the water molecules, but no ions are formed. How do we know when a dissolved substance forms ions in solution? Since ions are charged, they possess certain electrical properties. One of these properties is that oppositely charged entities attract one another. Suppose we had a solution containing **cations** (positive ions) and **anions** (negative ions) and we placed into this solution a piece of inert metal (such as platinum) that carried a negative charge. This negatively charged metal would attract the positive ions in the solution, and, thus, the cations would migrate toward the metal and ultimately form a layer of ions around the metal. If a positively charged piece of metal were placed in the solution, the negative ions would be attracted to the metal and would form a layer of ions around the metal.

How would it be possible to obtain the two charged pieces of metal? This could be accomplished by connecting the two pieces of metal through metal wires to the two terminals of a **battery** or a **generator,** as shown in Figure 10-3. The battery or generator serves as a device for pumping electrons from one piece of metal to the other. This ease of movement of electrons arises from the nature of the structure of metals. Since one piece of metal connected to the battery has an excess of electrons, it will be negatively charged and the piece that is deficient in electrons will be positively charged. The electron pump (battery)

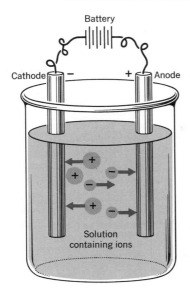

Figure 10-3
An electrolysis
apparatus. From
*Introduction to
Chemistry* by T. R.
Dickson, John Wiley &
Sons, New York, 1971.

serves to maintain what is called a **potential difference** between the two pieces of metal. If these two pieces of metal connected to the electron pump are immersed in a solution containing ions, the cations are attracted toward the negative piece of metal and the anions toward the positive metal. In this situation, the pieces of metal are called **electrodes.** The electrode that attracts the cations is called the **cathode** and the electrode that attracts the anions is called the **anode.**

An interesting phenomenon occurs if the potential difference between the electrodes is great enough. The battery or generator provides a driving force that can result in a chemical reaction that involves the gaining of electrons by some species at the cathode surface and the loss of electrons by another species at the surface of the anode. That is, since the battery or generator tends to pump electrons, a situation arises in which some species in solution lose electrons at the anode and others gain electrons at the cathode. In this manner, a complete electrical circuit is set up, in which electrons are pumped through the metal to the cathode where electrons are gained by some species in solution and, simultaneously, some species in solution lose electrons to the anode which provides more electrons to be pumped to the cathode. The fact that, under the correct conditions, ions can migrate in the solution and electrons are lost and gained at the electrodes, means that a solution containing ions will conduct electricity. However, electrical conductivity in a solution is not the same as that which occurs in a metal. It is not possible for electrons to flow through the solution, but electrical conductivity in solution occurs as a result of the movement of ions and the electrode reactions involving the loss and gain of electrons. The species that react at the cathode and anode depend on the nature of the species present in solution. Some species react readily at the electrodes while others do not react. The process of subjecting a solution to the conditions which will produce electrode reactions is called **electrolysis.**

10-5 ELECTROLYSIS OF SODIUM CHLORIDE SOLUTIONS

As an example of an electrolytic process that illustrates electrical conductivity by a solution, let us consider what happens when a rather concentrated sodium chloride (brine) solution is placed in an electrolytic cell. When the electrodes are connected to the external source of electricity (electron pump), the chloride ions are attracted to the anode where they can lose electrons and form molecular chlorine. This reaction can be represented as

$$2 \text{ Cl}^- \text{ (aqueous)} \longrightarrow \text{Cl}_2 \text{ (gas)} + 2 \text{ e}^- \text{ (anode)}$$

Since molecular chlorine is diatomic, two chloride ions must react to produce one chlorine molecule. A reaction in which a species loses or gains electrons at an electrode and is converted to a new species is called an **electrode reaction.** An electrode reaction, such as that given above, only occurs when a simultaneous reaction involving the gain of electrons occurs at the other electrode in the system. In this example, we would expect the sodium ions to migrate toward the cathode. However, sodium ions cannot gain electrons in aqueous solution to form sodium metal, because sodium metal cannot exist in water. Some other species must react at the cathode. The only other species present is water, which in this case will gain electrons according to the electrode reaction

$$2 \text{ e}^- \text{ (cathode)} + 2 \text{ H}_2\text{O} \longrightarrow \text{H}_2 \text{ (gas)} + 2 \text{ OH}^- \text{ (aqueous)}$$

These two electrode reactions occur simultaneously when the solution of sodium chloride is subjected to electrolysis. For every two electrons gained at the cathode, two electrons are lost at the anode. These electrode reactions are actually chemical reactions involving electron transfer. The overall result of the electron transfer process occurring during the electrolysis of the sodium chloride solution can be represented by the equation

$$2 \text{ Na}^+ + 2 \text{ Cl}^- + 2 \text{ H}_2\text{O} \xrightarrow{\text{electrolysis}} \text{Cl}_2 \text{ (gas)} + \text{H}_2 \text{ (gas)} + 2 \text{ Na}^+ + 2 \text{ OH}^-$$

Thus, when a rather concentrated solution of sodium chloride is subjected to electrolysis, chlorine gas is produced at the anode and hydrogen gas is produced at the cathode. Furthermore, for every two chloride ions which react, two hydroxide ions are produced. See Figure 10-4 for an illustration of this electrolysis process.

Figure 10-4
The electrolysis of a sodium chloride solution. From *Introduction to Chemistry* by T. R. Dickson, John Wiley & Sons, New York, 1971.

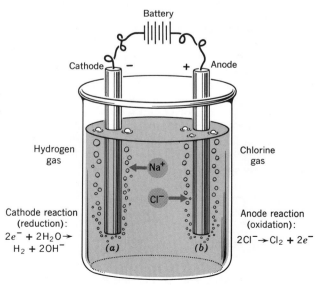

Battery

Cathode   +   Anode

Hydrogen gas   Chlorine gas

Na$^+$

Cl$^-$

Cathode reaction (reduction):
$2e^- + 2H_2O \rightarrow$
$H_2 + 2OH^-$

(a)   *(b)*

Anode reaction (oxidation):
$2Cl^- \rightarrow Cl_2 + 2e^-$

Overall reaction: $2Cl^- + 2H_2O \rightarrow Cl_2 + H_2 + 2OH^-$

10-6 CHLOR–ALKALI PLANTS AND MERCURY IN THE ENVIRONMENT

The electrolysis of brine is an important industrial process carried out in numerous **chlor–alkali plants** in the United States. These plants produce chlorine gas and sodium hydroxide solutions called **alkali** solutions or **caustic soda.** The hydrogen is vented to the atmosphere. Chlor–alkali plants are a good example of an unexpected source of environmental pollution. Most plants use an electrolysis process in which mercury metal serves as the electrode at which the hydroxide ion and hydrogen are produced. After electrolysis, the used brine is flushed out of the cell and replaced. This waste contains some mercury as does the caustic soda produced. Furthermore, the hydrogen vented to the atmosphere contains some mercury vapor.

Many chlor–alkali plants have been operating in this country for decades. In 1970, Norvald Fimreite, a student at the University of Western Ontario, reported dangerous levels of mercury in fish from Lake St. Clair. Fishing in the lake was banned and an investigation was begun by the Canadian government. The investigation revealed that much of the mercury came from the wastes of a chlor–alkali plant located on the shores of the lake. (See Section 4-11 for a discussion of mercury in the environment.) Similar investigation in the United States showed that most chlor–alkali plants were releasing from 2 to 30 kilograms of mercury into the environment each day. The industries claimed that they knew of the mercury losses but thought that the mercury would be deposited in the rivers and lakes and cause no problems. Unfortunately, the metallic and inorganic mercury compounds released by the plants were converted to methyl mercury compounds which entered the food chain. The investigations of 1970 showed higher than expected mercury levels in the waterways of 33 states. The chlor–alkali industries instituted certain control methods and were able to reduce mercury discharges in the waste water by over 85 percent and cut down on atmospheric mercury emission. However, these plants remain a source of mercury in the environment.

10-7 ELECTROLYTES AND NONELECTROLYTES

The conduction of electricity by a solution involves the movement of ions and certain electrode reactions. Consequently, observing that a solution will conduct electricity indicates that ions are present in the solution. A substance that forms an aqueous solution that conducts electricity is called an **electrolyte.** Of course, an electrolyte must form ions in solution when it dissolves. Most soluble ionic substances are electrolytes, as are some molecular substances. A substance that forms an aqueous solution that does not conduct electricity is called a **nonelectrolyte.** Many molecular substances (e.g., sugar and alcohol) are nonelectrolytes. The fact that a substance is a nonelectrolyte indicates that it does not form ions when it dissolves. Whether or not a substance is an electrolyte can be determined by preparing an aqueous solution of the substance and then experimentally observing the conductivity of the solution.

10-8 CHEMICAL EQUILIBRIUM

Suppose we had a water solution containing a certain concentration of hydrogen chloride, HCl, and another solution of the same concentration of acetic acid, $HC_2H_3O_2$. When we test the conductivity of these solutions, they both are found to conduct electricity, so we classify hydrogen chloride and acetic acid as electrolytes. However, the hydrogen chloride solution is found to be a much stronger conductor of electricity than the acetic acid solution. This difference in conductivity indicates that the acetic acid solution has fewer ions than the

Figure 10-5a
Fish analogy for
dynamic equilibrium.

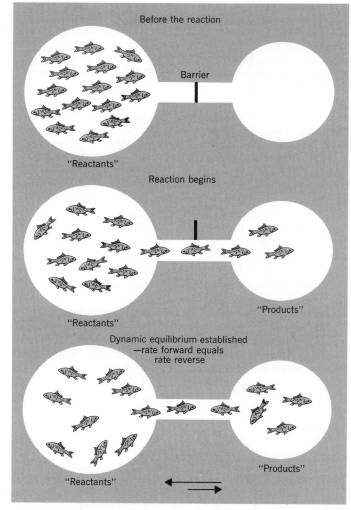

Fish anology for dynamic equilibrium. Note that once dynamic equilibrium
is established there is no net change in the "concentration" of the reactants or products, but there is a continuous interchange between the two
sides.

hydrogen chloride solution. Since they are both electrolytes and we are comparing solutions of the same concentration, we might expect the solutions to
contain the same number of ions, but they do not. The difference in the number
of ions is a result of the extent to which these two substances enter into chemical reaction with water to form ions.

When some molecular substances, such as hydrogen chloride, dissolve in
water, they enter into a chemical reaction which consumes the substance to
form ions. The reaction between hydrogen chloride and water is

$$HCl + H_2O \longrightarrow H_3O^+ + Cl^-$$

On the other hand, some molecular substances dissolve in water and enter into
a chemical reaction which produces relatively few ions. For example, when
acetic acid is dissolved in water, the reaction is

$$HC_2H_3O_2 + H_2O \rightleftharpoons H_3O^+ + C_2H_3O_2^-$$

Figure 10-5b
An idealized view
of dynamic
equilibrium.

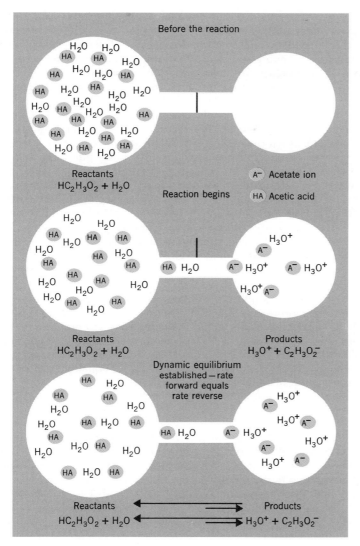

Dynamic equilibrium involved in the reaction of acetic acid with water to form acetate ion and hydronium ion.

This reaction is an example of a **reversible chemical reaction** involving chemical equilibrium (note the double arrow). In previous discussions we have given chemical equations showing the reactants and products separated by a single arrow. Since many reactions actually involve chemical equilibrium, let us consider the nature of chemical equilibrium.

Figure 10-5 illustrates **chemical equilibrium** using an analogy involving fish in tanks. In Figure 10-5a the fish in the left-hand tank represent the initial reactants. When the barrier between tanks is removed, some of the fish make their way to the right-hand tank and become products. As they accumulate on the right, some will swim back to the left-hand tank. After a certain time, they will distribute between the two tanks so that there is a certain "concentration" of reactants and products. But, the fish will continuously swim back and forth between the tanks changing places with other fish. The idea is that even though they are swimming back and forth, the relative "concentrations" in both tanks remain constant. When this occurs a state of **dynamic equilibrium** is said to exist. The term **dynamic** means that some of the "reactants" are forming "products" and some of the "products" are forming "reactants." The term **equilibrium** means that the rates of the two changes are the same and no net change in the "concentrations" of the "reactants" and "products" is apparent. That is, the relative numbers of fish in the two tanks remain constant even though the fish are continuously moving back and forth between tanks.

Now let us consider the reversible reaction between acetic acid and water in terms of chemical equilibrium. (See Figure 10-5 b.) The $HC_2H_3O_2$ enters into reaction with H_2O to form H_3O^+ (hydronium ion) and $C_2H_3O_2^-$ (acetate ion). When these two ions begin to accumulate, they enter into a reaction with one another which reforms the acetic acid and water. After a while, a point is reached when the rate of the reaction forming the ions equals the rate of the reaction of the ions, and a state of dynamic chemical equilibrium exists. That is, acetic acid and water are continuously reacting to give hydronium ion and acetate ion, and the ions are continuously reacting to give the original reactants; but, since the rates of the two competing reactions are the same, the relative concentrations of the reactants and products becomes fixed at certain levels. In fact, the equilibrium reaction in the case of acetic acid produces low concentrations of ions.

A state of dynamic equilibrium in a reaction is denoted by use of a double arrow in the equation for the reaction. The relative sizes of the arrows used indicate which species exist at greater concentrations at equilibrium. The larger arrow points in the direction of the species present at greater concentrations. The equilibrium is said to favor this direction. For example, the reversible reaction between acetic acid and water (see the above equation and note arrows) favors the formation of water and acetic acid. Another way of stating this is that water and acetic acid enter into a slight chemical reaction to form hydronium ion and acetate ion.

Reversible chemical reactions involving chemical equilibrium are quite common, and many of the reactions we have discussed are reversible reactions. Usually, it is only necessary to refer to the equilibrium to explain certain chemical phenomena. For instance, in Section 8-6 the reaction of carbon monoxide with the hemoglobin in the blood was discussed. Fortunately, this is a reversible reaction so that when a person breathes air having a lower level of carbon monoxide, the carboxyhemoglobin will decompose, releasing the carbon monoxide ($HbCO \leftrightarrows Hb + CO$). When we want to emphasize the reversibility or state of equilibrium of a reaction, we include the double arrow in the equation. When we are not concerned with denoting equilibrium, we use a single arrow between the reactants and products in the same way that we denote a nonequilibrium reaction.

10-9 ACIDS AND BASES

As you probably know, vinegar has a distinct sour taste. However, you may not know that when you drop a piece of chalk in a vinegar solution, carbon dioxide will bubble off as the chalk dissolves; or that a piece of freshly sandpapered zinc, when placed in vinegar, will slowly dissolve accompanied by the evolution of hydrogen gas. Vinegar is a water solution of acetic acid.

It is possible to classify certain compounds according to similarities in chemical properties, that is, similarities in the kinds of chemical reactions the substances undergo. Chemists noted long ago that certain substances, which became known as **acids,** were characterized as having sour tastes, being able to dissolve certain metals, changing the color of the vegetable dye called litmus from blue to red, and reacting with chemicals called bases. The substances which became known as **bases** had bitter tastes, felt slippery to the touch, would change the color of blue litmus to red, and could react chemically with acids. As a deeper understanding of the nature of chemicals developed, it became apparent that there were many substances which could be classified as acids or bases. It is now possible to give more definite chemical definitions for these classes of compounds.

A Swedish chemist, Svante Arrhenius, in 1884, proposed the first significant definitions. Several are possible. He defined an acid as a substance that forms hydrogen ions (H^+) in water solution and a base as a substance that forms hydroxide ions (OH^-) in water solution. The **Arrhenius theory** was significant, since it provided a basis for the description of ions in aqueous solutions. However, for acids and bases in aqueous solutions, the best definitions come from the **Brønsted–Lowry theory** of acids and bases. According to this theory, the terms acid and base should be defined as follows:

> **Acid**—A chemical species that can donate hydrogen ions or protons (H^+) in a chemical reaction.

> **Base**—A chemical species that can accept hydrogen ions or protons in a chemical reaction.

According to these definitions, an acid is any hydrogen-containing species in which the covalent bond holding the hydrogen can be broken so that the hydrogen ion can be lost. A base is a species which is capable of forming a new covalent bond with a proton donated by an acid. Of course, when an acid loses a proton a base must be present to accept it. Thus, an acid can react with a base in a proton transfer or **acid–base reaction.** Such a reaction can be generally represented as

$$\underset{\text{acid}}{H–A} + \underset{\text{base}}{B} \longrightarrow HB^+ + A^-$$

As an example of an acid–base reaction, consider the reaction between hydrogen chloride gas and water:

$$\underset{\text{acid}}{HCl} + \underset{\text{base}}{H_2O} \longrightarrow \underset{\underset{\text{hydronium ion}}{}}{H_3O^+} + \underset{\underset{\text{chloride ion}}{}}{Cl^-}$$
$$\text{hydrochloric acid}$$

Note that the hydrogen chloride is an acid which loses a proton to water, which is a base. Hydrogen chloride is quite soluble in water and produces a water solution called **hydrochloric** acid or muriatic acid. Table 10-2 lists a few more common acids.

As another example of an acid–base reaction consider the reversible reaction of ammonia with water:

$$\underset{\text{base}}{NH_3} + \underset{\text{acid}}{H_2O} \rightleftharpoons \underset{\underset{\text{ion}}{\text{ammonium}}}{NH_4^+} + \underset{\underset{\text{ion}}{\text{hydroxide}}}{OH^-}$$

Ammonia is quite soluble in water but only reacts to a slight extent. Nevertheless, ammonia reacts as a base (proton acceptor) and water reacts as an acid

Acid	Formula	Acid	Formula
Acetic acid	$HC_2H_3O_2$	Nitric acid	HNO_3
Boric acid	H_3BO_3	Phosphoric acid	H_3PO_4
Hydrochloric acid	HCl	Sulfuric acid	H_2SO_4

Table 10-2
Some Common Acids

Table 10-3
Some Common Bases

Base	Formula	Base	Formula
Ammonia	NH_3	Sodium carbonate	Na_2CO_3
Calcium hydroxide	$Ca(OH)_2$	Sodium hydroxide	NaOH
Magnesium hydroxide	$Mg(OH)_2$	Sodium phosphate	Na_3PO_4
Potassium hydroxide	KOH		

(proton donor). A water solution of ammonia is called **aqua ammonia**. Table 10-3 lists some typical bases. Note that water is an acid in this reaction, but it behaved as a base in the previous reaction with hydrogen chloride. Water can be an acid in some cases and a base in other cases. A species that can act as an acid or a base is called **amphiprotic**. Since water is amphiprotic, a proton transfer can occur between water molecules:

$$H_2O + H_2O \rightleftharpoons H_3O^+ + OH^-$$

This reversible reaction does not occur to a great extent, but does indicate that in pure water equal but small concentrations of hydronium ion and hydroxide ion are present. Pure water or a water solution which contains neither an acid nor a base will have equal concentrations of hydronium and hydroxide ions and are called **neutral solutions**. When an acid is mixed with water it reacts to form more hydronium ion. A water solution which has a concentration of hydronium ion that is greater than pure water is called an **acidic solution**. When a base is mixed with water it reacts to form more hydroxide ion. A water solution which has a concentration of hydroxide ion that is greater than pure water is called a **basic solution**.

10-10 SALTS

When an acid solution is mixed with a base solution an acid–base reaction can occur which **neutralizes** the acid and the base. For instance, when hydrochloric acid is mixed with a sodium hydroxide solution, the reaction is

$$H_3O^+ + Cl^- + Na^+ + OH^- \longrightarrow \underline{Na^+ + Cl^- + 2\,H_2O}$$

sodium chloride solution

Whenever an acid solution and a base solution are mixed, the result is a solution of a positive ion and a negative ion. If the water is removed by evaporation, the resulting ionic compound is called a **salt**. All salts are ionic compounds containing some metal ion (or ammonium ion, NH_4^+) combined with some negative ion other than hydroxide ion. Sometimes the word salt is used to refer to sodium chloride, NaCl, but the term is used in a general sense to refer to certain ionic compounds. Many salts are chemically manufactured for various uses. A few important salts and their uses are listed in Table 10-4.

10-11 GASES IN WATER

You have undoubtedly noticed gaseous carbon dioxide bubbling out of a bottle of carbonated beverage. The carbon dioxide is dissolved in the beverage under a slight pressure. Removal of the cap releases the pressure, and much of the carbon dioxide bubbles out. Tap water contains some dissolved air. Heating of tap water will cause some of this air to bubble out of the water. This is why, as

Table 10-4
Some Typical Salts

Salt	Formula	Common uses
Sodium chloride	NaCl	Industrial, de-icing of roads
Sodium carbonate	Na_2CO_3	Soda ash, industrial
Sodium bicarbonate	$NaHCO_3$	Baking soda, industrial
Sodium nitrate	$NaNO_3$	Fertilizers, explosives
Sodium thiosulfate	$Na_2S_2O_3$	Photographic hypo
Sodium tetraborate decahydrate	$Na_2B_4O_7 \cdot 10H_2O$	Borax
Potassium chloride	KCl	Fertilizers
Potassium bromide	KBr	Medicine and photography
Potassium nitrate	KNO_3	Fertilizers, explosives
Ammonium nitrate	NH_4NO_3	Fertilizers, explosives
Ammonium sulfate	$(NH_4)_2SO_4$	Fertilizers
Calcium phosphate	$Ca_3(PO_4)_2$	Fertilizers
Calcium sulfate dihydrate	$CaSO_4 \cdot 2H_2O$	Gypsum, plaster board

you heat a pan of water, small air bubbles form on the sides and bottom of the pan.

Natural waters contain variable amounts of **dissolved oxygen,** O_2, nitrogen, N_2, and carbon dioxide, CO_2. As shown in Figure 10-6, these gases dissolve in water from the atmosphere and can be produced or consumed in the biological processes that occur within the natural waters. Carbon dioxide gas is found in natural waters as dissolved CO_2, but some carbon dioxide may exist in the form of hydrogen carbonate (bicarbonate) ion:

$$CO_2 + 2 H_2O \rightleftharpoons H_3O^+ + HCO_3^-$$

The solubilities of oxygen, nitrogen, and carbon dioxide are listed in Table 10-5. Note that the solubilities are less at higher temperatures. Thus, warmer water will contain smaller concentrations of these gases than cooler water. Although these three gases are the most important of the dissolved gases, other gases are occasionally found in water. Table 10-6 lists some of these.

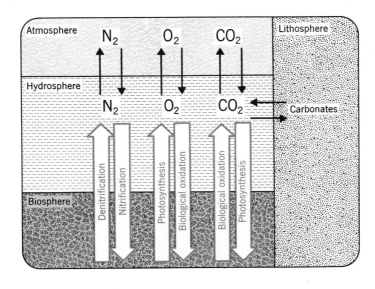

Figure 10-6
Sources of dissolved gases in the hydrosphere.

Table 10-5
Comparative
Solubilities of Gases
in Pure Water

Gas	Solubility (parts per million) at			
	0°C	20°C	40°C	50°C
CO_2	3,480	1,690	970	760
N_2	29	19	15	12
O_2	70	43	30	27

Table 10-6
Some Other Gases
Found in Natural
Waters

Gas	Formula	Source
Ammonia	NH_3	Industrial and sewage wastes
Chlorine	Cl_2	Industrial and sewage wastes
Hydrogen sulfide	H_2S	Natural
Methane	CH_4	Natural
Nitrogen dioxide	NO_2	Industrial wastes and air pollution
Sulfur dioxide	SO_2	Industrial wastes and air pollution

10-12 NATURAL WATERS

Recall from Section 10-3 that in the water cycle water moves from the oceans to the land by evaporation and precipitation. Then it returns to the oceans, passing through streams, rivers, and lakes. Water passing through this cycle comes into intimate contact with the atmosphere and the rocks, minerals, and salts of the lithosphere. As a result of dissolving processes, these natural waters become solutions containing various ions and molecules. The composition of natural waters is discussed below.

OCEAN WATER The ocean is the habitat of tremendous amounts of animal and plant life. The ocean is actually a vast solution of ions and other substances in which these plants and animals exist. Most of the dissolved constituents of the ocean are ions. Studies of sea water from many sources have shown that there are apparently only small variations in the relative amounts of these ions. The major constituents of typical sea water in grams per kilogram of sea water are:

Constituent	Concentration (grams per kilogram)	Constituent	Concentration (grams per kilogram)
Sodium ion, Na^+	10.76	Chloride ion, Cl^-	19.353
Magnesium ion, Mg^{2+}	1.294	Sulfate ion, SO_4^{2-}	2.712
Calcium ion, Ca^{2+}	0.413	Hydrogen carbonate ion, HCO_3^-	0.142
Potassium ion, K^+	0.387	Bromide ion, Br^-	0.067
Strontium ion, Sr^{2+}	0.008	Fluoride ion, F^-	0.001
		Iodide ion, I^-	0.00006
Boric acid, H_3BO_3	0.004		
Nitrogen, N_2	0.010		
Oxygen, O_2	0.007		
Carbon dioxide, CO_2	0.6		

Ocean water contains about 3.5 percent by mass dissolved ions. As can be seen in the above listing, sodium ion and chloride are the predominant ions. It is obvious why ocean water tastes salty. Many other elements, called trace elements, are present in ocean water in very small concentrations. When a sample of sea water is evaporated, a mixture of ionic compounds remains.

The **salinity** of sea water as defined by oceanographers is the mass in grams of the solids in 1 kilogram of sea water evaporated to a constant mass at 480°C. Since the relative amounts of dissolved substances is somewhat invariant, the salinity of sea water can be directly related to the chlorinity of the seawater. **Chlorinity** is defined as the number of grams of chloride ion, bromide ion, and iodide ion contained in 1 kilogram of sea water. In the experimental determination of chlorinity, the bromide ions and iodide ions are assumed to be replaced by chloride ions for calculation purposes. The experimentally observed relationship between salinity and chlorinity is

salinity (grams per kilogram of sea water)
$$= 1.805 \text{ chlorinity (grams per kilogram of sea water)} + 0.030$$

Some of the major constituents found in sea water are extracted commercially. Millions of tons of salt (NaCl) are obtained from the ocean every year by solar evaporation. Over 100,000 tons of bromine are obtained from the ocean annually, using chlorine to convert the bromide ion to bromine:

$$Cl_2 + 2\ Br^- \longrightarrow 2\ Cl^- + Br_2 \text{ (bromine)}$$

A certain amount of magnesium metal is obtained from the magnesium ion in ocean water by the following series of reactions:

$$Mg^{2+} + 2\ OH^- \longrightarrow Mg(OH)_2$$

$$Mg(OH)_2 + 2\ H_3O^+ + 2\ Cl^- \longrightarrow MgCl_2 + 4\ H_2O$$

$$MgCl_2 \xrightarrow{\text{electrolysis}} Mg + Cl_2$$

Some day it may be economically possible to recover some of the trace elements such as gold or uranium found in the ocean.

LAKE AND RIVER WATER   Ocean water has a relatively constant salinity, but river and lake waters have variable composition. This is to be expected, since the rivers and lakes often contain water which has come into contact with various geological formations. This water may have flowed long or short distances over the land, dissolving minerals and substances of decaying plant life along the way. In addition, the waters may contain materials contributed by humans.

Some land-locked lakes are known as salt lakes, since they have accumulated vast amounts of dissolved mineral salts. Except for the waters of salt lakes, the natural waters of lakes and rivers are not salty and are called **fresh water.** Of course, such water is usually neither fresh nor pure, but is a solution of dissolved ions and molecules. The most common positive ions in such water are calcium ion, Ca^{2+}, magnesium ion, Mg^{2+}, and sodium ion, Na^+. The most common negative ions are hydrogen carbonate (bicarbonate) ion, HCO_3^-, and sulfate ion, SO_4^{2-}. There are many other substances present in lake and river water.

Since surface water, along with well water, is for public use and consumption, national chemical standards have been established for drinking water. The U.S. Public Health Service **chemical standards for drinking water** are listed below.

Chemical species	Maximum allowable concentration in milligrams per liter (parts per million)
Arsenic (ionic)	0.05
Barium ion	1.0
Cadmium ion	0.01
Chloride ion	250
Chromium (ionic)	0.05
Copper (ionic)	1
Cyanide ion	0.2
Fluoride ion	about 2.0
Iron (ionic)	0.3
Lead (ionic)	0.05
Linear alkyl sulfonate (detergent)	0.5
Manganese (ionic)	0.05
Nitrate ion plus nitrite ion	10 (as N)
Selenium (ionic)	0.01
Silver ion	0.05
Sulfate ion	250
Synthetic organic chemicals (carbon–chloroform extract)	0.15
Total dissolved solids	500
Zinc ion	5

These standards along with bacteriological standards serve as guides to suppliers of public water to maintain health safety, color, appearance, taste, and odor of drinking water.

In 1970, a Public Health Service survey revealed that 41 percent, on an average basis, of public water systems surveyed in the United States were of inferior quality, and many of these supplied water that violated the recommended water quality standards. These results did not mean that the public water was unfit to drink, but was of a poor quality in terms of color, taste, and odor, or contained too much bacteria or dissolved chemicals. However, some of the small water systems serving less than 100,000 people were found to be supplying potentially hazardous water (water that violates any of the mandatory Public Health Service drinking water standards). Some of this water contained bacteriological contamination and some contained excessive amounts of lead, cadmium, arsenic, or selenium. The Public Health Service warned that such water has to be closely watched so that the quality does not further degrade.

10-13   HARD WATER AND WATER SOFTENING

Hard water is water containing calcium ion and magnesium ion in concentrations greater than about 60 parts per million. These two positive ions react with the negative ions of soap (see Section 11-6) to form soap scums:

$$Ca^{2+} + 2 \text{ soap}^- \longrightarrow Ca(soap)_2 \left. \begin{array}{l} \text{solid} \\ \text{insoluble} \\ \text{soap scums} \end{array} \right.$$
$$Mg^{2+} + 2 \text{ soap}^- \longrightarrow Mg(soap)_2$$

Thus, if water is too hard, it interferes with the use of soap. **Synthetic detergents** can be used more readily in hard water (see Section 11-6). Hard water can also produce scales in pipes, boilers, and even pots and pans. Such scales are a result of the formation of solid calcium and magnesium carbonates:

$$Ca^{2+} + CO_3^{2-} \longrightarrow CaCO_3 \Big]~\text{scale-like}$$
$$Mg^{2+} + CO_3^{2-} \longrightarrow MgCO_3 \Big/~\text{solids}$$

Water can be softened by removal of calcium and magnesium ions. Water from which most of the calcium and magnesium ions have been removed is called **soft water.** Some waters are naturally soft, while others have varying degrees of hardness. One common means of softening water is by use of **ion exchange.** The ion exchange method is illustrated in Figure 10-7. Hard water is passed through a bed of special plastic called an ion exchange resin. As the water passes through the resin, the calcium and magnesium ions are attracted to the resin where they are exchanged for sodium ions. Thus, the sodium ions replace most of the calcium and magnesium in the water. This exchange can be represented as

$$(\text{resin}^{2-})~Na^+ + Ca^{2+} \Longleftrightarrow (\text{resin}^{2-})~Ca^{2+} + 2~Na^+$$
$$Na^+$$

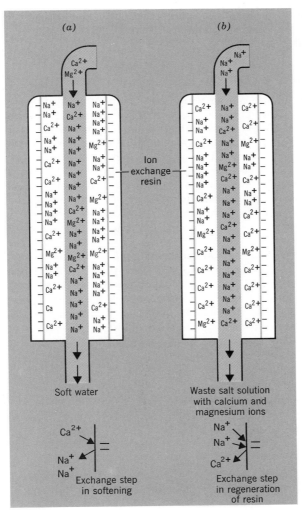

Figure 10-7
Water softening.

Water softening by ion exchange.

Regeneration of ion exchange resin.

Note that this is a reversible equilibrium reaction. As the hard water passes over the resin, the reaction proceeds to the right as written to soften the water. When most of the sodium ion has been exchanged, the resin no longer reacts. However, it can be regenerated so that it once more contains sodium ions. Such a regeneration takes advantage of a chemical principle called the **Le Chatelier principle,** which may be expressed as follows:

> When a system at equilibrium is upset by a change in any factor affecting the equilibrium, the system will change in a manner that tends to restore the original conditions.

Since the above exchange reaction is a chemical equilibrium, the principle can apply. As long as hard water is passed over the resin and there are sodium ions to be exchanged, the reaction reaches an equilibrium in which the rate of the forward reaction equals the rate of the reverse reaction and no net exchange occurs. To regenerate the resin, we pass a concentrated sodium chloride solution over the resin bed. The presence of large amounts of sodium ion upsets the equilibrium, and the reaction proceeds in a manner which tends to restore the equilibrium. This causes the sodium ion to replace the calcium ion on the resin. Once the resin is replenished, the excess salt solution with the calcium and magnesium ions is washed away, and the resin can again be used for water-softening purposes.

10-14 ACIDITY AND pH

Recall that a solution can be acidic, basic, or neutral. Acidic solutions have concentrations of hydronium ion, H_3O^+, that are higher than pure water, and basic solutions have concentrations of hydronium ion that are less than pure water. In certain biological and chemical reactions that occur in water solutions, the concentration of hydronium ion, or **acidity,** is important. For instance, the acidity of human blood must be maintained at a certain level. Any deviation of the hydronium ion concentration in blood outside a certain range of this level can lead to sickness and death. Fish and plants cannot survive in water that has too high a concentration of hydronium ion. When discussing acidity it is customary to express concentrations of hydronium ion in terms of moles per liter or molarity of hydronium ion.

Actually, the hydronium ion concentration in water can vary widely. A solution of an acid could contain 1 mole of hydronium ion per liter, while a solution of a base could contain 10^{-13} mole of hydronium ion per liter. Pure water at 25°C contains 10^{-7} mole hydronium ion per liter. Since these concentrations of hydronium ion include a very wide range, a special scale has been devised to express such concentrations in aqueous solutions. This is called the pH (read: pee-āch) scale and is based on the following definition: **pH** is the negative logarithm of the hydronium ion concentration,

$$pH = -\log[H_3O^+]$$

The $[H_3O^+]$ represents the concentration of hydronium ion in moles per liter. The pH allows for the expression of low concentrations of hydronium ion as a simple number, as shown below.

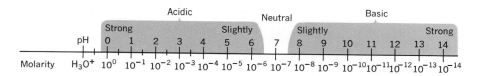

Note that there is a tenfold difference between subsequent pH units. For instance, pH 6 is 10 times more acid than pH 7, and pH 5 is 100 times more acid then pH 7, and so on. For solutions with $[H_3O^+]$ ranging from 1 M to 10^{-14} M, the pH ranges from 0 to 14. Solutions with pH below 7 are acidic, while solutions with pH above 7 are basic (sometimes called alkaline). A neutral solution at 25°C has a pH of 7. The pH is merely used as a means of expressing the hydronium ion concentration in a solution. If the $[H_3O^+]$ is an integral power of 10, the pH is easily determined as

$$[H_3O^+] = 10^{-pH}$$

That is, the pH is the numerical value of the power of ten. Thus, a solution with a $[H_3O^+]$ of 10^{-4} M has a pH of 4, and a solution with a $[H_3O^+]$ of 10^{-9} M has a pH of 9. When the pH is not an integral power of 10, the logarithm of the concentration must be taken. For example, a solution with $[H_3O^+]$ of 2×10^{-3} M has a pH of

$$pH = -\log(2 \times 10^{-3})$$
$$pH = -\log 2 - \log 10^{-3}$$
$$(\log 2 = 0.30 \text{ and } \log 10^{-3} = -3)$$
$$pH = -0.30 - (-3) = -0.30 + 3 = 2.7$$

The log of 10 raised to a whole number power is the power ($\log 10^{-3}$ is -3). The log of a number between 1 and 10 can be found in a table of logarithms or from the relationship shown below.

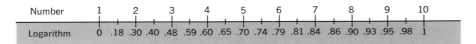

Number	1	2	3	4	5	6	7	8	9	10
Logarithm	0 .18	.30 .40	.48 .59	.60 .65	.70 .74	.79 .81	.84 .86	.90 .93	.95 .98	1

Hydronium ion is present in many common solutions and mixtures. Gastric juices, citric fruit juices, soft drinks, vinegar, urine, milk, and some natural fresh water are acidic in nature. Pure water and saliva are neutral, and blood is slightly basic. Table 10-7 lists the pH of some common solutions.

In certain biological processes that occur in solution, the pH must be maintained within certain limits. Human blood normally has a pH of about 7.4 at room temperature. Any deviation in the pH of blood above 7.9 or below 7.0 will lead fairly quickly to death.

A special instrument called a **pH meter** can be used to measure the pH of an aqueous solution. Actually, the pH meter measures an electrical property of the solution which is proportional to the pH. By careful calibration using solutions

pH	Solution	pH	Solution
1	Volcanic waters Gastric juice	7	Pure water Blood
2	Lemon juice Vinegar	8	Sea water
3	Orange juice Wine	9	Alkaline lakes
4	Tomatoes	10	Soap solutions
5	Coffee, black	11	Household ammonia
6	Urine Saliva	12	

Table 10-7
The pH Values of
Typical Solutions

of known pH, the pH meter can be used to determine whether a solution is acidic, basic, or neutral.

10-15   ACID RAINS AND ACID MINE DRAINAGE

The acidity of natural waters is relevant in two notable environmental problems. One problem is that of **acid rains**. Normally, the pH of rain water is around 5.7. It is slightly acidic as a result of dissolved carbon dioxide from the atmosphere. Studies of rains in European countries have shown an increasing trend to higher acidities of rains with pH values ranging between 3 and 5. This increased acidity results from the absorption by the rain of sulfuric acid and nitric acid formed in the atmosphere from sulfur oxide and nitrogen oxide air pollutants. Similar, but not as extensive, studies in New England states have shown that rains are about 10–100 times more acidic than normal. Again, the acid rains are attributed to air pollutants from industrial and metropolitan areas in the eastern United States. The health effects of acid rains are not known, but it is known that the rains have reduced forest growth and have caused corrosion damage to bridges and buildings.

Another problem associated with acidity is that of **acid mine drainage**. This problem is associated with waters draining from old and abandoned coal mines. In the mines, iron sulfides or pyrites are oxidized to sulfuric acid. The sulfuric acid dissolves in water passing through the mines. The water becomes very acidic and dissolves minerals as it drains from the mines. This acidic solution containing dissolved minerals is called acid mine drainage. When the drainage enters waterways it causes water pollution. In 1969, acid mine drainage was blamed for the pollution of an estimated 16,000 kilometers of waterways in the United States, mostly in Pennsylvania, West Virginia, and Kentucky. Prevention of acid mine drainage can sometimes be accomplished by flooding or sealing abandoned mines or by diverting water which flows through the mines.

BIBLIOGRAPHY   ARTICLES AND PAMPHLETS

"Drinking Water: Is it Drinkable?" *Environmental Science and Technology* (Oct. 1970).
Likens, G. E. "Acid Rains." *Environment* **14** (2) (1972).
Wood, J. M. "A Progress Report on Mercury." *Environment* **14** (1) (1972).

QUESTIONS AND PROBLEMS

1. Describe the distribution of water in the hydrosphere.
2. What is ground water? What is an aquifer?
3. Describe the water cycle.
4. Define the following terms:
    (a) electrolysis   (b) electrolyte   (c) nonelectrolyte
5. Describe dynamic equilibrium and chemical equilibrium.
6. Define the following terms:
    (a) acid   (b) base
7. Give an example of an acid.

8. Give an example of a base.

9. What are the most common dissolved gases found in natural waters? Why?

10. Give definitions of the following oceanographic terms:
    (a) chlorinity   (b) salinity

11. Is drinking water "pure" water? Explain your answer.

12. What is hard water?

13. Why is it inconvenient to use soap in hard water?

14. Describe the ion exchange method of water softening.

15. What is pH?

16. What are acid rains?

17. Describe acid mine drainage.

18. A household cleaning solution is found to have a hydronium ion concentration of 10^{-11} M. What is the pH of the solution? Is the solution acidic or basic?

19. A soft drink is found to have a hydronium ion concentration of 10^{-4} M. What is the pH? Is the drink acidic or basic?

20. Do you feel that the owner of an abandoned mine or the owner of property which includes an abandoned mine should be responsible for the prevention of acid mine drainage from the property? Explain your answer.

11-1 INTRODUCTION

Usable water is among our most important resources. Drinkable water is fundamental to life, and civilizations have always developed around adequate water supplies. Modern civilizations have developed techniques to transport water over long distances and to manage the waters so that they can be used properly and reused. However, since water is very plentiful in certain regions it has become misused and neglected in ways that have polluted the waters.

Actually, water pollution has been a problem ever since people have congregated in cities. Throughout the centuries open sewers and contaminated water supplies were common sources of disease and death. Most public water supplies in the United States are much safer than those that existed in the past and are superior to those that exist in most foreign countries. Of course, industrialization and high population densities generate new aspects to water pollution problems.

In 1887, A. R. Leeds of Stevens Institute wrote an article entitled "The Monstrous Pollution of the Water Supply of Jersey City and Newark." Leeds documented the sources and extent of pollution in the Passaic River which runs through New Jersey. Since then, certain practices, such as allowing dead animal carcasses to rot in streams and locating cemeteries on river shores have been curtailed, but the use of rivers for disposal of industrial wastes and sewage continues. Today, nearly every stream, river, estuary, and lake in the United States is polluted to some degree. In this chapter we will consider what water pollution is, the types and sources of pollutants, the effects of pollution, and the methods of water treatment.

In most regions of the United States water will always be available. There is no problem of running out of water. However, if this water is not treated properly, it can be degraded in quality so that it will be very expensive to clean up. An adequate water supply is often taken for granted, but there are forces which threaten this supply. It is imperative that we make an effort to conserve and protect our waters.

11-2 FEDERAL WATER POLLUTION CONTROL ACT

The most far-reaching Federal water pollution legislation are the 1972 Water Pollution Control Act Amendments to the Water Pollution Control Act of 1965. The amendments nobly proclaim that it is the national goal that the discharge of pollutants into navigable waters of the country be eliminated by 1985. To accomplish this the U.S. Environmental Protection Agency (EPA) is to implement the law through a variety of activities including the following:

1. Establish national water quality standards.
2. Solicit water quality improvement plans for interstate waters from various states, and accept, reject, or alter these plans.
3. Establish the acceptable pretreatment levels of pollutants in waste waters and determine the possible degree to which pollutant levels can be decreased by secondary treatment.
4. Establish standards for effective waste water treatment.
5. List industries for which water treatment standards apply.
6. Specify toxic water pollutants and establish standards for allowable levels of toxic pollutants in waste waters.
7. Regulate the granting of federal aid for local waste water treatment facilities.
8. Establish guidelines for, and achieve, the reduction of water pollutants in waste water to the levels which the best technological methods permit.
9. Report to Congress on the social, economic, and environmental effects of achieving or not achieving the reduction of water pollution.

Many controversies and problems are to be expected when the EPA attempts to carry out the law. One problem is to have some uniformity among states concerning water pollution control plans. Many industries may claim that they cannot economically meet the required reduction in water pollutant levels in industrial waste waters. In the near future, decisions will have to be made concerning the feasibility and desirability of some aspects of the law.

11-3 WATER USE

An industrial society uses tremendous amounts of water. In the United States, an estimated 1.4 billion cubic meters (370 billion gallons) of water are used each day. Actually, very little of this water is consumed in a manner in which it is chemically converted to other substances. Most of the water is used in crop irrigation, as a medium in certain industrial processes, and to carry away our domestic and industrial wastes. Around 40 percent of the water is used in agricultural irrigation. Over 50 percent is used by industry, including steam-generating electric power plants which account for about three-fifths of industrial use. Approximately 10 percent is used for municipal public water supplies.

Unlike water used in industry, agricultural water is used only once before it evaporates or reenters the environmental waters. The industrial uses of water other than steam–electric plants are shown in Figure 11-1. Around 40,000 gallons of water are used in the production of 1 ton of steel, and 360,000 gallons are used in the production of 1 ton of aluminum. It requires about 770 gallons of water in the refining of 1 barrel of crude oil. Around 200,000 gallons are used to make 1 ton of viscose rayon, and 600,000 gallons for 1 ton of synthetic rubber. The Federal Water Pollution Control Act Amendments of 1972 will call for more strict controls of the contents of industrial waste waters. Consequently, some industries are designing processes in which water can be recycled numerous times within a plant before being cleaned up and released to the environment. In petroleum and coal industries, water is used on the average more than five

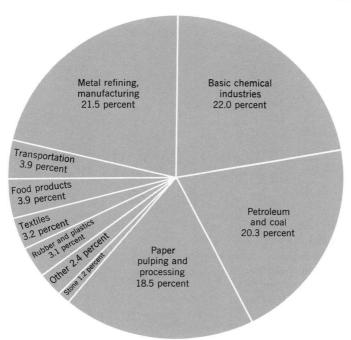

Figure 11-1
Various industrial uses
of water.

Metal refining,
manufacturing
21.5 percent

Basic chemical
industries
22.0 percent

Transportation
3.9 percent

Food products
3.9 percent

Textiles
3.2 percent

Rubber and plastics
3.1 percent

Other 2.4 percent

Stone 1.2 percent

Paper
pulping and
processing
18.5 percent

Petroleum
and coal
20.3 percent

times before being discarded. Of course, not all industries are attempting to recycle water and clean up waste water effluents.

Only a small percentage of our water is used domestically, but domestic use is increasing at a faster rate than the population growth. A typical American uses well over 50 gallons of water each day. A bath requires 25 gallons, a garbage disposal uses 2 gallons per day, and a dishwasher uses around 10 gallons per day (half as much is used if dishes are washed by hand). The watering of a typical lawn might use 1,000 gallons. A flush toilet uses 3–5 gallons of water to carry away less than 1 quart of wastes.

We obtain about 80 percent of our water from surface sources such as lakes and rivers, and well water from aquifers supplies the rest. The use of ground water poses no problems as long as the water table is adequate and stable. In many regions of the country the water table level is dropping as a result of excessive water use. In the West and Southwest, the problem of ground water depletion is most serious. Water is being withdrawn at a rate faster than the aquifers are being recharged. Some farming regions of Texas and New Mexico have been abandoned as a result of a drastic drop in the table. In some regions of the country very deep ground water sources are being tapped. Some of these sources are of ancient origin and are only replenished by natural processes over hundreds or thousands of years. As an example, consider the case of the Mohave electric power plant located in southern Nevada. This power plant burns around 14,000 tons of coal daily. The coal is strip mined from Black Mesa and sent 273 miles west to the power plant in a slurry (water–coal mixture) pipeline. The pipeline is the longest slurry pipeline in the world and uses around 3,000 gallons of water per minute. In this indian region of little rainfall, the water is obtained from wells tapping deep aquifers 2,000 feet below the land surface. The power generated by the Mohave plant is sent long distances to metropolitan areas like Las Vegas and Southern California.

To prevent or alleviate water shortages in certain parts of the country, we have built vast complexes for storage and transport of water from region to region.

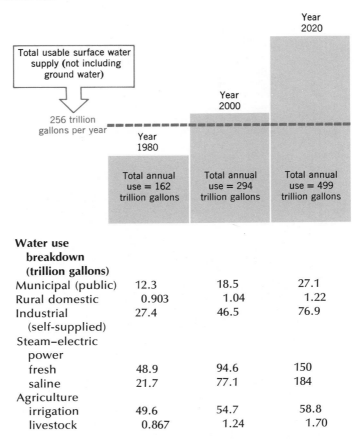

Water use breakdown (trillion gallons)			
Municipal (public)	12.3	18.5	27.1
Rural domestic	0.903	1.04	1.22
Industrial (self-supplied)	27.4	46.5	76.9
Steam–electric power			
fresh	48.9	94.6	150
saline	21.7	77.1	184
Agriculture			
irrigation	49.6	54.7	58.8
livestock	0.867	1.24	1.70

The California Aqueduct transports millions of gallons of water each year over hundreds of miles from northern California to the semi-arid Los Angeles area. As water needs increase and water quality decreases, water transport could cause conflicts between water-rich regions and water-deficient regions.

As can be seen in Figure 11-2, it is expected that the water use in the United States will exceed the usable surface supply by the year 2000. This does not mean we will run out of water, but it does mean that we will have to reuse and recycle water and attempt to preserve the quality of the water. The reuse of water has been carried out for many years. In certain river basins in the United States, water is reused 10–20 times before it reaches the ocean. The problem of reuse is that the water must be purified before it can be used again. Some users purify their waste waters before returning them to the source. Others do not attempt much purification, which tends to degrade the quality of the source so that much effort in purification must be expended by those desiring to use the waters again. It is apparent that the users of water, whether industrial or municipal, must be more concerned with the purity of the waste waters that are returned to the environment. The purification of waste water involves much chemistry and the expenditure of energy. The methods of water treatment will be discussed in Sections 11-10 and 11-11.

11-4 WATER POLLUTANTS

As we have seen, natural waters are dilute solutions containing many chemicals. Natural water also contains a variety of suspended matter and colloidal particles. Table 11-1 lists the composition of typical fresh water. The presence of most of

Source	Suspended	Colloidal dispersion	Solution	
Lithosphere (minerals and rocks)	Sand Clays Soil	Clays Soil	CO_2 from carbonates Na^+ K^+ Ca^{2+} Mg^{2+} Fe^{2+}	Cl^- SO_4^{2-} HCO_3^- NO_3^- PO_4^{3-} F^-
Atmosphere	Dusts Particulates	Dusts Particulates	O_2 N_2 CO_2	H_3O^+ HCO_3^-
Biosphere	Algae, other aquatic plants and animals Bacteria	Organic macromolecules Viruses	Organic molecules CO_2 O_2 N_2 H_2S CH_4 NH_4^+ NO_3^- SO_4^{2-}	

Table 11-1
Components of
Natural Waters

these impurities results from natural processes and cannot be avoided. The quality of fresh water can be adversely affected by the addition of other impurities by humans.

Water has a variety of uses. A body of water may be used for recreation purposes, to support fish and wildlife, for agricultural irrigation, for industrial purposes, or for a public water supply. Obviously, water of varying purity is required for these uses.

What is polluted water? Actually, there are varying degrees of water pollution, depending upon the intended use. In California and similarly in other states, **water pollution** is defined legally as follows:

> An impairment of the quality of water which unreasonably affects: (1) such water for beneficial uses, or (2) facilities which serve such beneficial uses.

Often, the taste, odor, and appearance of water indicate that it is polluted. In some cases, only precise chemical tests reveal the presence of dangerous pollutants. In any case, polluted water is water that is not fit for the intended use.

Let us now consider the nature of some **water pollutants.** Table 11-2 lists the major kinds of pollutants. Pollutants in liquid forms come from the discharge of municipal, agricultural, and industrial wastes into the waterways, as well as seep-

Type	Example	Type	Example
Chemical	Organic compounds Inorganic ions Radioactive material	Physical	Floating solids Suspended material Settleable material
Biological	Pathenogenic bacteria Viruses Algae Aquatic weeds		Foam Insoluble liquids Heat

Table 11-2
Common Physical,
Chemical, and
Biological Pollutants

age from septic tanks, animal feedlots, sanitary landfills, and acid mine drainage. These liquids contain dissolved minerals, human and animal wastes, man-made chemicals, and suspended and colloidal matter. Pollutants in solid forms include such material as sand, clay, soil, ashes, solid sewage, agricultural vegetable matter, grease, tars, garbage, paper, rubber, wood, metals, and plastics. Some of these solids or **physical pollutants** are of natural origin, but many are synthetic man-made substances which enter the waters as a result of human activities. Other physical pollutants include foams, scums, oil slicks, and heat (thermal pollution). Physical pollutants affect the appearance of water and, by settling out on the bottom or floating on top of the waters, they interfere with animal life.

Chemical pollutants include dissolved or dispersed organic and inorganic substances. Inorganic pollutants come from the domestic, agricultural, and industrial discharges which contain a variety of dissolved substances. These pollutants include soluble metallic salts such as chlorides, sulfates, nitrates, phosphates, and carbonates. Also included are waste acids, bases, and toxic dissolved gases, such as sulfur dioxide, ammonia, hydrogen sulfide, and chlorine. Acids can be deadly to aquatic life and can cause the corrosion of metals and concrete. **Organic pollutants** are carbon-containing compounds from domestic, agricultural and industrial wastes. These include organics from human and animal wastes, food processing and slaughterhouse wastes, industrial chemicals and solvents, oils, tars, dyes, and synthetic organic chemicals such as insecticides. As we shall see in Section 11-5, organic pollutants tend to deplete the dissolved oxygen in water. Some organics, such as oils and insecticides, can interfere with, or be toxic to, aquatic life.

Biological pollutants include disease-causing bacteria and viruses, algae, and other aquatic plants. Certain bacteria are harmless and others are involved in the decomposition of organic compounds in water. The undesirable bacteria and viruses are those which cause diseases like typhoid, dysentery, poliomye-

Figure 11-3
A decrease in the incidence of typhoid fever correlates with the growth of public water supplies and public sewers and sewage treatment. From *Advances in Environmental Sciences*, vol. 1, edited by J. N. Pitts and R. L. Metcalf, Wiley-Interscience, New York, 1969, p. 48.

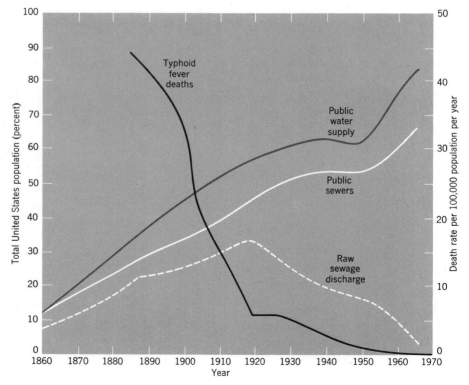

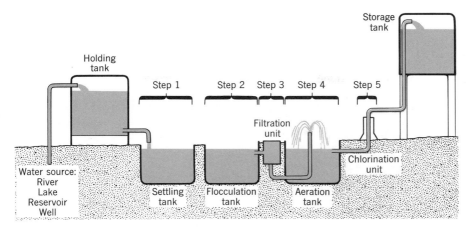

Figure 11-4
Five common
treatment steps for
public drinking water.

litis, hepatitis, and cholera. As can be seen in Figure 11-3, the development of public water supplies and public sewers has had a profound effect upon the incidence of typhoid in the United States. The control of viruses in water is difficult. Waterborne viral diseases must be watched closely, especially if water is to be reused.

The treatment of public waters in the United States before distribution is common practice. Figure 11-4 shows a typical treatment process including **settling tanks,** flocculation by aluminum sulfate, filtration, aeration, chlorination, and storage. In the **flocculation,** aluminum sulfate is added to the water, producing a fluffy gelatinous solid which entraps any suspended solids and aids in the settling process. The **filtration** removes any additional suspended solids. The **aeration** step, in which the water is sprayed in the air, and the **chlorination** step, in which chlorine gas is bubbled into the water, are carried out to kill by oxidation any bacteria that may be in the water. It is thought that chlorination may also help control viruses which may be present in the water. Not all public water systems include each of the purification techniques shown in Figure 11-4. The U.S. Public Health Service survey mentioned in Section 10-12 revealed that over 20 percent of the water systems surveyed had treatment deficiencies.

11-5 DEGRADATION IN WATER

Some pollutants are decomposed by chemical and biological processes occurring in water. These are called **degradable** or **biodegradable** pollutants. **Degradation** refers to breaking down into simpler substances. Most organic pollutants are degradable. However, some organics such as certain pesticides and detergents are nondegradable, or they decay very slowly in water. These are called **hard** or **refractory organics. Hard pesticides** include chlorinated pesticides such as DDT, chlordane, and endrin. In addition to hard organics, nondegradable chemical pollutants include nitrate ion, phosphate ion, sulfate ion, and various metal ions, the least desirable of which are mercury, lead, and cadmium ions.

The decomposition of organic materials in water occurs mainly through the action of bacteria and other organisms in the water. The bacteria use the organics as foods and utilize them as sources of energy in biological oxidation processes. In such bacterial decay, dissolved oxygen is utilized, and carbon dioxide, water, and various nondegradable ions are produced. The following are some general reactions showing the bacterial decay of organics in the presence of oxygen; such decay is called **aerobic** (Greek: air-life) decay:

$$CH \text{ (hydrocarbons)} + O_2 \xrightarrow{\text{bacteria}} CO_2 + H_2O$$

$$CH_2O \text{ (carbohydrates)} + O_2 \xrightarrow{\text{bacteria}} CO_2 + H_2O$$

$$\text{organic sulfur-containing compounds} + O_2 \xrightarrow{\text{bacteria}} CO_2 + H_2O + SO_4^{2-}$$

$$\text{organic nitrogen-containing compounds} + O_2 \xrightarrow{\text{bacteria}} CO_2 + H_2O + NO_3^-$$

$$\text{organic phosphorus-containing compounds} + O_2 \xrightarrow{\text{bacteria}} CO_2 + H_2O + PO_4^{3-}$$

All the reactions consume dissolved oxygen from water. Organics which enter into aerobic decay are called **oxygen-depleting** pollutants. When insufficient oxygen is present in water, bacterial decay of most organics can continue to occur. However, such decomposition in the absence of oxygen does not produce the same products. This **anaerobic** (without oxygen) decay is illustrated by the general equation

$$\begin{array}{l}\text{organic sulfur- and nitrogen-} \\ \text{containing compounds}\end{array} + H_2O \xrightarrow[\text{decay}]{\text{anaerobic}} CO_2 + H_2S + CH_4 + NH_4^+$$

Anaerobic decay produces gases which bubble from the water and which contribute offensive odors to the water. In fact, this decay is the major source of naturally produced methane and hydrogen sulfide present in the atmosphere. Once in the atmosphere, methane is converted to carbon dioxide, and hydrogen sulfide is converted to sulfates by atmospheric chemical reactions.

11-6 SOAPS AND DETERGENTS

Animal fats and vegetable oils can react with concentrated solutions of sodium hydroxide to form glycerol and organic ions called carboxylate ions:

$$\text{fat} + OH^- \longrightarrow \text{glycerol} + RCOO^- \quad (RCOO^- \text{ is a symbolic formula}$$
$$\text{for a carboxylate ion)}$$

This is called a **saponification** reaction. The organic ions can be precipitated out of solution by making the solution very concentrated in sodium ion by the addition of sodium chloride:

$$RCOO^- + Na^+ \longrightarrow RCOONa \text{ (soluble in water but not}$$
$$\text{concentrated salt solution)}$$

The resulting compound is called a **soap.** An example of the formation of soap is shown below.

$$\text{palm oil} \xrightarrow{\text{saponification}} \text{palmitate ion} + \text{oleate ion} + \text{other organic ions} + \text{glycerol}$$

$$\xrightarrow{+NaCl} \underbrace{\text{sodium palmitate} + \text{sodium oleate, etc.}}_{\text{soap}}$$

Soaps are utilized as cleansing agents because of the unique structure of these special organic ions. When an object is dirty, it usually is a result of adhering layers of grease and oil containing dust and foreign particles. When such an object is washed with water, much of the "dirt" is not washed away. However, when soap is placed in the water, it can dissolve to give carboxylate ions, as exemplified by the palmitate ion:

$$CH_3CH_2CH_2CH_2CH_2CH_2CH_2CH_2CH_2CH_2CH_2CH_2CH_2CH_2CH_2COO^-$$

long-chain hydrocarbon end   ionic end

The ionic end of these ions is very water-soluble, but the long-chain hydro-carbon ends have strong attraction for oil and grease molecules. These oil-attracting ends of the ion become embedded in the oil and grease layers, but the ionic ends remain dissolved in the water. As shown in Figure 11-5, this tends to pull the grease and oil particles into the solution so that they can be washed away. This kind of cleansing action is called **detergent action.** Around 500,000 tons of various types of soap are used in the United States each year. The trouble with soaps is that when they are used in hard water containing calcium and magnesium ion, soap scums can form, eliminating the soap:

$$RCOO^- + Ca^{2+} \text{ or } Mg^{2+} \longrightarrow (RCOO)_2Ca \text{ or } (RCOO)_2Mg$$
soap scum

Soaps do not work well in hard water, and when they are used in hard water, they leave soap scums on washed items.

It is possible to synthesize from petroleum chemicals other organic com-pounds that display detergent action. These compounds are called **syndets (syn-thetic detergents)** or simply **detergents.** Most syndets are sodium compounds of substituted benzene sulfonate, called **linear alkyl sulfonates** (LAS):

$$SO_3^-Na^+$$

hydrocarbon end

$$CH_3CHCH_2CH_2CH_2CH_2CH_2CH_2CH_2CH_2CH_3$$

The $-SO_3^-$ end is water-soluble, and the hydrocarbon end is oil-attracting. Syndets act as cleansing agents in the same manner as described above for soaps. The advantage of syndets is that they do not form scum in hard water. Today, synthetic detergents are more widely used than soaps. Around 3 million tons of syndets are used each year in the United States.

Detergents are used both in industry and in the home. However, since they are used in great quantities, they are a source of water pollution. Most syndets are broken down by bacteria; they are biodegradable. Consequently, only small amounts of these syndets end up in the water supplies. Unfortunately, some nonbiodegradable syndets are still used and become very inconvenient water pollutants. The presence of syndets in natural water can cause serious foaming problems.

11-7   PHOSPHATES IN DETERGENTS

One of the major dilemmas of detergent use is that commercial detergents must contain certain additives which can become serious water pollutants. Detergents contain small amounts of perfumes, bleaches, and optical brighteners. Optical brighteners are dyes which give clothes a clean appearance. The major additive in many detergents is sodium tripolyphosphate, $Na_5P_3O_{10}$. This is the so-called **phosphate** in detergents. In 1970, a typical heavy-duty laundry detergent con-tained 18 percent LAS, 50 percent sodium tripolyphosphate, 19 percent inert fill-ers, and the remainder a variety of ingredients. Since then, some manufacturers have cut down the percentage of phosphate. The nonionic detergents, which include a syndet other than LAS, contain very little, if any, phosphate.

The phosphate additive in detergent is called a **builder.** Phosphate builders function in three ways. First, as bases they make wash water basic (high pH) which is necessary for detergent action. Second, the phosphates react with the hard water ions, such as calcium ions and magnesium ions, in a manner which

Figure 11-5
Detergent action. (a)
Oil-attracting
hydrocarbon ends of
soap ions become
embedded in the oily
or greasy dirt
deposits. (b) The soap
ions surround
portions of the dirt
and pull the dirt from
the surface to which it
is adhering into the
water. (c) The dirt
particles are kept in
suspension by a
surrounding layer of
soap ions. The
negatively charged
ionic ends of the soap
ions keep the particles
in suspension due to
their water-attracting
tendency. The
negative charges
surrounding the
particles also prevent
them from clumping
together before they
are rinsed away.

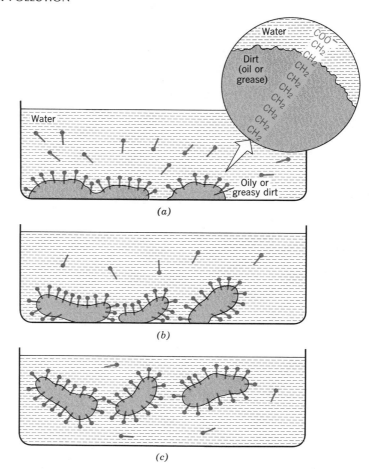

(a)

(b)

(c)

prevents these ions from interacting with the detergent. Third, phosphates help keep fats and dirt in suspension so that they can be washed away.

Once a detergent has been used, the phosphates end up in the sewage. Most sewage treatment plants are not designed to remove phosphates and, thus, the phosphates pass directly into the aquatic environment in sewage effluent. It is estimated that around 50 percent of the phosphates in sewage come from detergents. The remainder come from the phosphorus compounds in human and animal wastes and phosphate fertilizers. The problem with excess phosphates as well as nitrates in natural waters is that they appear to act as nutrients for algae and other aquatic plants. As explained in Section 11-9, rapid plant growth can aid in the degradation of natural waters.

Much effort is being expended in an attempt to find a substitute for phosphates in detergents. Some nonphosphate detergents, involving detergents other than LAS, are in use, but a satisfactory safe substitute for phosphates has not been developed. One approach to reduce the phosphate level in waste waters is to convince consumers to use smaller amounts of detergents for washing purposes. Furthermore, it is technically feasible to remove much of the phosphates in sewage by special chemical sewage treatment methods mentioned in Section 11-11.

11-8 BIOCHEMICAL OXYGEN DEMAND

One of the most remarkable properties of natural water is the capacity of the water to reduce organic pollution by bacterial action. This property, along with the fact that when wastes are dumped into natural waters they are greatly diluted, has enabled us to use our waterways for disposal of vast amounts of waste. However, as the amounts of wastes increase, we begin to strain the capacity of the waters to handle the pollution load. Recall that the aerobic decay of organics utilizes dissolved oxygen. As the dissolved oxygen is used up, it is replaced by more oxygen from the atmosphere. This replacement of dissolved oxygen occurs more readily in a moving, turbulent body of water. As long as there is sufficient dissolved oxygen in the water, aerobic decay will continue. If the level of organic pollution is high, the oxygen is used up more rapidly than it is replaced. In such a situation, the dissolved oxygen in the water is soon depleted. Once the oxygen is depleted, anaerobic decay begins and the waters become fouled by the products of this kind of decay. Furthermore, the depletion of oxygen can result in the death of fish and other aquatic life. The decay products of these animals can add to the organic level increasing the amount of anaerobic decay. Of course, anaerobic conditions do not always occur, but the chances increase as the amount of degradable organics increases. The **biochemical oxygen demand (BOD)** serves as a quantitative measure of the level of organic oxygen demanding wastes in water. In a water sample, the BOD expresses the number of milligrams of dissolved oxygen per liter used up as the organic wastes are consumed by bacteria in the water. Biochemical oxygen demand is usually expressed in terms of parts per million oxygen and is determined by measuring the decrease over a period of five days of the dissolved oxygen in a water sample maintained at 20°C. Water of high BOD requires large amounts of oxygen. Thus, the BOD of water can indicate the organic pollution load of the water. **Potable** (drinkable) water normally has a BOD of 0.75–1.5 parts per million (ppm) oxygen. Relatively unpolluted water has a BOD of around 1–3 parts per million. Water is considered to be polluted if the BOD exceeds around 5 parts per million. Untreated municipal sewage has a BOD range of 100–400 parts per million. Some industrial and agricultural wastes have BOD levels in the thousands of parts per million range. Of course, such BOD levels are diminished upon dilution of the wastes, but in some cases, very drastic pollution problems are caused by these wastes. It is possible to decrease the BOD level of wastes by the sewage treatment discussed in Section 11-10.

One of the dangers of polluted waters is the possible presence of disease-causing bacteria. One test for bacteria is accomplished by measuring the amount of the common coliform bacteria in water. **Coliform bacteria** are a harmless bacteria which reside in the large intestines of humans. These bacteria are quite plentiful in human feces. These fecal bacteria become part of sewage wastes. They do not thrive in the aquatic environment, and a coliform bacteria count gives an indication of how recently and to what extent sewage pollution has occurred. The coliform count is assumed to be related to the amounts of any disease-causing bacteria. A high coliform count is an indication of human sewage contamination with the possible presence of dangerous bacteria.

11-9 EUTROPHICATION

Perhaps you have seen a stream or lake clogged with masses of green algae or other aquatic plants. Certain wastes act as nutrients or fertilizers for aquatic plants. Phosphates and nitrates in water can stimulate plant growth. This is especially true in the relatively calm waters of lakes which support algae. When lakes receive large amounts of nutrients from sewage wastes and agricultural runoff, this can cause a rapid increase in algae growth. This rapid growth of algae is termed an **algae bloom** and becomes visible as large masses of floating algae and clumps of algae on the beaches. Other aquatic weeds can grow rapidly and may clog the waters with masses of plant material. When the algae and plants die, their decomposition consumes the dissolved oxygen. This can lead to the death of fish and other aquatic life and produce anaerobic conditions.

The accumulation of nutrients in natural waters causes a hastening of eutrophication. **Eutrophication** (Greek: well-nourished) is a natural aging process in which overnourished lakes accumulate large amounts of decaying plant material on the bottom. This tends to fill the lake, and it becomes more shallow, warmer, and accumulates more nutrients. Plants take over the lake basin as it fills, and it slowly is converted to a marsh and finally a meadow or a forest. This is a natural lake-aging process which may occur over a period of hundreds of years. The presence of nutrients in man-made wastes very often greatly exceeds the amounts of natural nutrients. As a result of these wastes, eutrophication can occur over a shorter period, resulting in very rapid aging of a lake or other body of water. In fact, in a matter of several decades wastes have had a definite aging effect on some lakes, such as Lake Erie. Actually, natural eutrophication is not a water pollution problem. Problems arise when man-made nutrients cause rapid eutrophication, resulting in algae blooms, fish kills, and anaerobic conditions.

Several chemical elements in dissolved or ionic forms are needed for algae growth. The major nutrients include carbon, nitrate ion, and phosphate ion. It is not known which of these nutrients is the growth-limiting factor. It is thought that when the level of growth-limiting nutrient decreases, rapid plant growth ceases. Studies of algae growth reveal that, depending upon the kind of algae and other conditions in the waters, carbon is the limiting nutrient in some cases, phosphate is in other cases, and nitrate in still other cases. Some scientists claim that algal growth can be curtailed by reducing phosphates in wastes. Others hold that reduction of phosphate will not reduce algal growth as long as dissolved organics are present as sources of carbon dioxide.

The Canadian government has ordered detergent manufacturers to cut down the phosphate levels in detergents and is considering a total ban of phosphates. In the United States, no action has been taken at the federal level. Scientists point out that eutrophication is often a localized problem, and a total phosphate ban may not be needed. State and regional control of phosphates in detergents could be exercised to reduce eutrophication. Furthermore, it is possible to remove much of the phosphate in municipal sewage by chemical treatment methods. It is of interest to note that phosphorus is one of the vital elements needed for food crop growth and fertilizers containing phosphorus as phosphate are in wide use. (See Section 15-6 for a discussion of chemical fertilizers.) However, the sources of phosphates are limited and could some day be diminished to a point which might curtail food production. In view of this, the use of phosphates in laundry detergents and the subsequent loss of the phosphates into the environment is a waste of one of our important resources.

11-10   SEWAGE TREATMENT

As can be seen from Table 11-3, large amounts of sewage wastes are generated by industry and individuals each year. In addition to this, there are large amounts of agricultural wastes and irrigation water runoff. Most of the domestic and industrial wastes are treated in some way to decrease the BOD level and to remove dangerous components. A portion of the industrial sewage is fed into municipal sewage lines. However, many industries treat their own waters or dispose of wastes in special ways. Unfortunately, some industries and municipalities are responsible for adding large amounts of untreated or partially treated sewage to our waters. These wastes often have high BOD levels, and some contain toxic or refractory chemicals.

The treatment of domestic sewage is usually directly controlled by municipal or regional agencies. In our highly urbanized society well over 60 percent of the population is served by public sewers. Over 30 percent utilize septic tanks or cesspools, while less than 10 percent use nonwater outdoor facilities. Presently,

Industry (cooling water not included)	Waste water (billion gallons)	Settleable and suspended solids (million tons)	Table 11-3 Estimated Annual Amounts of Sewage Wastes
Food and kindred products	800	3.5	
Textile mills	160	?	
Paper and paper products	2,200	1.6	
Chemical and allied products	4,300	1.0	
Petroleum and coal	1,500	0.24	
Rubber and plastics	200	0.03	
Primary metals	5,000	2.5	
Machinery	170	0.03	
Electrical machinery	100	0.01	
Transportation equipment	280	?	
All other manufacturing	520	0.49	
Total industry	15,230	9.4	
Domestic served by sewers	6,300	4.8	

Adapted from Federal Water Pollution Control Administration data.

around 10 percent of the sewage handled by public sewers receives no treatment. This raw sewage is dumped into various bodies of water. The vast majority of domestic sewage does receive some treatment in sewage disposal plants before being released into environmental waters. Some of this treatment is adequate, while some is marginal. Many domestic sewage systems are tied in with storm drain systems. This causes problems during rainy seasons when large amounts of storm water wash untreated sewage directly through the plant into the environment.

As shown in Figure 11-6, there are three potential steps in the treatment of domestic sewage. These three steps are called primary, secondary, and tertiary treatment. The main purpose of sewage treatment is to lower the BOD level of the waste water. Around 30 percent of domestic sewage receives only primary treatment, while around 60 percent is subject to secondary treatment. Very little tertiary treatment is being applied to domestic sewage. Secondary treatment can cost twice as much as primary treatment. Tertiary treatment, at the present, is quite costly. Let us consider each of these treatment steps. (See Figure 11-6.)

PRIMARY TREATMENT   As the waste water enters the plant it is passed through a bar screen to screen out large objects. Some plants have large grinders which pulverize the large objects so that they can pass through the treatment step. The sewage then flows slowly through a large chamber where the sand, pebbles, and other heavy material settle out. From this chamber the sewage flows into a large settling tank. The suspended solids settle to the bottom (as sewage sludge) and the greases and oils float to the top as scum. The water between the sludge and scum is drained off and released to the environment or is passed to secondary treatment. Sometimes the effluent is first chlorinated to kill bacteria before it is released to the environment. Primary treatment removes around 60 percent of the suspended solids and around 35 percent of the BOD.

SECONDARY TREATMENT   The most common type of secondary treatment is **activated sludge** treatment. Sewage flows from primary treatment to an **aeration tank** where air (or pure oxygen in a few plants) is bubbled through it. The aeration of the sewage results in rapid growth of bacteria and other microorganisms. The bacteria utilize the oxygen to decompose the organic wastes in the sewage. Suspended solids mixed with bacteria form a sludge called activated sludge.

Figure 11-6
Sewage treatment
steps.

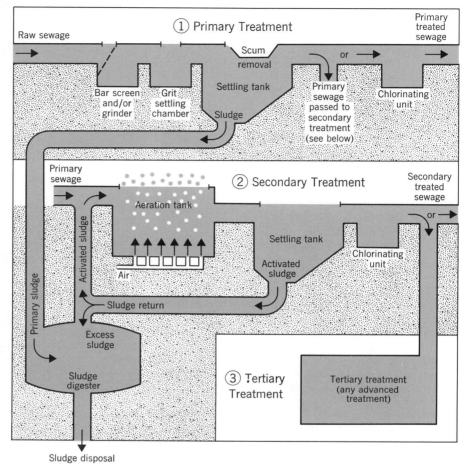

Some plants use a device called a **trickling filter** rather than the activated sludge process. In this method the sewage water is sprayed on a bed of stones about 6 feet deep. As the water trickles over and around the stones it comes into contact with bacteria which decompose the organic pollutants. The bacteria in turn are consumed by a variety of other organisms residing in the filter.

Treated sewage from the aeration tank or trickling filter is passed to another large settling tank where the activated sludge settles to the bottom. The activated sludge which accumulates on the bottom is transferred back to the aeration tank and mixed with incoming sewage. Excess sludge is collected, treated, and disposed of. The disposal of sewage sludge is a major problem. A large treatment plant produces appreciable amounts of sludge which must be disposed of as solid wastes. Some is used for fertilizer but most has to be buried in a landfill or dumped in the ocean. Primary treatment followed by secondary treatment removes up to 90 percent of the suspended solids and around 90 percent of the BOD. After secondary treatment the waste water is usually subjected to chlorination before being released to the environment.

TERTIARY TREATMENT Any kind of treatment beyond the secondary step is termed tertiary treatment. The purpose of tertiary treatment is to remove organic pollutants, nutrients such as phosphate ion and nitrate ion, or excessive mineral salts. The major objective of tertiary treatment is to make the waste water as pure as feasible before returning it to the environment. There are several kinds of tertiary treatment. Precipitation, sedimentation, and filtration can be used for nutrient removal. Carbon adsorption is used for removal of organics. Techniques such as reverse osmosis, ion exchange, and electrodialysis are used for demineralization (the removal of inorganic ions). Some of these advanced treatment processes are discussed in Section 11-11.

11-11 ADVANCED TREATMENT METHODS

The tertiary treatment methods discussed here are undergoing intensive research and are being tested in a few pilot plants throughout the nation. None of these methods are in wide use, but some have good potential of supplementing secondary treatment.

NUTRIENT REMOVAL The most common method for phosphate ion removal from sewage involves the addition of chemicals which form solid precipitates with phosphate ion. These solid precipitates are coagulated, allowed to settle, and filtered from the treated water. Other methods of phosphate removal, including biological removal and selective ion exchange, are being investigated.

Nitrogen occurs in sewage mostly as nitrate ion and ammonium ion. Bacteria convert organic nitrogen to ammonium ion, and other bacteria convert ammonium ion to nitrate ion. Nitrogen nutrients are more difficult to remove than phosphates. One method is the use of denitrifying bacteria to convert nitrate ion to nitrogen gas. In this biodenitrification method water is slowly passed through a column containing the bacteria, and the nitrogen is released to the atmosphere. Another method in use removes nitrogen which is in the form of ammonium ion. In this ammonia-stripping method the waste water is treated with a base to convert the ammonium ion to ammonia:

$$NH_4^+ + OH^- \longrightarrow NH_3 + H_2O$$

Air is then bubbled through the water to strip out the ammonia as ammonia gas. Other methods under investigation are oxidation of ammonium ion to nitrogen gas by electrolysis, and ion exchange of ammonium ion and nitrate ion.

CARBON ADSORPTION The most successful tertiary method for dissolved organic removal is activated carbon adsorption. In this method the waste water is passed over a series of columns packed with granular activated carbon. The carbon is very porous and has a high surface area. Organic pollutants adhere to the surface of the carbon, which also serves to filter out suspended solids. This method is especially good for removal of refractory organics. Carbon adsorption removes most of the residual organics and nearly all of the suspended solids from the water. The carbon can be regenerated for reuse by heating in a furnace to burn off the absorbed organic material.

REVERSE OSMOSIS The osmosis process utilizes a membrane which is permeable only to water. See Section 9-9 for a discussion of osmosis. When water that contains mineral ions is separated from pure water by an osmotic membrane, the pure water passes through and dilutes the water that contains ions. This is the normal osmosis process. Osmosis can be reversed by applying pressure to the solution side of the membrane. (See Figure 11-7). The pressure forces the water from the solution side to the pure water side, leaving the ions behind. This is called reverse osmosis. Figure 11-7 illustrates several forms of the reverse osmosis technique. This method is not only useful for **demineralization** but also is effective in removing dissolved organics and nutrients. Reverse osmosis can achieve around 90 percent demineralization with loss of about 25 percent of the

Figure 11-7
Techniques of reverse
osmosis. Parts c–e
from *ABC's of
Desalting,* U.S.
Department of the
Interior.

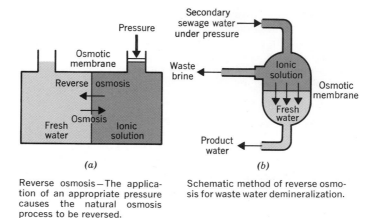

(a)

Reverse osmosis—The application of an appropriate pressure
causes the natural osmosis
process to be reversed.

(b)

Schematic method of reverse osmosis for waste water demineralization.

feed water as water brine. This appears to be one of the more promising of the tertiary treatment methods.

ION EXCHANGE The ion exchange method for removing excessive positive and negative inorganic ions is based upon the same principle of ion exchange that is used for water softening. (See Section 10-13.) The difference is that different kinds of resins are utilized. The ion exchange method is illustrated in Figure 11-8. The water is first passed through a bed of **cation exchange resin** where positive ions in the water are exchanged for hydrogen ions (H^+) in the resin. The water leaving this resin contains a mixture of anions and hydronium ions (H_3O^+) which form by combination of the hydrogen ions and water. This water passes into a bed of **anion exchange resin** where negative ions in the water are exchanged for hydroxide ions (OH^-) in the resin. The hydroxide ions combine with the hydronium ions to form water:

$$H_3O^+ + OH^- \longrightarrow 2\ H_2O$$

Thus, by use of ion exchange, the inorganic mineral ions are replaced by water.

Figure 11-7c

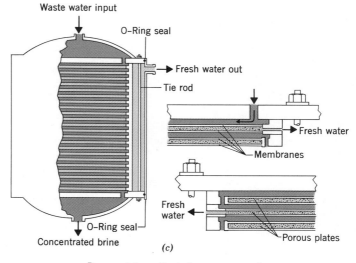

(c)

Porous plate method of reverse osmosis.

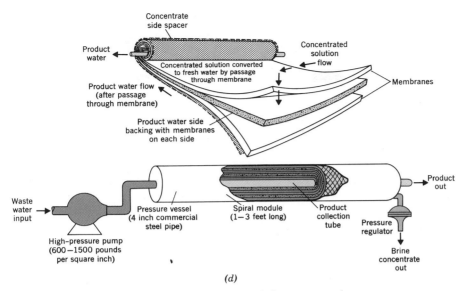

Concentrate side spacer

Product water

Concentrated solution converted to fresh water by passage through membrane

Concentrated solution flow

Membranes

Product water flow (after passage through membrane)

Product water side backing with membranes on each side

Waste water input

High-pressure pump (600–1500 pounds per square inch)

Pressure vessel (4 inch commercial steel pipe)

Spiral module (1–3 feet long)

Product collection tube

Pressure regulator

Product out

Brine concentrate out

(d)

Spiral-wound membrane method of reverse osmosis.

Ion exchange can be used to accomplish up to 90 percent demineralization of water. The resins are regenerated by passing sulfuric acid through the cation resin and an ammonia solution through the anion resin.

ELECTRODIALYSIS When an electrical potential is applied between electrodes immersed in a solution of ions, the positive ions move toward the negative electrode and the negative ions move toward the positive electrode. As shown in

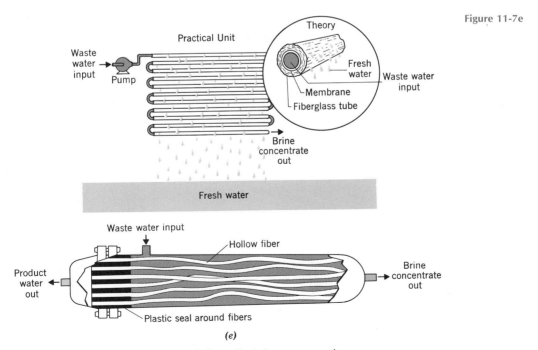

Figure 11-7e

Practical Unit

Theory

Waste water input

Pump

Fresh water

Waste water input

Membrane

Fiberglass tube

Brine concentrate out

Fresh water

Waste water input

Hollow fiber

Product water out

Brine concentrate out

Plastic seal around fibers

(e)

Hollow fiber or tubular method of reverse osmosis.

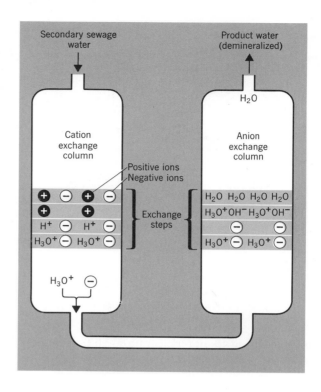

Figure 11-8
Demineralization of water by ion exchange.

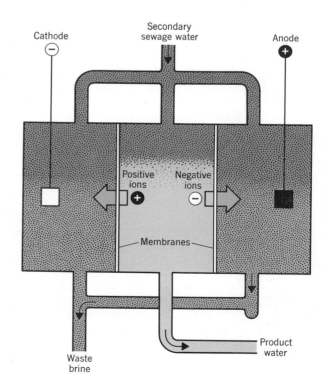

Figure 11-9
Electrodialysis.

Figure 11-9, the electrodialysis demineralizing method uses an electrical cell in which special selective dialyzing membranes are placed in front of the electrodes. The cation-permeable membrane allows passage of positive ions, and the anion-permeable membrane allows passage of negative ions. As waste water flows through the electrodialysis cell, anions are swept to one side of the cell and cations are swept to the other side, under the influence of the electrical potential. The water leaving the central compartment of the cell will contain fewer ions than the water initially entering the cell. Electrodialysis can achieve approximately 35 percent demineralization with loss of about 16 percent of the feed water as waste brine.

11-12   OCEAN POLLUTION

The ocean is the ultimate sink for many of our wastes. Rivers wash polluted waters into the oceans. Coastal communities release untreated or partially treated sewage into coastal waters. Solid wastes and chemicals are dumped into the oceans in vast amounts. Shipping operations pour large quantities of sewage and oil into the ports and open oceans of the world.

According to a report prepared for the U.S. Bureau of Solid Waste Management in 1970, over 50 million tons of wastes are dumped into the oceans around the United States each year. These wastes are carried to sea in ships and barges and do not include wastes from sewage outfalls. The dumping occurs in over 100 dumping sites along the Atlantic and Pacific coasts and in the Gulf of Mexico. These dumping sites range from 15 to 100 miles from shore. The wastes which are dumped are composed as follows:

Type	Percentage of total
Dredging materials	80
Industrial wastes	10
Sewage sludge	9
Miscellaneous wastes (refuse, garbage, explosives, etc.)	1

Amounts of industrial wastes have been increasing rapidly in recent years. Industrial wastes consist of waste acids, petroleum refinery wastes, pesticide wastes, paper mill wastes, and a variety of miscellaneous materials. The major industries involved are petroleum refineries, steel manufacturers, paper mills, pesticide manufacturers, chemical manufacturers, oil well drilling operations and metal finishing and plating operations.

The oceans seem infinite in size. They cover some 363 million square kilometers, which is about 71 percent of the surface of the earth. The volume of the oceans is roughly 1.5 billion cubic kilometers. This is indeed a large amount of water which can have a significant diluting factor on wastes that are dumped into the oceans. However, we must keep in mind that the oceans are the habitat of a vast amount of animal and plant life which exist in a delicate ecological balance. The U.S. Food and Drug Administration (FDA) estimates that around 500,000 different kinds of chemical substances are being dumped into the oceans. Some countries dispose of radioactive wastes by dumping sealed containers of such wastes in deep waters of the oceans. The fact is, we do not know what the effect of large amounts of ocean dumping might be. There are reports that certain coastal waters are becoming highly polluted. Finland and Poland have complained of pollution in the shallow waters of the Baltic Sea caused by industrial wastes. Recreation beaches in southern France and Italy are periodically closed due to the sewage problem. Many harbors and bay areas of the United States suffer from intermittent pollution episodes.

Insecticides (apparently from agricultural uses) have been found to be widely distributed in the animal life of the oceans. Furthermore, mercury, possibly from industrial wastes, has been found in some fish at alarming levels. In 1971, most swordfish showed mercury levels ranging from 0.9 to 2.4 parts per million, which well exceeds the FDA safe level of 0.5 part per million. Of course, a certain amount of ocean dumping has to take place, since there is no alternative. However, more control and better records of dumping activities is necessary. Unfortunately, many countries do not have legal jurisdiction over dumping which takes place in international waters.

11-13 OIL POLLUTION

A certain amount of ocean pollution comes from the wastes of millions of surface ships which sail the seas. Probably the most important of these wastes is oil. A quarter of a century ago Thor Heyerdahl made the voyage of the Kon-Tiki across the Pacific Ocean. He wrote of the beauty and purity of the oceans. In 1970, he made a similar voyage across the Atlantic in the Ra expedition. He reported that on 43 of the 57 days of the voyage the ocean was visibly polluted with globs of solidified oil clumps, oil slicks, and other debri. On one occasion in the mid-Atlantic, he observed that the ocean changed from clear blue to a grayish-green color which he compared to the water around sewage outfalls along some coastal areas.

World-wide annual oil production is roughly 2.5 billion tons. Of this, some 60 percent is transported across the oceans. It is estimated that 0.1 percent, or over 2 million tons of this oil, is lost in the loading and unloading operations. It is common practice for oil tankers to use ocean water for ballast and to dump this oil-contaminated water into the seas. Other ships pump waste oil into the seas as bilge wastes. Ships dump over 3.5 million tons of oil from these two sources alone. Oil pollution also comes from oil and gas drilling in coastal waters, underwater pipeline leakage, and occasional disasters such as the wreckage of the oil tanker Torrey Canyon or the San Francisco Bay disaster involving the wreckage of the oil tanker Oregon Standard. Around 1.1 billion gallons of automobile lubricants and 1.5 billion gallons of industrial lubricants are used in the United States each year. It is estimated that about 1 billion gallons of these lubricating oils are not chemically consumed and are disposed of in some way. Some is burned, but the remainder appear to be disposed of by dumping on the ground or into sewers. Apparently, a portion of this waste oil ends up in the ocean. **Re-refining** of waste oil is difficult due to the additives and impurities which accumulate during use. Re-refining cannot compete economically with virgin oil. The conversion of used oil to fuel oil is being investigated. However, it appears that most used oils contain high levels of various metals, such as lead, which can contribute to air pollution. The loss of 1 billion gallons of oil each year, much of which is disposed of in unknown ways, indicates that more control over used oil disposal is needed.

Oil pollution of the ocean can have an effect on aquatic animal and plant life. Once oil is spilled in the ocean it spreads out in a thin film called an **oil slick.** Oil is a mixture of various hydrocarbons, some of which are highly volatile. These volatile components evaporate from the slick, leaving behind the less volatile hydrocarbons. The remaining oil mixes with the water to form an oily conglomeration which floats about in tar-like clumps. Some portions of the oil sink to produce an oil coat on aquatic life on the ocean bottom. Sometimes an oil slick will wash ashore to coat the shoreline. Apparently, some of the oil in an oil slick is consumed by bacteria and microorganisms in the water. However, there is a possibility that a portion of any spilled oil will remain in the ocean or on the bottom for a long period of time.

The dramatic but significant effect of oil spills includes the coating of beaches and rocks with oil, which interferes with and is toxic to the plant and animal life of the tide regions. Vast numbers of water fowl are often coated with oil as they try to feed in the slick-covered waters. Such oil coatings are fatal to most birds. There is evidence that the oil which reaches the ocean bottom has a devastating effect on some of the underwater plants and animals. It is also thought that some chemical components in oil may interfere with chemicals which the sea animals use as **pheromones** (chemicals secreted by animals for communication and life processes). These chemicals involve functions such as escaping from predators, homing, selection of habitat, sexual attraction, and feeding. These processes are often governed by small amounts of these secreted chemicals present in ocean water. If oil components interfere with or substitute for some of these chemicals, aquatic life processes could be greatly upset.

Oil slicks and oil pollution might also interfere with the production of phytoplankton in the ocean. **Phytoplankton** consists of microscopic plant life which grows in surface areas of the ocean penetrated by sunlight. These plants are food sources for some animal life and actually occupy a place in the beginning of oceanic food chains. Decreases in the amounts of phytoplankton could have a drastic effect on the populations of numerous animal species in the oceans. Some geologists claim that a drastic decrease in phytoplankton occurred around 300 million years ago. As a result, nearly 20 percent of existing fish and mammal species are thought to have become extinct.

The major problem is that the amount of oil shipped across the oceans is increasing, and most oil spills are a result of human error. A greater effort has to be expended toward designing transportation, loading, and unloading methods that are safer and less polluting. Furthermore, off-shore drilling and production methods must be improved to prevent unnecessary spills.

BOOKS

BIBLIOGRAPHY

American Chemical Society. *Cleaning Our Environment: The Chemical Basis for Action.* section 2. "The Water Environment." Washington, D.C.: ACS Special Issues Sales, 1969.

Behrman, A. S. *Water is Everybody's Business.* Garden City, N.Y.: Doubleday, 1968.

Carr, D. E. *Death of the Sweet Waters.* New York: Berkeley Publishing, 1971.

Hood, D. W., ed. *Impingement of Man on the Oceans.* New York: Wiley, 1971.

Marx, W. *The Frail Ocean.* New York: Sierra Club–Ballantine, 1967.

River of Life. Water: The Environmental Challenge. U.S. Department of the Interior Conservation Yearbook No. 6. Washington, D. C.: U.S. Government Printing Office, 1970.

ARTICLES AND PAMPHLETS

American Chemical Society. "Chemistry and the Oceans." *Chemical and Engineering News.* Reprint.

Crossland, J., and J. McCaull. "Overfed." *Environment* **14** (9) (1972).

"The Great Phosphorus Controversy." *Environmental Science and Technology* (Sept. 1971).

Hickel, W. J., C. L. Klein, and D. D. Dominick. *A Primer on Waste Water Treatment.* Washington D.C.: U.S. Government Printing Office, 1969.

Keller, E. "Fish Kills." *Chemistry* **41** (9), 8 (1968).

Kushner, L. M., and J. I. Hoffman. "Synthetic Detergents." *Scientific American* 26–30 (Oct. 1951).

McCaull, J. "The Tide of Industrial Waste." *Environment* **14** (10) (1972).

"The Ocean." *Scientific American* **221** (Sept. 1969).

"Ocean Dumping Poses Growing Threat." *Environmental Science and Technology* (Oct. 1970).

Othmer, D. F. "Water and Life." *Chemistry* **12** (Nov. 1970).

Rey, G., et al. "Industrial Water Reuse." *Environmental Science and Technology* (Sept. 1971).

U.S. Department of the Interior. *The A-B-Seas of Desalting.* Office of Saline Water, Washington, D.C.: U.S. Government Printing Office, 1968.

U.S. Department of the Interior. *Clean Water for the 1970's—A Status Report.* Washington, D.C.: Federal Water Quality Administration, 1970.

Washburn, C. A. "Clean Water and Power." *Environment* **14** (7) (1972).

Wolff, A. "Showdown at Four Corners." *Saturday Review: The Society* June 3, (1972).

QUESTIONS AND PROBLEMS

1. What are the three main uses of water in the United States? What activity uses the most water?

2. Will we ever run out of water? Explain.

3. What is meant by polluted water?

4. Describe the three major types of water pollutants.

5. What does the term "biodegradable" mean?

6. What are hard or refractory organic compounds?

7. Describe the aerobic decay of degradable organic compounds in natural waters.

8. What is anaerobic decay?

9. What is the difference between a soap and a detergent?

10. Why are phosphates added to commercial detergents?

11. What is biochemical oxygen demand? What does BOD measure?

12. What is the purpose of a coliform bacteria count in natural water?

13. Describe the process of eutrophication.

14. How are nutrients in man-made wastes involved in eutrophication?

15. Describe primary sewage treatment.

16. Describe secondary sewage treatment.

17. Do you feel that all domestic sewage in the United States should be subjected to secondary treatment? Why?

18. What is tertiary sewage treatment?

19. Give three examples of tertiary treatment methods and describe each of the methods.

20. Do you feel that ocean dumping is necessary? Explain.

21. Is oil pollution of the oceans unavoidable? Why?

22. Do you feel that an oil company should be liable for any oil spills that are caused by the drilling or transportation activities of the company? Why?

23. What do you think happens to the motor oil in your car or your family car when you get an oil change? Is there any way you can check on the disposal of this oil?

24. Would you favor the granting of tax advantages to companies that re-refine waste oils? Why?

25. Rank order the following (1 = most agreement, and so on) choices corresponding to your feelings concerning the use and depletion of water from ancient aquifers:

    _____ This use is of no concern to me.
    _____ This use is very short-sighted and more effort should be made to find alternate water sources.
    _____ People have the right to obtain water where it is available and convenient to use.

ORGANIC CHEMISTRY 12

12-1 SYNTHETIC ORGANIC CHEMICALS

For centuries chemists have isolated and investigated chemical compounds found in nature. They classified these compounds into two groups: Inorganic compounds were those isolated from mineral sources; organic compounds were those isolated from plant and animal sources. A wide variety of compounds were found to be manufactured by plants and animals during life processes. Upon investigation, it was found that the one thing that organic compounds had in common was that they all were carbon-containing compounds. Today, organic compounds are considered as a special group of carbon-containing compounds. Inorganic compounds are composed of a wide variety of the other elements and, except for the carbonates, most inorganic compounds do not contain carbon.

Much effort was expended in determining the composition of the variety of compounds found in nature. Chemists began investigating organic compounds and the ability of carbon to form bonds. They found that they were able to synthesize some chemical compounds which were chemically equivalent to naturally occurring compounds. Moreover, they learned that it was possible to synthesize numerous chemical compounds which were not found in nature. That is, they were able to create new combinations of elements which were not known to exist. These achievements contributed to the development of a vast technology in which usable synthetic chemicals are produced.

Several decades ago, **synthetic organic** substances were not common in consumer products. Today, man-made substances surround us and are incorporated into many of the things we use. Synthetic organic chemicals are prepared from naturally occurring molecules (usually derived from petroleum) by industrial chemical processes which convert the natural products to more useful molecules.

The polyester shirt or blouse you might be wearing is made of a synthetic fiber, and the plastic buttons and colored dyes used on the garment are synthetics. The vinyl tiles or carpet on the floor are synthetic plastics, and the rubber heels on your shoes are synthetic rubber. The aspirin or vitamin tablets you may have taken today contain synthetic pharmaceuticals. The bread you ate for breakfast came from a plastic wrapped container and the bread might have contained synthetic food additives as preservatives. The apple you had for lunch may have carried a small amount of synthetic pesticide into your body.

Figure 12-1
Synthetic organic
chemical production.

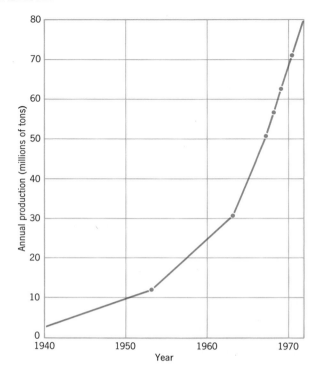

Figure 12-1
Synthetic organic chemical production.

The increase in the annual amounts of synthetic organic chemicals produced in the United States has been phenomenal. As shown in Figure 12-1, annual synthetics production has increased from 5 million tons in 1943 to 80 million tons in 1973. These synthetics include plastics, plasticizers, paints, pesticides, preservatives, and pharmaceuticals, among others.

Special disposal and pollution problems are associated with some synthetics. Some are resistant to bacterial decay in water and soil and persist in the environment. Some plastics, when buried in landfills, may not degrade for hundreds or thousands of years. When detergents came into wide use in the 1950s, the synthetic detergent used in them was not easily biodegradable. The detergent levels increased drastically in environmental waters, causing serious foaming problems. In 1965, most commercial detergent preparations were changed so that they contained a more biodegradable detergent. Some pesticides and refractory industrial organics are highly toxic to animals and have been associated with fish kills, and others give bad tastes or odors to fish and shellfish.

12-2 CARBON CHEMISTRY

A branch of chemistry called **organic chemistry** is devoted to the study of the compounds of carbon. The most important property of carbon is the ability of carbon atoms to form chemical bonds with one another and with a variety of other elements. Carbon forms four covalent bonds in organic compounds. Thus, carbon is said to be **tetravalent.** This tetravalency or tendency to form four bonds correlates with the fact that carbon has four valence electrons:

When a carbon atom forms covalent bonds with four other atoms, the four pairs of electrons are **tetrahedrally** distributed about the carbon atom as shown in Figure 12-2. For writing convenience, the four possible covalent bonds of carbon may be represented in a plane as

$$-\overset{\displaystyle |}{\underset{\displaystyle |}{C}}-$$

Of course, other atoms are bonded to carbon through these bonds.

One of the reasons there are numerous organic compounds is that carbon atoms can form strong bonds with other carbon atoms while at the same time they can form strong bonds with other nonmetals. This means that it is possible to have chains of carbon atoms bonded to one another and to other types of atoms. It is possible for two carbon atoms to be linked by a single covalent bond:

$$-\overset{\displaystyle |}{\underset{\displaystyle |}{C}}-\overset{\displaystyle |}{\underset{\displaystyle |}{C}}-$$

A given carbon atom can form more than one single bond with other carbon atoms. This gives rise to a very large number of possible sequences of carbon atoms bonded to one another.

As illustrated in Figure 12-3, carbon atoms can bond together in a variety of ways much like the links of a chain. In fact, the bonded carbon sequence in an organic compound is sometimes called the carbon chain. There are literally billions of possible sequences of bonded carbon atoms. This accounts for the fact that there are millions of known organic compounds. Fortunately, most of these compounds are rare or unimportant, and the others can be studied by grouping them into categories of compounds having similar structures. Organic chemistry involves the study of the various groups of compounds. In this chapter, twelve groups of the most common types of organic compounds will be described.

12-3   BONDING IN ORGANIC COMPOUNDS

Organic compounds involve bonded carbon sequences in which some of the carbon bonds are used in bonding to atoms of other elements. Hydrogen is the

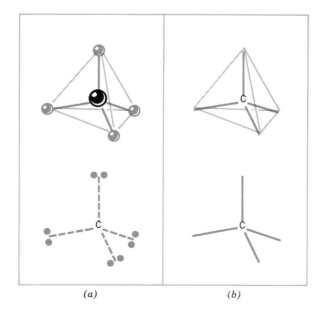

(a)   (b)

Figure 12-2
Tetrahedral bonds of carbon. (a) When carbon forms bonds with other atoms, the bonding electrons are oriented in a tetrahedral arrangement about the center of the carbon atom. (b) When carbon is bonded to four other atoms, the bonds are arranged in a tetrahedral fashion about the carbon.

Figure 12-3
The ability of carbon atoms to link together in a variety of ways is similar to the formation of chains from individual links. Bonded carbon sequences in molecules are sometimes called carbon chains. When molecules include carbon sequences linked to the main chain they are sometimes called branched chains, and the side chains are called branches. Carbons sometimes link together in a cyclic fashion in which the end carbon bonds to the first carbon. These are called cyclic or ring molecules.

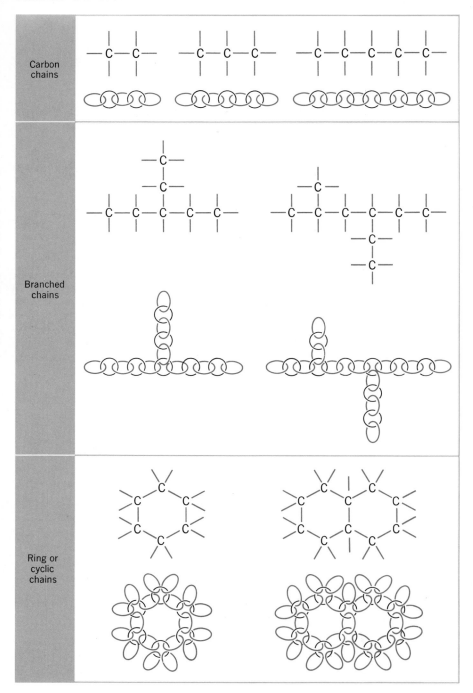

element most often found bonded to carbon in organic compounds. A hydrogen atom can share one pair of electrons with a carbon atom:

$$-\overset{\displaystyle |}{\underset{\displaystyle |}{C}}-H$$

The formula above does not represent a compound, but just indicates how a

● Hydrogen
◐ Carbon

Figure 12-4
Three representations
of a methane
molecule.

carbon–hydrogen bond can be represented. In a compound, all the bonds of carbon must be used. The compound methane, CH_4, can be represented as (also see Figure 12-4)

$$H-\overset{\displaystyle H}{\underset{\displaystyle H}{C}}-H$$

This formula, which indicates the atoms that are bonded, is called a **structural formula.** The formula CH_4, which only indicates the type and number of atoms comprising a molecule, is called a **molecular formula.** Structural formulas are used to indicate which atoms are bonded to which in the molecule. On the other hand, molecular formulas are used to indicate the composition of the molecules of the compounds.

Butane, C_4H_{10}, has the following structural formula:

$$H-\overset{\displaystyle H}{\underset{\displaystyle H}{C}}-\overset{\displaystyle H}{\underset{\displaystyle H}{C}}-\overset{\displaystyle H}{\underset{\displaystyle H}{C}}-\overset{\displaystyle H}{\underset{\displaystyle H}{C}}-H$$

Rather than writing the complete structural formula of a compound like butane, it is possible to represent the compound by a **condensed structural formula** which indicates the bonding sequence without showing all the bonds. For example, the condensed structural formula for butane is

$$CH_3CH_2CH_2CH_3$$

Such formulas should be interpreted as indicating that the carbons are bonded to one another in sequence and each carbon is bonded to the hydrogens (or other atoms) that are next to it in the formula. Condensed formulas are more convenient to write than full structural formulas.

Two carbon atoms are capable of sharing two pairs of electrons with one another to form a **double covalent bond (double bond):**

$$\overset{}{\underset{}{C}}=\overset{}{\underset{}{C}}$$

The double bond uses two bonds for each carbon, leaving two other bond positions on each carbon available for bonding to other atoms. For example, the compound ethylene, C_2H_4, involves a double bond:

$$\overset{H}{\underset{H}{}}C=C\overset{H}{\underset{H}{}} \qquad or \qquad CH_2{=}CH_2$$

Sometimes two carbon atoms actually share three pairs of electrons to form a **triple covalent bond (triple bond)** as shown on the next page.

$$-C\equiv C-$$

When two carbon atoms are bonded by a triple bond, each carbon can form one other bond with a different atom. For example, acetylene, C_2H_2, involves a triple bond:

$$H-C\equiv C-H \quad \text{or} \quad CH\equiv CH$$

Oxygen can bond to carbon in two ways. Oxygen can form two covalent bonds (divalent) so that it is possible for oxygen to bond to a carbon by a double bond:

$$\underset{\diagup C \diagdown}{\overset{\overset{\displaystyle O}{\|}}{}}$$

This leaves two bond positions on the carbon that are used for bonding to other atoms. The structural formula of formaldehyde, CH_2O, is

$$\underset{H \quad\quad H}{\overset{\overset{\displaystyle O}{\|}}{C}}$$

Carbon and oxygen can be bonded by a single bond:

$$-\overset{|}{\underset{|}{C}}-O-$$

This leaves one bond position on the oxygen and three on the carbon. The structural formula of ethyl alcohol, C_2H_5OH, is

$$H-\overset{\overset{\displaystyle H}{|}}{\underset{\underset{\displaystyle H}{|}}{C}}-\overset{\overset{\displaystyle H}{|}}{\underset{\underset{\displaystyle H}{|}}{C}}-O-H$$

Usually nitrogen and carbon are bonded by a single bond:

$$-\overset{|}{\underset{|}{C}}-\overset{|}{N}-$$

This leaves two bond positions on the nitrogen and three on the carbon. The structural formula of methylamine, CH_3NH_2, is

$$H-\overset{\overset{\displaystyle H}{|}}{\underset{\underset{\displaystyle H}{|}}{C}}-\overset{}{\underset{\underset{\displaystyle H}{|}}{N}}-H$$

Carbon can form single bonds with the halogens (F, Cl, Br, I):

$$-\overset{|}{\underset{|}{C}}-Cl$$

The structural formula of chloroform, $CHCl_3$, is

$$H-\overset{\overset{\displaystyle Cl}{|}}{\underset{\underset{\displaystyle Cl}{|}}{C}}-Cl$$

The knowledge of the manner in which carbon forms bonds with atoms serves as a foundation for the discussion of organic compounds.

12-4 ALKANES AND ISOMERISM

Organic compounds containing only carbon and hydrogen are called **hydrocarbons.** Hydrocarbons that involve only single bonded carbons (no double or triple bonds) are called **saturated hydrocarbons.** It is important to examine some of these hydrocarbons, since they serve as a basis for the nomenclature and structural formulas of a large number of organic compounds. There are numerous possible hydrocarbons. The simplest hydrocarbon is methane, CH_4:

$$CH_4$$

Second is ethane, C_2H_6:

$$CH_3CH_3$$

Third is propane, C_3H_8:

$$CH_3CH_2CH_3$$

Notice that these compounds differ by $-CH_2-$ units.

$$CH_3{-}CH_2{-}CH_3$$

The structural formulas of a large number of hydrocarbons can be written which differ by $-CH_2-$ units. Any group of compounds in which the members differ in this manner is called an **homologous series.** The saturated hydrocarbons comprise an homologous series corresponding to the general formula C_nH_{2n+2} ($CH_4, n = 1$; $C_2H_6, n = 2$; $C_3H_8, n = 3$). These compounds are called the **alkanes.** Eight of the most important alkanes are listed in Table 12-1.

Let us consider the five-carbon alkane, C_5H_{12}. In such a compound, the sequence of carbons can be

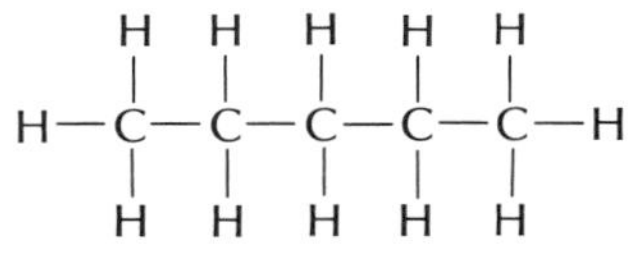

But notice below that it is possible to have other branched carbon sequences with the same molecular formula:

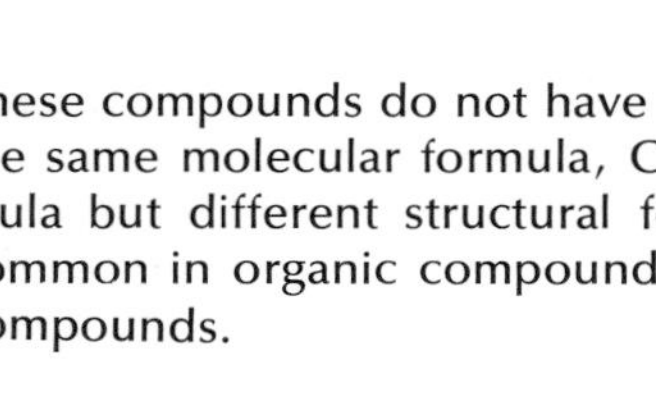

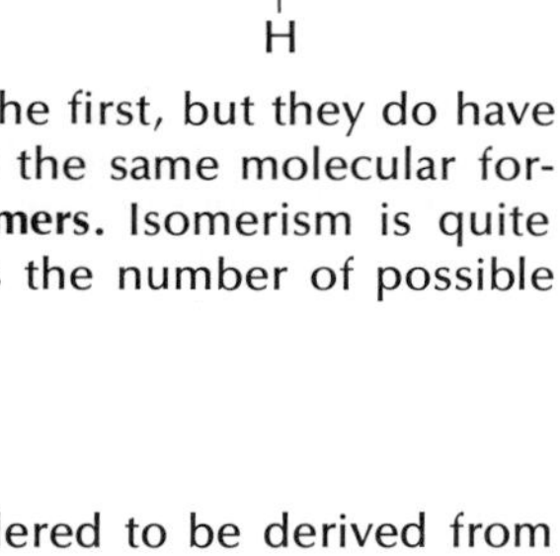

These compounds do not have the same structure as the first, but they do have the same molecular formula, C_5H_{12}. Compounds with the same molecular formula but different structural formulas are called **isomers.** Isomerism is quite common in organic compounds and greatly increases the number of possible compounds.

12-5 NOMENCLATURE AND GROUPS

A large number of organic compounds can be considered to be derived from the replacement of one or more hydrogens of alkanes by another atom or group

Table 12-1
Some Alkanes

Name	Molecular formula	Structural formula	Condensed structural formula
Methane	CH_4		CH_4
Ethane	C_2H_6		CH_3CH_3
Propane	C_3H_8		$CH_3CH_2CH_3$
Butane	C_4H_{10}		$CH_3CH_2CH_2CH_3$ or $CH_3(CH_2)_2CH_3$[a]
Pentane	C_5H_{12}		$CH_3CH_2CH_2CH_2CH_3$ or $CH_3(CH_2)_3CH_3$[a]
Hexane	C_6H_{14}		$CH_3CH_2CH_2CH_2CH_2CH_3$ or $CH_3(CH_2)_4CH_3$[a]
Heptane	C_7H_{16}		$CH_3CH_2CH_2CH_2CH_2CH_2CH_3$ or $CH_3(CH_2)_5CH_3$[a]
Octane	C_8H_{18}		$CH_3CH_2CH_2CH_2CH_2CH_2CH_2CH_3$ or $CH_3(CH_2)_6CH_3$[a]

From *Introduction to Chemistry* by T. R. Dickson, John Wiley & Sons, New York, 1971.
[a] The $(CH_2)_n$ notation refers to n CH_2 units in a row.

of atoms. For instance, the compound

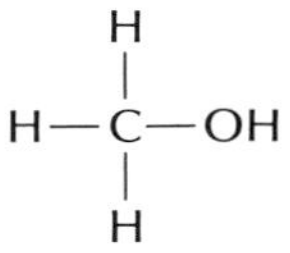

$$H-\underset{\underset{H}{|}}{\overset{\overset{H}{|}}{C}}-OH$$

can be viewed as being derived from the replacement of an H of methane with a
−OH group. Another view of this compound is that it is composed of a

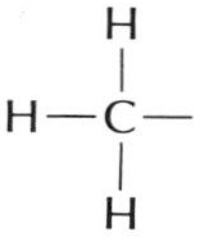

$$H-\underset{\underset{H}{|}}{\overset{\overset{H}{|}}{C}}-$$

group bonded to a −OH. This is a good way to view such compounds, since it
simplifies nomenclature.

Groups corresponding to the alkanes can be considered to be formed by
removing one hydrogen to leave a bonding position. Such groups formed from
the alkanes are called **alkyl groups.** They are named by using the name of the
alkane from which they are derived, with the -ane ending changed to -yl. For example,

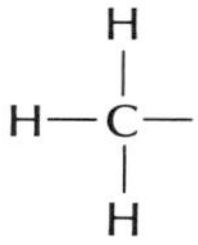

$$H-\underset{\underset{H}{|}}{\overset{\overset{H}{|}}{C}}-\text{ is methyl group} \quad \text{and} \quad H-\underset{\underset{H}{|}}{\overset{\overset{H}{|}}{C}}-\underset{\underset{H}{|}}{\overset{\overset{H}{|}}{C}}-\text{ is ethyl group}$$

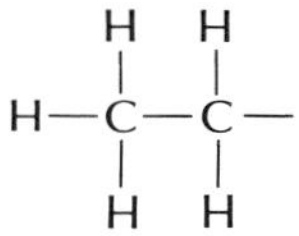

These groups can also be shown in condensed, more easily written formulas as
CH_3- (methyl) and CH_3CH_2- (ethyl). Table 12-2 lists some common alkyl groups.
A special symbol, R−, is used to represent any alkyl group. A few examples of
compounds containing alkyl groups are:

CH_3OH	methyl alcohol
CH_3CH_2Cl	ethyl chloride
$(CH_3)_2CHNH_2$	isopropyl amine

Formula	Name	
CH_3-	methyl group	**Table 12-2**
CH_3CH_2-	ethyl group	Some Alkyl Groups
$CH_3CH_2CH_2-$	propyl group	
$\underset{CH_3CH-}{\overset{\overset{CH_3}{\|}}{}}$ or $(CH_3)_2CH-$	isopropyl group	
$CH_3CH_2CH_2CH_2-$	butyl group	
$\underset{CH_3CHCH_2-}{\overset{\overset{CH_3}{\|}}{}}$ or $(CH_3)_2CHCH_2-$	isobutyl group	
$CH_3-\underset{\underset{CH_3}{\|}}{\overset{\overset{CH_3}{\|}}{C}}-$ or $(CH_3)_3C-$	tertiary butyl group	
R−	general symbol for any alkyl group	

Generally, there are two ways to name organic compounds. Historically, common names were developed for many compounds. However, in order to make nomenclature more systematic, the International Union of Pure and Applied Chemistry (IUPAC) has proposed some systematic nomenclature rules called the IUPAC rules. Some compounds can be named both by a common name and a systematic IUPAC name. In this chapter compounds will be named almost exclusively by common names.

12-5 ALKENES

Hydrocarbons that have a double bond between any two carbons are called **unsaturated hydrocarbons,** or **olefins.** The simplest olefin is ethylene:

$$\begin{array}{c}
\text{H} \\ \diagdown \\ \text{H}
\end{array}
\text{C=C}
\begin{array}{c}
\diagup \text{H} \\ \\ \diagdown \text{H}
\end{array}$$

The compounds formed from this compound by adding longer carbon sequences are members of the homologous series called the **alkenes.** The general formula for alkenes is C_nH_{2n} ($CH_2{=}CH_2, n = 2$; $CH_3CH{=}CH_2, n = 3$; etc.).
Some important alkenes and substituted alkenes are listed below.

Formula	Name
$CH_2{=}CH_2$	ethylene
$CH_3{-}CH{=}CH_2$ or $CH_3CH{=}CH_2$	propylene
$CH_2{=}C\begin{smallmatrix}\diagup\text{H}\\\diagdown\text{Cl}\end{smallmatrix}$ or $CH_2{=}CHCl$	vinyl chloride
$\begin{smallmatrix}\text{F}\\\diagdown\end{smallmatrix}C{=}C\begin{smallmatrix}\diagup\text{F}\\\diagdown\text{F}\end{smallmatrix}$ or $CF_2{=}CF_2$	tetrafluoroethylene

Whenever, a double bond occurs in a sequence of carbons, it is called an **unsaturated linkage.** Molecules of **polyunsaturated vegetable oils** involve carbon sequences with unsaturated linkages. However, vegetable oils are not classified as alkenes, since they contain oxygen as well as carbon and hydrogen. Vegetable oils are classified as lipids and are discussed in Section 13-3. Compounds with double bonds are generally quite chemically reactive in comparison with alkanes. The typical reactions involving alkenes are **addition reactions** in which the double bond is broken and other atoms become bonded to the carbons of the original double bonds. Two examples of addition reactions are

$$\begin{array}{c}\text{H}\\\diagdown\\\text{H}\end{array}\text{C=C}\begin{array}{c}\diagup\text{H}\\\diagdown\text{H}\end{array} + \text{HCl} \longrightarrow \text{H}{-}\overset{\overset{\displaystyle\text{H}}{|}}{\underset{\underset{\displaystyle\text{H}}{|}}{\text{C}}}{-}\overset{\overset{\displaystyle\text{H}}{|}}{\underset{\underset{\displaystyle\text{H}}{|}}{\text{C}}}{-}\text{Cl} \qquad \text{(ethyl chloride)}$$

$$CH_3{-}CH{=}CH_2 + Cl_2 \longrightarrow CH_3\overset{\overset{\displaystyle\text{Cl}}{|}}{CH}{-}\overset{\overset{\displaystyle\text{Cl}}{|}}{CH_2} \qquad \text{(propylene dichloride)}$$

12-7   POLYMERIZATION AND PLASTICS

Some alkenes can react under specific conditions in the presence of a catalyst, so that the individual alkene molecules add to one another. In this reaction, the double bonds are broken and many molecules (hundreds or thousands) link together to form very large molecules called **polymers.** The original alkene used to prepare the polymer is called the **monomer.** These reactions are called **polymerization** reactions. (See Figure 12-5.) The polymer molecules form compounds that are known as **plastics** and **elastomers,** which are used to make many of the materials and objects we use in everyday life. Table 12-3 lists some of the products that are made of polymers. Many substances that occur in nature contain polymers. Cotton, wool, silk, and natural rubber are polymers. In fact, as we shall see in Sections 13-2 and 13-4, many important biological molecules are polymers.

In an equation for a polymerization reaction, it is not possible to give the exact formula of the polymer, since its individual molecules, made up of hundreds or thousands of monomer units, vary in chain length. We could represent the polymerization of ethylene as

$$ n\ CH_2{=}CH_2 \xrightarrow[\text{heat pressure}]{\text{catalyst}} H-\underset{\underset{H}{|}}{\overset{\overset{H}{|}}{C}}-\underset{\underset{H}{|}}{\overset{\overset{H}{|}}{C}}-\underset{\underset{H}{|}}{\overset{\overset{H}{|}}{C}}-\underset{\underset{H}{|}}{\overset{\overset{H}{|}}{C}}-\underset{\underset{H}{|}}{\overset{\overset{H}{|}}{C}}-\underset{\underset{H}{|}}{\overset{\overset{H}{|}}{C}}-\underset{\underset{H}{|}}{\overset{\overset{H}{|}}{C}}-\cdots $$

but this is not convenient. However, notice that the polymer could be represented by the repeating sequence

$$ -\underset{\underset{H}{|}}{\overset{\overset{H}{|}}{C}}-\underset{\underset{H}{|}}{\overset{\overset{H}{|}}{C}}- $$

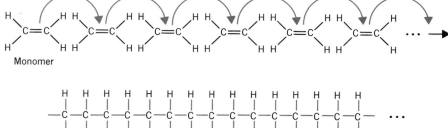

Monomer

Polymer

(a)

Links

Chain

(b)

Figure 12-5
Polymerization. (a) In a polymerization reaction, monomer molecules are linked together to form large molecules called polymers. (b) The formation of a polymer can be likened to the forming of a chain. Monomer units are linked together one after another to form the chain-like polymer molecule. Polymerization reactions are sometimes called chain reactions and the bonded atoms comprising the polymer are sometimes called the polymer chain.

Table 12-3
Polymers—Plastics,
Resins, and Rubbers

Name	Use
Polyethylene	Electrical insulation, packaging material (sandwich bags, plastic wrap), molded toys and utensils, milk carton coatings
Polypropylene	Molded containers, bottles, hospital utensils (sterilizable), washing machine parts, automobile interior parts
Polyvinyl chloride	Electrical insulation, toys, garden hoses, automobile seat covers, washable wallpaper, packaging material, bottles (shampoo), "patent leather," vinyl flooring, water pipes
Polytetrafluoroethylene	Teflon, electrical insulator, chemically inert material, nonstick coatings (pots, pans, and tools)
Polyvinylacetate	Paints, adhesives for textiles, paper, and wood, sizing for textiles
Polyvinyl alcohol	Emulsifiers in cosmetics, water-soluble packaging materials
Polymethylmethacrylate (acrylic)	Plexiglas, Lucite, paints, light fixtures, signs, airplane windows, helicopter bubbles, dentures
Polystyrene	Styrofoam, packaging material, bottle caps, refrigerator interiors, toys, containers, kitchen utensils, foam insulation
Polyvinyldiene chloride	Saran, packing material, coating material
Nylon (polyamides)	Machine parts (gears, cams), boil-in-the bag containers, filaments used in brushes, surgical sutures, fishing lines, carpets, women's stockings
Cellulose acetate	Photographic film, toothbrushes, combs
Phenol–formaldehyde resins	Bakelite, electrical insulators
Melamine–formaldehyde resins	Formica
Polyester resins	Molding compounds, fiberglass resins, Mylar film, synthetic fibers, paints, permanent-press clothing
Epoxide resins	Epoxy glues, coatings
Polyurethanes	Foams for packaging, insulation, furniture
Silicones	Polishes, lubricants, high-temperature rubber
cis-Polyisoprene	Natural rubber
Styrene–butadiene rubber	Synthetic rubber
Polychloroprene	Neoprene synthetic rubber

From *Introduction to Chemistry* by T. R. Dickson, John Wiley & Sons, New York, 1971.

This sequence repeats along the polymer except at the ends, but the ends are a very small part of the entire polymer. Polymers are often represented by a characteristic repeating sequence. Therefore, the polymer of ethylene could be represented as

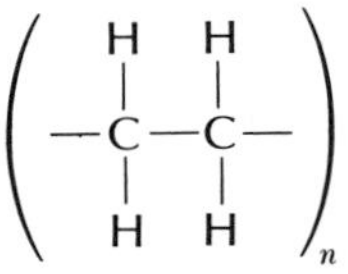

$$\left(\begin{matrix} & H & H \\ & | & | \\ -C & -C- \\ & | & | \\ & H & H \end{matrix} \right)_n$$

where *n* is some large number corresponding to the number of monomer units comprising the polymer. Now the equation for the polymerization of ethylene can be written as

$$n \; CH_2{=}CH_2 \xrightarrow[\text{heat pressure}]{\text{catalyst}} \left(\begin{array}{c} H \;\; H \\ | \;\;\; | \\ -C-C- \\ | \;\;\; | \\ H \;\; H \end{array} \right)_n \text{(polymer)}$$

(monomer)

The polymer of ethylene is called polyethylene and is used to make plastic bottles, toys, and other products. A few additional polymerization reactions are given below:

$$n \; CH_3{-}CH{=}CH_2 \longrightarrow \left(\begin{array}{c} H \;\; CH_3 \\ | \;\;\; | \\ -C-C- \\ | \;\;\; | \\ H \;\; H \end{array} \right)_n \text{(polypropylene)}$$

(propylene)

Polypropylene is used to make synthetic fibers, films, and other products.

$$n \; CF_2{=}CF_2 \longrightarrow \left(\begin{array}{c} F \;\; F \\ | \;\;\; | \\ -C-C- \\ | \;\;\; | \\ F \;\; F \end{array} \right)_n \text{(polytetrafluoroethylene, or Teflon)}$$

(tetrafluoroethylene)

Teflon is a chemically inert plastic used for nonstick coatings on machines, tools, and cooking utensils.

Large quantities of polymers are produced each year in the United States in the forms of molded plastics, films, and fibers, and a variety of synthetic rubbers. The many uses of plastics and rubbers are listed in Table 12-3. The annual production of a few plastics is given in Table 12-4. Synthetic fibers are in wide use in carpets, clothing, and other fabrics. In fact, as shown in Figure 12-6, synthetic fibers are now used in greater quantities than natural cotton and wool fibers.

Plastic use involves potential pollution problems during manufacture and disposal. Manufacture of plastics and synthetic fibers include industrial processes in which air and water pollution can occur. Many of the monomers used for polymers are gases or volatile liquids which can be lost to the atmosphere during monomer or polymer preparation. The manufacturing and processing of synthetic fibers uses large quantities of water which is used to carry away wastes. The most bothersome wastes are lint fibers. The synthetic fibers are of low biodegradation levels and cannot be removed from waste water by bacterial action. As a result, these fibers can accumulate in treatment plants and can be carried into environmental waters.

The low biodegradability of plastics in general makes the disposal of plastics difficult. Most plastic material which is buried in landfills or dumped in the

Material	Production (thousands of tons)	Material	Production (thousands of tons)
Acrylics	260	Polystyrene	1,150
Nylons	750	Polyvinyl chloride	
Polyesters	900	(and copolymers)	1,600
Polyethylene	2,900	Synthetic rubber	960
Polypropylene	520		

Table 12-4 Estimated Annual Production of Selected Plastics and Rubbers

Figure 12-6
Annual textile fiber
use in the United
States.

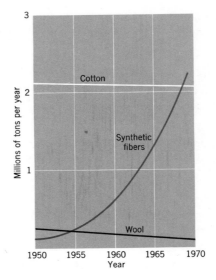

oceans will not degrade for hundreds of years. The incineration of certain chlorine-containing plastics, like polyvinyl chloride (PVC), produces hydrogen chloride gas. This gas can condense with water to produce hydrochloric acid which can damage incinerators or become a hazardous air pollutant. Unfortunately, the recycling of plastics and synthetic fibers is not a common practice. Of course, one of the major barriers to the reuse of plastics is that they become part of the complex mixture of materials which makes up domestic and industrial solid wastes.

12-8   ALCOHOLS

In this section an important group of organic compounds is discussed. Compounds are grouped according to similarities. The alkanes are saturated hydrocarbons, and the alkenes are hydrocarbons with a double bond. Compounds can be grouped according to some structural characteristic that sets them apart from other compounds.

The group of compounds involving an –OH (hydroxy group) bonded to a R– (alkyl group) are called **alcohols** (R–OH). The **hydroxy** group is characteristic of alcohols, some of which have familiar common names. A few typical alcohols are described below.

$$CH_3OH \quad \text{methyl alcohol, methanol, or wood alcohol}$$

Around 2.5 million tons of methyl alcohol are manufactured

$$CO + 2 H_2 \xrightarrow{\text{catalyst}} CH_3OH$$

in the United States each year. It is used in the manufacture of numerous chemical products such as formaldehyde, jet fuel, and antifreeze. (DANGER: Methyl alcohol or wood alcohol is very poisonous.)

$$\overset{\displaystyle OH}{\underset{\displaystyle |}{CH_3CHCH_3}} \quad \text{or} \quad (CH_3)_2CHOH \quad \text{isopropyl alcohol}$$

Isopropyl alcohol is used as a disinfectant and in rubbing alcohol.

CH_3CH_2OH or C_2H_5OH ethyl alcohol   or   ethanol

Millions of tons of ethyl alcohol are produced annually in the United States. It is used in numerous manufacturing processes and in the preparation of alcoholic beverages. Ethanol can be produced by the fermentation of sugars.

Fermentation is a chemical process in which complex organic molecules are broken down into simpler compounds like ethanol. This process is catalyzed by certain enzymes, which are complex chemical catalysts produced by living cells. The sugars used for fermentation are often formed by enzymatic decomposition of starches from corn, potatoes, rice, or grain. The fermentation reactions which produce ethyl alcohol are shown in the following equations:

$$\text{starch} \xrightarrow[\text{from malt}]{\substack{\text{enzyme} \\ \text{diastase}}} n\ C_{12}H_{22}O_{11} \qquad \text{(maltose)}$$

$$\underset{\text{maltose}}{C_{12}H_{22}O_{11}} \xrightarrow[\text{from yeast}]{\substack{\text{enzyme} \\ \text{maltase}}} 2\ C_6H_{12}O_6 \qquad \text{(glucose)}$$

$$\underset{\text{glucose}}{C_6H_{12}O_6} \xrightarrow[\text{yeast}]{\substack{\text{several enzymes} \\ \text{also in}}} 2\ CO_2 + 2\ C_2H_5OH \qquad \text{(ethyl alcohol)}$$

In addition to fermentation, over 1 million tons of ethyl alcohol are manufactured each year by the following industrial method:

Step 1: $CH_2{=}CH_2 + H_2SO_4 \longrightarrow \overset{\displaystyle OSO_3H}{\overset{|}{CH_2CH_3}}$

Step 2: $\overset{\displaystyle OSO_3H}{\overset{|}{CH_3CH_2}} + 2\ H_2O \longrightarrow CH_3CH_2OH + H_3O^+ + HSO_4^-$

Alcohols that have more than one hydroxy group attached to a carbon sequence are called polyhydroxy alcohols. Two important polyhydroxy alcohols are described below.

$\overset{\displaystyle OH \quad\ OH}{\underset{\displaystyle CH_2{-}CH_2}{\overset{|\qquad\ |}{}}}$ or CH_2OHCH_2OH ethylene glycol

Ethylene glycol is used as an antifreeze and engine coolant.

$\overset{\displaystyle OH \quad\ OH \quad\ OH}{\underset{\displaystyle CH_2{-}CH{-}CH_2}{\overset{|\qquad\ |\qquad\ |}{}}}$ or $CH_2OHCHOHCH_2OH$ glycerol   or   glycerin

Glycerol is used in the manufacture of plastics, drugs, cosmetics, inks, food products, and nitroglycerin.

12-9 ALDEHYDES AND ACIDS

Alcohols that have the hydroxy group attached to a carbon, which is in turn bonded to only one other carbon ($R{-}CH_2OH$), are called **primary alcohols.** When a primary alcohol reacts with certain substances called oxidizing reagents, the $-CH_2OH$ grouping can be converted to an **aldehydo group** (see next page).

$$-C\overset{\displaystyle O}{\underset{\displaystyle H}{\diagup}}$$

This chemical reaction can be represented as

$$RCH_2OH + \text{oxidizing reagent} \longrightarrow R-C\overset{\displaystyle O}{\underset{\displaystyle H}{\diagup}}$$

The compounds that contain an aldehydo group and correspond to formula

$$R-C\overset{\displaystyle O}{\underset{\displaystyle H}{\diagup}} \quad \text{or} \quad RCHO$$

are called **aldehydes.** A few typical aldehydes are described below.

$$\overset{\displaystyle H}{\underset{\displaystyle H}{>}}C{=}O \quad \text{or} \quad CH_2O \quad \text{formaldehyde}$$

Formaldehyde is used for the manufacture of plastics such as Formica. Water solutions of formaldehyde, called formalin, are used as disinfectants and to preserve tissue.

$$CH_3-C\overset{\displaystyle O}{\underset{\displaystyle H}{\diagup}} \quad \text{or} \quad CH_3CHO \quad \text{acetaldehyde}$$

Acetaldehyde is used in the manufacture of plastics and for some medical purposes.

Aldehydes react with certain oxidizing reagents to convert the aldehydo group to a

$$\overset{\displaystyle O}{\overset{\displaystyle \|}{-C-OH}}$$

called a **carboxylic** group:

$$R-C\overset{\displaystyle O}{\underset{\displaystyle H}{\diagup}} + \text{oxidizing reagent} \longrightarrow R-C\overset{\displaystyle O}{\underset{\displaystyle OH}{\diagup}}$$

The compounds that include a carboxylic group and correspond to the formula

$$R-C\overset{\displaystyle O}{\underset{\displaystyle OH}{\diagup}} \quad \text{or} \quad RCOOH$$

are called the **carboxylic acids.** Some carboxylic acids are

$$HC\overset{\displaystyle O}{\underset{\displaystyle OH}{\diagup}} \qquad \text{or} \qquad HCOOH \qquad \text{formic acid}$$

$$CH_3C\overset{\displaystyle O}{\underset{\displaystyle OH}{\diagup}} \qquad \text{or} \qquad CH_3COOH \qquad \text{acetic acid}$$

Vinegar is a dilute aqueous solution of acetic acid.

$$CH_3CH_2C \underset{OH}{\overset{O}{\diagup}} \qquad or \qquad CH_3CH_2COOH \qquad propionic\ acid$$

$$CH_3CH_2CH_2C \underset{OH}{\overset{O}{\diagup}} \qquad or \qquad CH_3CH_2CH_2COOH \qquad butyric\ acid$$

12-10 ESTERS

A carboxylic acid can react with an alcohol to form a product called an **ester**:

$$R-C \underset{OH}{\overset{O}{\diagup}} + ROH \xrightarrow{catalyst} R-C \underset{O-R}{\overset{O}{\diagup}} + H_2O$$

$$(ester)$$

Name	Structure	Source or flavor
Ethyl formate	$CH_3CH_2-O-\overset{O}{\overset{\|}{C}}-H$	Rum
Isobutyl formate	$\underset{}{CH_3}\!\!\overset{CH_3}{\underset{}{CHCH_2}}-O-\overset{O}{\overset{\|}{C}}-H$	Raspberries
Ethyl acetate	$CH_3CH_2-O-\overset{O}{\overset{\|}{C}}CH_3$	Used in lacquers
n-Pentyl acetate (n-amyl acetate)	$CH_3(CH_2)_4-O-\overset{O}{\overset{\|}{C}}CH_3$	Bananas
Isopentyl acetate (isoamyl acetate)	$\overset{CH_3}{\underset{}{CH_3CHCH_2CH_2}}-O-\overset{O}{\overset{\|}{C}}CH_3$	Pears
n-Octyl acetate	$CH_3(CH_2)_7-O-\overset{O}{\overset{\|}{C}}CH_3$	Oranges
Ethyl butyrate	$CH_3CH_2-O-\overset{O}{\overset{\|}{C}}CH_2CH_2CH_3$	Pineapples
n-Pentyl butyrate	$CH_3(CH_2)_4-O-\overset{O}{\overset{\|}{C}}CH_2CH_2CH_3$	Apricots
"Waxes"	$CH_3(CH_2)_n-\overset{O}{\overset{\|}{C}}-O-(CH_2)_nCH_3$	$n = 23–33$, carnauba wax $n = 25–27$, beeswax $n = 14–15$, spermaceti

Table 12-5
Some Esters

From *Introduction to Chemistry* by T. R. Dickson, John Wiley & Sons, New York, 1971.

The RO– grouping of the alcohol replaces the –OH of the acid. For example,

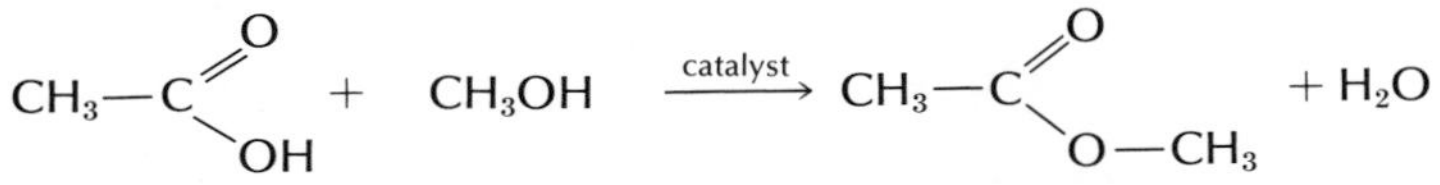

$$CH_3-C\!\!\begin{array}{c}\nearrow^{O}\\\searrow_{OH}\end{array} \quad + \quad CH_3OH \quad \xrightarrow{\text{catalyst}} \quad CH_3-C\!\!\begin{array}{c}\nearrow^{O}\\\searrow_{O-CH_3}\end{array} \quad + H_2O$$

    acetic acid   methyl alcohol   methyl acetate (an ester)

Many esters occur naturally and are often responsible for some of the tastes and odors of fruits. Table 12-5 lists some typical esters of this type. Esters can also be formed between noncarboxylic acids and alcohols. It is possible to have phosphate esters derived from phosphoric acid:

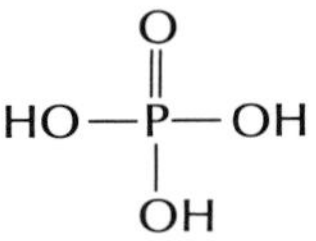

$$\begin{array}{c}O\\\|\\HO-P-OH\\|\\OH\end{array}$$

Two examples of such phosphate esters are

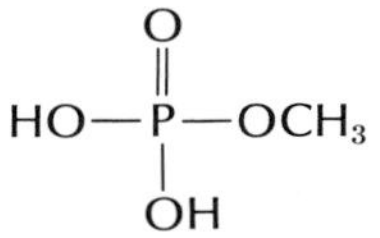

$$\begin{array}{c}O\\\|\\HO-P-OCH_3\\|\\OH\end{array} \quad \text{methyl phosphate}$$

used as a gasoline additive to control the ignition process in the engine and

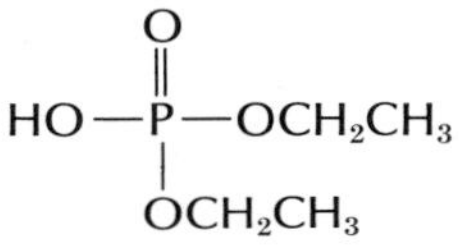

$$\begin{array}{c}O\\\|\\HO-P-OCH_2CH_3\\|\\OCH_2CH_3\end{array} \quad \text{diethyl phosphate}$$

12-11 AMINES

The structural formula of ammonia is

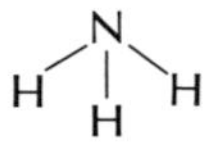

$$H\!\!\begin{array}{c}^{N}\\|\\H\end{array}\!\!H$$

Compounds in which an –H of ammonia is replaced by an alkyl group, –R, are called **amines.**

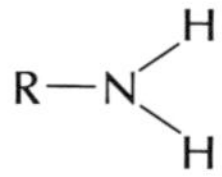

$$R-N\!\!\begin{array}{c}\nearrow^{H}\\\searrow_{H}\end{array} \quad \text{or} \quad R-NH_2 \quad \text{or} \quad RNH_2$$

The –NH$_2$ group is called the **amino** group. Amines with one alkyl group are called **primary amines.** Amines with two alkyl groups (which need not be alike) are called **secondary amines,** and those with three alkyl groups are called **tertiary amines.**

$$\begin{array}{c}R-N-H\\|\\R\end{array} \qquad\qquad \begin{array}{c}R-N-R\\|\\R\end{array}$$

     secondary amine    tertiary amine

Some typical amines are

$$CH_3CH_2NH_2 \quad \text{ethylamine}$$

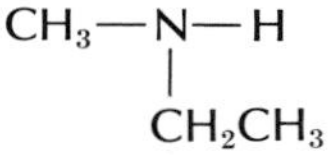

$$\begin{array}{c}CH_3-N-H\\|\\CH_2CH_3\end{array} \quad \text{methylethylamine}$$

$$CH_3—N—CH_3 \quad \text{trimethylamine}$$
$$\overset{|}{CH_3}$$

A fourth alkyl group on the nitrogen occurs in ions called **quaternary ammonium ions:**

$$\overset{\displaystyle R}{\underset{\displaystyle R}{R—\overset{|}{\underset{|}{N}}—R}} \quad +$$

A quaternary ammonium ion is analogous to ammonium ion, NH_4^+. Two quaternary ammonium ions that are important in some biological processes are

$$\overset{+}{CH_3}—\overset{\displaystyle CH_3}{\underset{\displaystyle CH_3}{\overset{|}{\underset{|}{N}}}}—CH_2CH_2OH \quad \text{choline}$$

$$\overset{+}{CH_3}—\overset{\displaystyle CH_3}{\underset{\displaystyle CH_3}{\overset{|}{\underset{|}{N}}}}—CH_2CH_2—O—\overset{\displaystyle O}{\overset{||}{C}}—CH_3 \quad \text{acetyl choline}$$

Note that acetyl choline is an ester of choline and acetic acid. These two ions are involved in life processes such as growth, metabolism, and nerve impulse transmission.

12-12 AMIDES

Under certain conditions, ammonia and some amines can combine with carboxylic acids in a manner which is quite similar to the formation of an ester:

$$R—\overset{\displaystyle O}{\overset{||}{C}}\boxed{—OH \quad H}—\overset{|}{\underset{|}{N}}—H \longrightarrow R—\overset{\displaystyle O}{\overset{||}{C}}—\overset{|}{\underset{|}{N}}—H + H_2O$$
$$\qquad\qquad\qquad\quad H \qquad\qquad\qquad H$$
$$\text{amide}$$

The product of such a reaction is called an **amide;** the linking of an amine and an acid is called an **amide linkage.** As we shall see in Section 13-4, amide linkages are involved in the formation of proteins. A few examples of amides are

$$CH_3—\overset{\displaystyle O}{\overset{||}{C}}—NH_2 \quad \text{acetamide}$$

$$CH_3CH_2CH_2\overset{\displaystyle O}{\overset{||}{C}}—NH_2 \quad \text{butyramide}$$

$$CH_3\overset{\displaystyle O}{\overset{||}{C}}—\overset{\displaystyle H}{\overset{|}{N}}—CH_3 \quad \textit{N}\text{-methylacetamide}$$

The *N*-preceding the methyl indicates that the methyl group is attached to the nitrogen of the amide.

12-13 AMINO ACIDS

A class of compounds that are important in biochemistry is the amino acids. The common **amino acids** are carboxylic acids with an amino group bonded to the carbon next to the carboxylic group carbon:

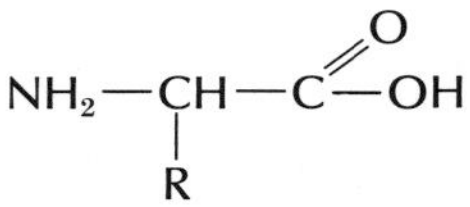

$$NH_2-CH-C\overset{O}{\underset{R}{\diagup}}OH$$

The –R group may be an alkyl group (i.e., –CH$_3$) or a substituted alkyl group (i.e., –CH$_2$OH). Various amino acids differ only in the structure of the –R group. The common amino acids that are found in nature have been given common names. For example,

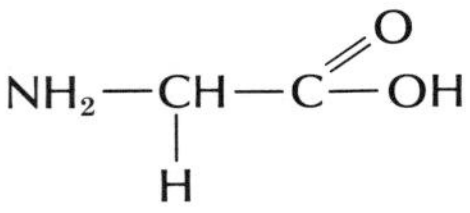

$$NH_2-CH-C\overset{O}{\diagup}OH \qquad \text{glycine}$$
$$\underset{H}{|}$$

$$NH_2-CH-C\overset{O}{\diagup}OH \qquad \text{alanine}$$
$$\underset{CH_3}{|}$$

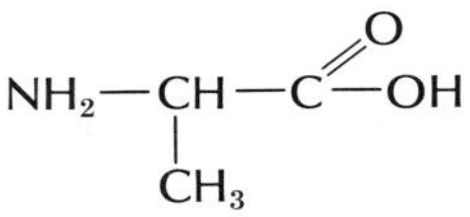

Table 12-6
Common Amino
Acids

All the amino acids except proline and hydroxyproline have the general formula

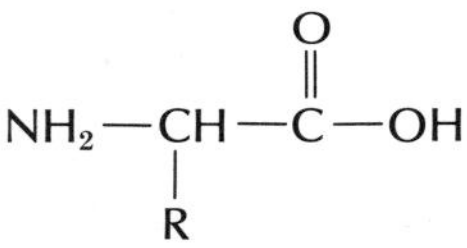

$$NH_2-CH-\overset{O}{\overset{\|}{C}}-OH$$
$$\underset{R}{|}$$

in which R is the characteristic group for each acid. The R groups, names, and abbreviations[a] are as follows.

1. Glycine	—H	Gly
2. Alanine	—CH$_3$	Ala
3. Serine	—CH$_2$OH	Ser
4. Cysteine	—CH$_2$SH	Cys
5. Cystine	—CH$_2$—S—S—CH$_2$—	Cys—S—S—Cys[b]

6. Threonine[c]	—CH—CH$_3$
	    |
	   OH
		Thr

7. Valine[c]	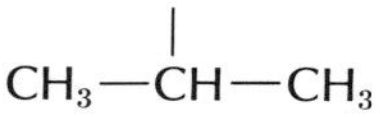 CH$_3$—CH—CH$_3$	Val

8. Leucine[c]	—CH$_2$—CH—CH$_3$
	       |
	     CH$_3$
		Leu

9. Isoleucine[c]	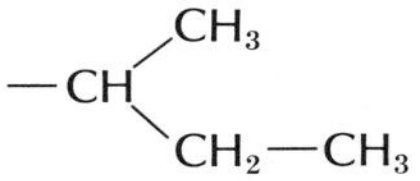 —CH$\overset{\diagup CH_3}{\diagdown CH_2-CH_3}$	Ile
10. Methionine[c]	—CH$_2$—CH$_2$—S—CH$_3$	Met
11. Aspartic acid	—CH$_2$CO$_2$H	Asp
12. Glutamic acid	—CH$_2$—CH$_2$—CO$_2$H	Glu
13. Lysine[c]	—CH$_2$—CH$_2$—CH$_2$—CH$_2$—NH$_2$	Lys
14. Arginine	—CH$_2$—CH$_2$—CH$_2$—NHCNHNH$_2$	Arg

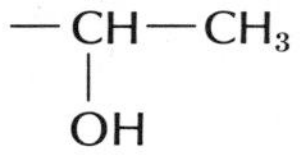

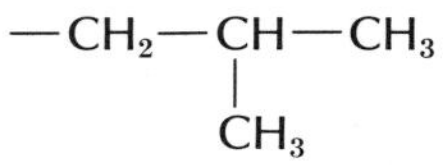

The names and structures of some other amino acids are given in Table 12-6. As we shall see in Section 13-4, the amino acids are the building units of proteins found in living organisms. Since an amino acid can act either as an acid or an amine, amino acids can link together through amide linkages.

12-14   BENZENE AND CYCLIC COMPOUNDS

In some hydrocarbons, the carbon sequence is one in which the carbons form a ring. These hydrocarbons are called **cyclic,** or **ring,** compounds. For example, cyclohexane can be represented by the formula

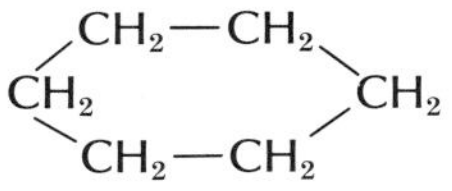

$$
\begin{array}{c}
CH_2\!-\!CH_2 \\
CH_2 \qquad\quad CH_2 \\
CH_2\!-\!CH_2
\end{array}
$$

Another example is cyclopentane:

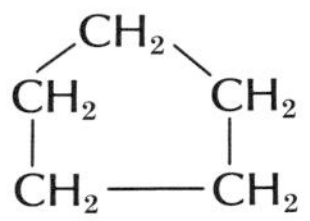

$$
\begin{array}{c}
CH_2 \\
CH_2 \qquad CH_2 \\
CH_2\!-\!CH_2
\end{array}
$$

Often, these cyclic compounds are represented for convenience by a geometrical figure representing the carbon sequence as shown on the next page.

15. Phenylalanine[c]	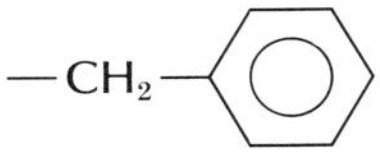 $-CH_2-$ ◯	Phe
16. Tyrosine	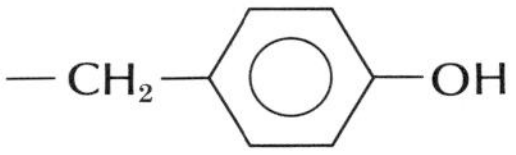 $-CH_2-$ ◯ $-OH$	Tyr
17. Tryptophan[c]	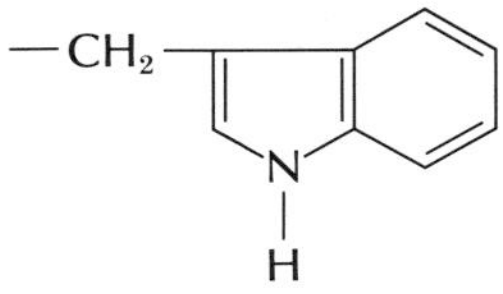 $-CH_2-$ (indole ring) N—H	Trp
18. Histidine	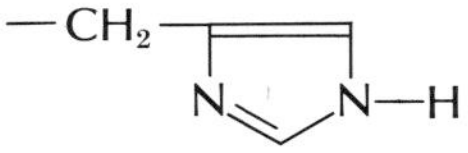 $-CH_2-$ (imidazole ring) N N—H	His
19. Proline	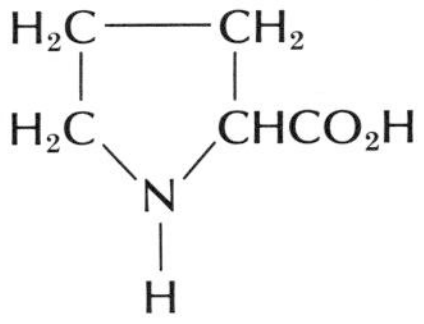 $H_2C\!-\!CH_2$ $H_2C\quad CHCO_2H$ N—H	Pro
20. Hydroxyproline	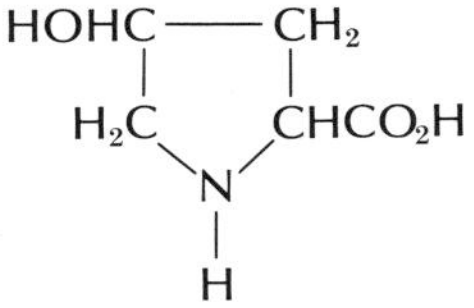 $HOHC\!-\!CH_2$ $H_2C\quad CHCO_2H$ N—H	Hyp

Table 12-6
(*Continued*)

From *Introduction to Chemistry* by T. R. Dickson, John Wiley & Sons, New York, 1971.
[a] These are the official IUPAC symbols for amino acids.
[b] Cystine involves two cysteine units joined by disulfide linkage (—S—S—).
[c] Essential amino acids needed in the diet of humans.

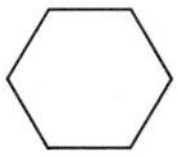

represents $\underset{\underset{CH_2-CH_2}{\diagdown}}{\overset{\overset{CH_2-CH_2}{\diagup}}{CH_2}}\; CH_2$ cyclohexane

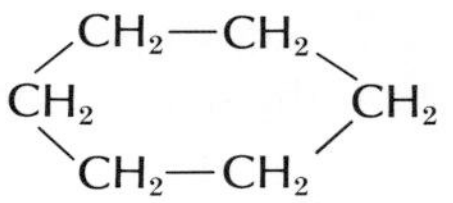

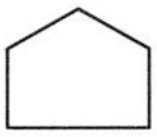

represents $\underset{\underset{CH_2-CH_2}{\vert}}{\overset{\overset{CH_2}{\diagup\diagdown}}{CH_2}}\; CH_2$ cyclopentane

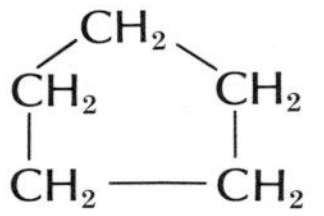

Substituted ring compounds can be represented using the figure with the group which replaced an –H connected to the figure. For instance, the alcohol cyclohexanol, $C_6H_{11}OH$, can be represented as

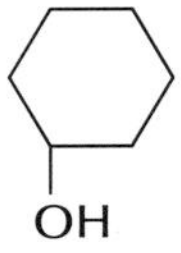

A very prevalent and important ring compound is **benzene**, C_6H_6, which can be represented by the structural formulas

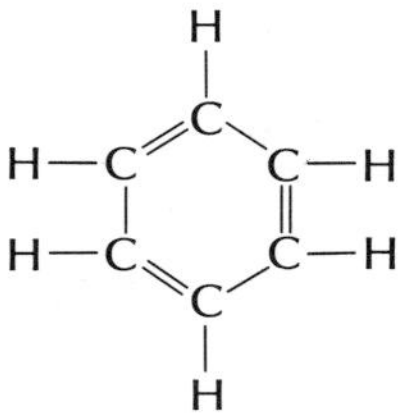

and

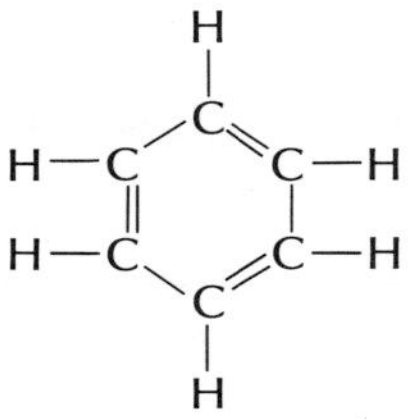

Benzene is an unusually stable ring compound, and many important substances are formed by substituting another group or other groups for hydrogens on the benzene ring. Notice that the structural formulas given above indicate that benzene has alternating single and double bonds around the ring. Either of the two arrangements shown could be used to represent benzene. Actually, these two formulas are different. One shows the double bonds in one possible arrangement, and the other shows the double bonds in another arrangement. However, studies have found that all the carbon–carbon bonds in benzene are the same. Thus, neither of the structures represents the actual bonding electron distribution in the benzene ring. The actual structure is something in between these two extremes. The actual structure is said to be a resonance hybrid between these two contributing structures. Then, how should we represent benzene? One possible way is to draw the structure to indicate that the carbon–carbon bonds are neither completely single nor completely double:

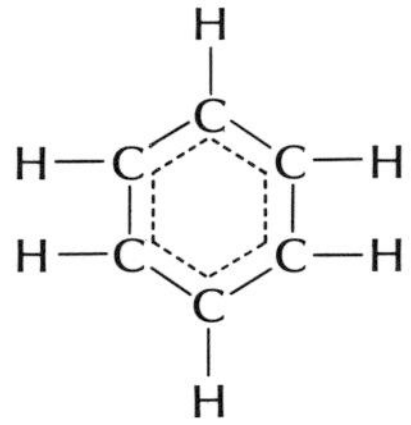

However, since it is inconvenient to write the structure every time we want to represent benzene, it is possible to use a geometrical figure as was done with the cyclic alkanes. The symbol used to represent benzene as a resonance hybrid is shown on the next page.

This symbol should be interpreted as representing the structure of benzene, C_6H_6. Sometimes, benzene is represented by similar symbols corresponding to the resonance contributing forms:

Benzene is used as a starting material to prepare many useful compounds. Over 4.5 million tons of benzene are produced annually in the United States. Compounds involving a benzene ring are sometimes called **aromatic compounds.** Substituted benzene compounds can be represented by showing the substituted group attached to the ring. For example,

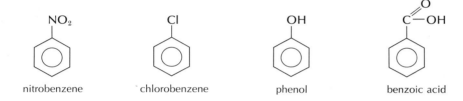

| nitrobenzene | chlorobenzene | phenol | benzoic acid |

When more than one substituted group appears on the ring, the positions on the ring are distinguished by numbering the carbons

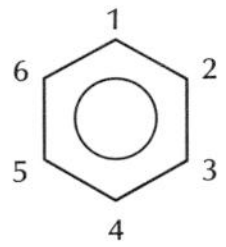

or by using the following notation:

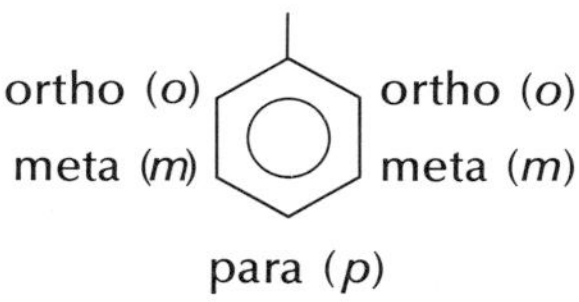

ortho (*o*)   ortho (*o*)

meta (*m*)   meta (*m*)

para (*p*)

To illustrate, consider the following compounds:

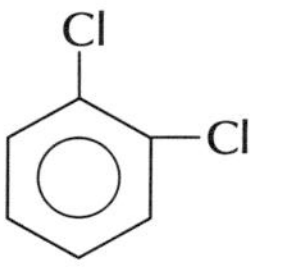

ortho-dichlorobenzene	meta-dichlorobenzene	para-dichlorobenzene
o-dichlorobenzene	*m*-dichlorobenzene	*p*-dichlorobenzene
1,2-dichlorobenzene	1,3-dichlorobenzene	1,4-dichlorobenzene

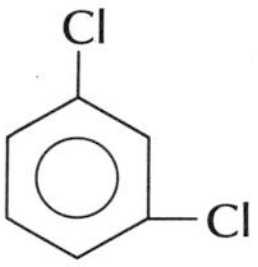

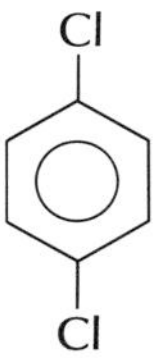

A few typical aromatic compounds are listed in Table 12-7.

Table 12-7
Some Typical
Aromatic Compounds

Formula	Name	Comment
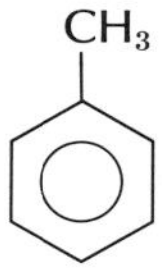 CH_3 (on benzene ring)	Toluene	Used to prepare TNT, solvent
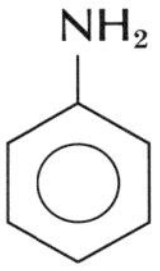 NH_2 (on benzene ring)	Aniline	Used in dyes and drugs
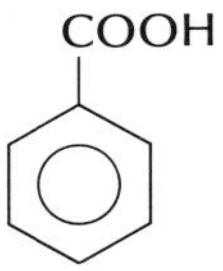 COOH (on benzene ring)	Benzoic acid	Used in food preservatives
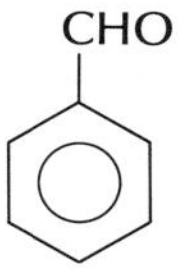 CHO (on benzene ring)	Benzaldehyde	Flavoring agent, "oil of almond"
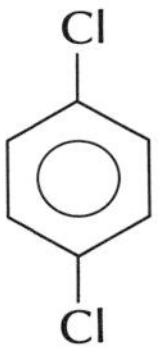 Cl (on benzene ring, Cl)	para-Dichlorobenzene	Moth balls
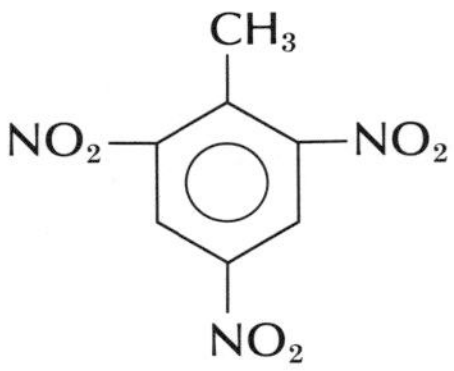 CH_3, NO_2—ring—NO_2, NO_2	2,4,6-Trinitrotoluene	TNT, an explosive

12-15 POLYCYCLIC AND HETEROCYCLIC COMPOUNDS

Compounds such as chlolesterol,

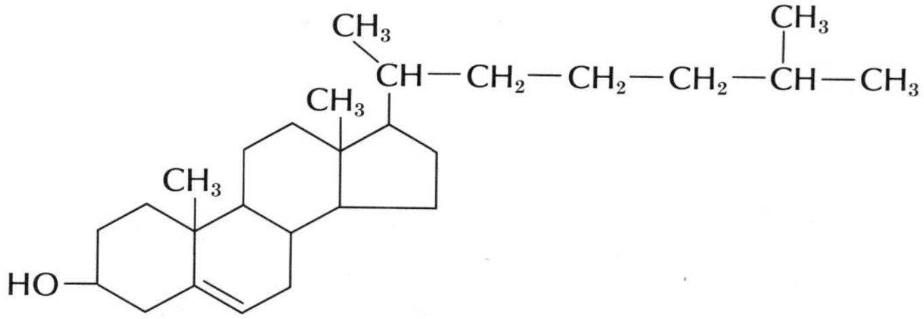

in which two or more rings mutually share carbons are called **polycyclic** (many-ring) compounds. Polycyclic compounds are common in biochemistry. Some

Table 12-7
(Continued)

Formula	Name	Comment
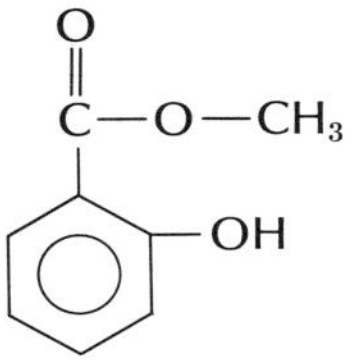	Methyl salicylate	Oil of wintergreen; used in rubbing ointments
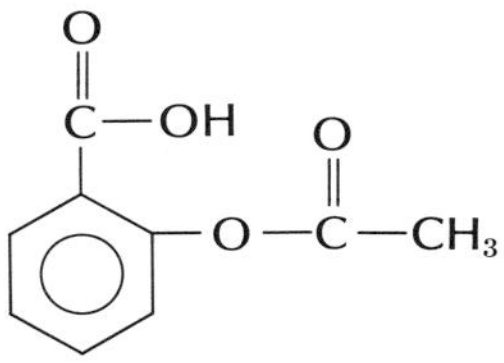	Acetylsalicylic acid	Aspirin; analgesic (pain reliever)
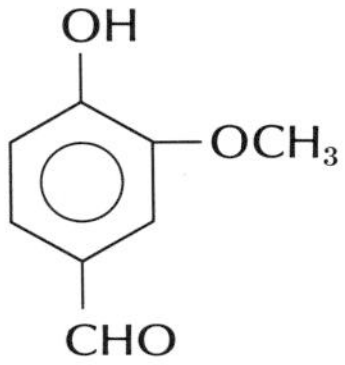	Vanillin	Vanilla flavoring
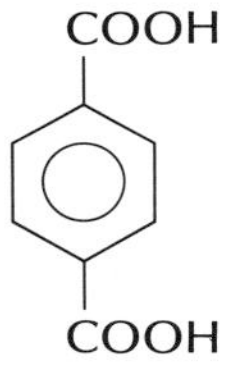	Terephthalic acid	Used in the manufacture of polyesters such as Dacron, Fortrel, and Mylar

From *Introduction to Chemistry* by T. R. Dickson, John Wiley & Sons, New York, 1971.

cyclic compounds such as

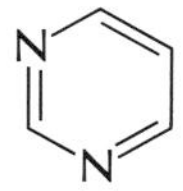

pyrimidine 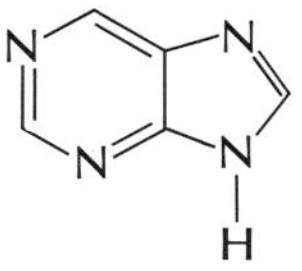   purine

have a few atoms other than carbon in the rings. Such cyclic compounds that include carbon and some other type of atom in the ring or rings are called **heterocyclic** compounds. Some important heterocyclic compounds are given below and on the next page.

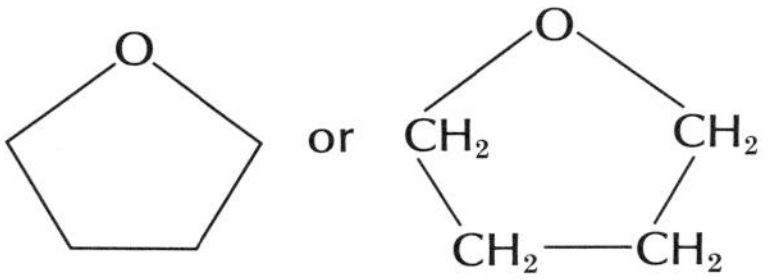

tetrahydrofuran 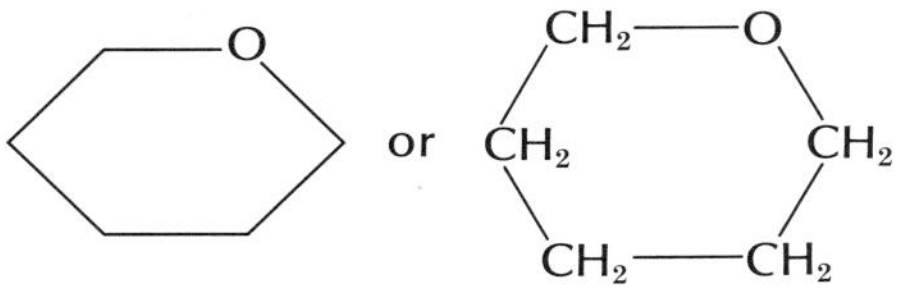   tetrahydropyran

cytosine   uracil   thymine   adenine   guanine

The above compounds are the heterocyclic bases or heterocyclic amines, discussed in Section 13-8, which are involved in biologically important ribonucleic acids (RNA) and deoxyribonucleic acids (DNA).

Only a few types of organic compounds have been discussed in this chapter. Organic chemistry includes the study of numerous other kinds of compounds and the chemical reactions they undergo. The organic chemistry considered in previous sections is summarized in Table 12-8.

12-16 SOME INDUSTRIAL CHEMICALS

Numerous chemicals are manufactured in our industrial society. Some are produced from raw materials found in the lithosphere and atmosphere, and others are produced from petroleum products. A few of the fundamental **industrial chemicals** are discussed below.

SULFUR Sulfur is among the most widely used industrial elements. Sulfur is found extensively in the crust of the earth both in the free elemental state and in

Table 12-8
Summary of Organic Chemicals

Group	General formula	Example
Alkane	C_nH_{2n+2}	$CH_3CH_2CH_3$, propane
Alkenes	C_nH_{2n}	$CH_2{=}CH_2$, ethylene
Alcohols	ROH	CH_3CH_2OH, ethyl alcohol
Aldehydes	RCHO	CH_3CHO, acetaldehyde
Acids	RCO_2H	CH_3CO_2H, acetic acid
Esters	RCO_2R	$CH_3CO_2CH_2CH_3$, ethyl acetate
Amines	RNH_2, R_2NH, R_3N	CH_3NH_2, methylamine
Amides	RCO_2NH_2	$CH_3CO_2NH_2$, acetamide
Amino acids	NH_2CHCO_2H $\mid$ R	$NH_2CH_2CO_2H$, glycine
Cyclic hydrocarbons	C_nH_{2n}	cyclohexane
Benzene	C_6H_6	
Aromatics	Substituted benzene	chlorobenzene
Heterocyclics	Element other than C in ring	pyridine

the combined state in minerals such as galena, PbS, pyrite, FeS_2 (fool's gold), and gypsum, $CaSO_4 \cdot 2\,H_2O$. It is also found as sulfur dioxide in sour natural gas and petroleum. Much sulfur is obtained in the United States by extracting free sulfur from deposits in which it is found as solid elemental sulfur. Sulfur is a brittle, yellow, solid nonmetal that melts at 113°C. In the solid state sulfur occurs in the form of S_8 molecules in which the sulfur atoms are bonded in a ring of the form

$$
\begin{array}{c}
\text{S} \quad \text{S} \quad \text{S} \\
\text{S} \quad \text{S} \\
\text{S} \quad \text{S}
\end{array}
$$

The most stable solid form of these molecules is the **rhombic** crystalline form, but they can also be arranged in the form of **monoclinic** crystals. (See Figure 12-7.) These two common crystalline forms of sulfur are called **allotropic** forms. Sulfur is used as a component of black gunpowder, as a fungicide, and in the manufacture of sulfuric acid, matches, and rubber.

Around 30 million tons of sulfuric acid are manufactured annually in the United States. It is used in the production of phosphate fertilizers, rayon, paper, detergents, dyes, plastics, paints, ammonium sulfate, and other chemicals. It is also used in the iron, steel, and petroleum industries. In the United States, most sulfuric acid is prepared by the **contact process** in which sulfur is reacted with oxygen to form sulfur dioxide gas:

$$ S + O_2 \longrightarrow SO_2 $$

The sulfur dioxide gas is made to react with more oxygen in the presence of a catalyst to produce sulfur trioxide gas:

$$ 2\,SO_2 + O_2 \xrightarrow{\;V_2O_5\;} 2\,SO_3 $$

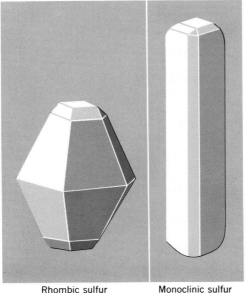

Rhombic sulfur
(diamond-shaped crystals)

Monoclinic sulfur
(needle-like crystals)

(a)   (b)

Figure 12-7
Two crystalline forms of sulfur. From *Introduction to Chemistry* by T. R. Dickson, John Wiley & Sons, New York, 1971.

The sulfur trioxide gas then combines with water to produce concentrated sulfuric acid:

$$SO_3 + H_2O \longrightarrow H_2SO_4$$

NITROGEN Large amounts of elemental nitrogen exist in the atmosphere. To make this nitrogen biologically and chemically useful, it is necessary to combine it with other elements. This combination process is called the fixation of the nitrogen. Some nitrogen is fixed by natural processes, but, in order to provide sufficient nitrogen-containing fertilizers, it is necessary to industrially fix large amounts of nitrogen. The most important manner in which atmospheric nitrogen is fixed is the **Haber** ammonia manufacturing process. In this process, elemental nitrogen and hydrogen are combined at high temperatures and pressures in the presence of a catalyst to form ammonia, NH_3. Over 13 million tons of ammonia are produced in the United States each year.

Ammonia is used in the manufacture of fertilizers, explosives, textiles, plastics, and in the manufacture of nitric acid. Nitric acid is produced by reacting ammonia with oxygen according to the equation

$$4\ NH_3 + 5\ O_2 \xrightarrow{\text{Pt}} 4\ NO + 6\ H_2O$$

The nitrogen oxide is then allowed to react with more oxygen to give nitrogen dioxide, which is in turn allowed to react with water to produce concentrated nitric acid:

$$2\ NO + O_2 \longrightarrow 2\ NO_2$$
$$3\ NO_2 + H_2O \longrightarrow 2\ HNO_3 + NO$$

Annual United States production of nitric acid is some 6.5 million tons. Nitric acid is used mainly in the manufacture of nitrate fertilizers and explosives.

ORGANICS One of the most widely used (12 million tons annually) organic chemicals in industry is ethylene, C_2H_4, which is manufactured from ethane, C_2H_6, obtained from petroleum. Ethylene is formed according to the dehydrogenation reaction given by the equation

$$C_2H_6 \longrightarrow C_2H_4 + H_2$$

The variety of substances produced from ethylene are indicated below.

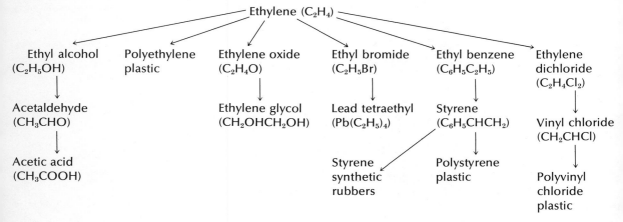

Another useful organic chemical is acetylene, C_2H_2, which is manufactured by heating CaO (lime) and C (coke) in an electric furnace to produce calcium carbide, CaC_2:

$$CaO + 3\ C \longrightarrow CaC_2 + CO$$

Calcium carbide is then combined with water to give acetylene gas:

$$CaC_2 + 2\ H_2O \longrightarrow Ca(OH)_2 + C_2H_2$$

The uses of acetylene are summarized below.

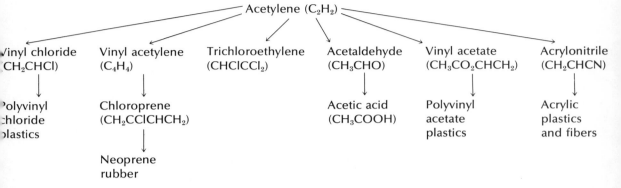

Acetylene (C_2H_2)

| Vinyl chloride (CH_2CHCl) | Vinyl acetylene (C_4H_4) | Trichloroethylene ($CHClCCl_2$) | Acetaldehyde (CH_3CHO) | Vinyl acetate ($CH_3CO_2CHCH_2$) | Acrylonitrile (CH_2CHCN) |

Polyvinyl chloride plastics

Chloroprene ($CH_2CClCHCH_2$)

Acetic acid (CH_3COOH)

Polyvinyl acetate plastics

Acrylic plastics and fibers

Neoprene rubber

BOOKS

BIBLIOGRAPHY

Briston, J. H. *Introduction to Plastics.* New York: Philosophical Library, 1968.
Holum, J. R. *Elements of General and Biological Chemistry.* 3rd ed. New York: Wiley, 1972.
Melville, H. *Big Molecules.* New York: Macmillan, 1958.
Pauling, L., and R. Hayward. *The Architecture of Molecules.* San Francisco: W. H. Freeman, 1964.
Read, J. *A Direct Entry to Organic Chemistry.* New York: Harper & Row, 1960.
Wendland, R. *Petrochemicals: The New World of Synthetics.* Garden City, N.Y.: Doubleday, 1969.

ARTICLES AND PAMPHLETS

"Can Plastics be Incinerated Safely?" *Environmental Science and Technology* (Aug. 1971).
Morton, M. "Big Molecules." *Chemistry* 13–17 (Jan. 1964).
Natta, G. "Precisely Constructed Polymers." *Scientific American* **205,** 33 (Aug. 1961).
Rossini, F. D. "Hydrocarbons in Petroleum." *Journal of Chemical Education* **37** (11), 554 (1960).
Smith, C. S. "Materials." *Scientific American* **217,** (Sept. 1967).
"Teflon—From Nonstick Frying Pans to Space Vehicles." *Chemistry* 21–22 (June 1965).

1. What is organic chemistry?

QUESTIONS AND PROBLEMS

2. What are synthetic organic chemicals?

3. How many bonds does carbon normally form in organic compounds?

4. What unique property of carbon results in the large variety of organic compounds?

5. What is an alkane? Give an example.

6. Define the term isomer.

7. What is an alkene? Give an example.

8. Describe a polymerization reaction using the terms monomer and polymer.

9. Give two examples of common plastics used in commercial products.

10. What are some of the environmental problems caused by plastics?

11. Describe the chemical nature and give an example of each of the following classes of organic compounds:

    (a) alcohol   (e) amine
    (b) aldehyde   (f) amide
    (c) carboxylic acid   (g) amino acid
    (d) ester

12. What is a cyclic hydrocarbon? Give an example.

13. What symbolic formula is used to represent benzene?

14. What is an aromatic compound? Give an example.

15. What is a heterocyclic compound?

16. Describe the chemistry of the manufacture of sulfuric acid.

17. What two common industrial chemicals are produced from nitrogen from the atmosphere?

18. Which two organic compounds are among the most widely used industrial chemicals?

19. Keep a list of the plastic and other synthetic organic chemicals which are involved in your life for one day. (See Table 12-3.)

13-1   BIOCHEMISTRY

Certain organic and inorganic compounds are constituents of animal and plant life. Many of these compounds are fundamental to the processes of life. The study of compounds involved in biological processes and the changes they undergo during life processes is called **biochemistry.** The field of biochemistry is one of the frontiers of science. Many scientifically exciting discoveries involving life processes are being made by biochemists. Such discoveries range from the establishment of a theory to explain the chemistry of heredity to the development of drugs and vaccines to cure disease.

Biochemists study the chemical aspects of living organisms. Living organisms are a complex mixture of various kinds of chemical compounds which act in concert to maintain the life of the organism. Considering the great diversity of plants and animals, it is surprising that many of the same chemical compounds and chemical processes are shared by all life forms. In this chapter, we will study some biochemically important substances and consider some of the life processes of humans.

Many chemical elements are involved in the biochemical compounds. However, since most of the biochemical compounds are organic, only a few elements make up most of these compounds. Combined hydrogen, oxygen, carbon, and nitrogen make up about 99 percent of the atoms of living organisms. Calcium, chlorine, magnesium, phosphorus, potassium, sodium, and sulfur make up most of the remaining 1 percent. Many other elements are present in small amounts as trace elements. (See Section 14-8.)

Before considering the details of some topics in biochemistry, let us take a broad overview of this chapter to establish some terminology.

Carbohydrates, or **sugars,** are the biochemical compounds produced mainly by plants. These compounds serve as food energy sources for humans. You are probably most familiar with table sugar (sucrose), but starch and cellulose are also carbohydrates. The chemical nature of carbohydrates is described in Section 13-2.

Lipids are another type of biochemical compound. Simple lipids make up animal fats and vegetable oils. Humans use lipids as a food source in such products as cooking oils, margarines, butter, and lard. Lipids will be discussed in Section 13-3.

Proteins are chemicals which serve as structural components of animal organisms and are involved in many life processes. Proteins are biological polymers composed of amino acids. Proteins are also an important food source for humans. The chemical nature of proteins is considered in Section 13-4.

Enzymes are biological catalysts which play an important part in biochemical reactions. The functions and naming of enzymes are considered in Sections 13-5 and 13-6.

The **nucleic acids** are biochemical polymers which play a fundamental part in the function of the cells that comprise living organisms. Nucleic acids are involved in the transmission of inherited characteristics passed on from the parents to the children. In Section 13-8, we will discuss these chemicals of heredity.

The remainder of the chapter deals with certain life processes involving blood, electrolytes, hormones, digestion, and metabolism.

13-2 CARBOHYDRATES

Carbohydrates, or sugars, are produced by green plants during photosynthesis. Carbohydrates contain only carbon, hydrogen, and oxygen. They occur in a variety of sizes ranging from simple sugars called **monosaccharides** to polymer molecules called **polysaccharides.** Most of the common monosaccharides are polyhydroxy aldehydes. For example, the most important monosaccharides are **ribose** and **glucose** (sometimes called dextrose):

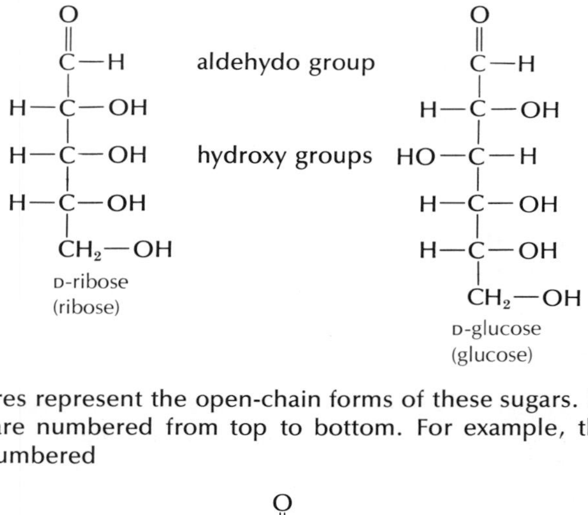

These structures represent the open-chain forms of these sugars. For reference, the carbons are numbered from top to bottom. For example, the carbons in glucose are numbered

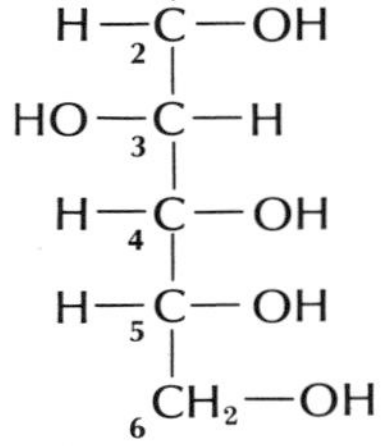

This monosaccharide very often occurs as a ring compound. These rings are usually formed by intramolecular reaction between the aldehydo group and the –OH group on the next-to-last carbon to form a ring. The two ring forms of

glucose can be pictured as

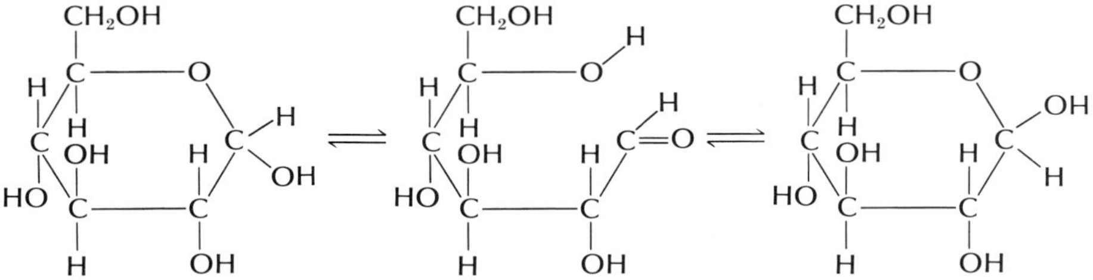

The –OH on carbon 1 is above the ring in one form and below the ring in the other form. These cyclic forms can be more conveniently written as a heterocyclic ring leaving the ring carbons out. Thus, the two forms of glucose can be represented as

β-D-glucose

[β(beta) form C-1—OH above ring]

α-D-glucose

[α(alpha) form C-1—OH below ring]

The biochemically important cyclic form of ribose involves a five-atom ring and can be represented as

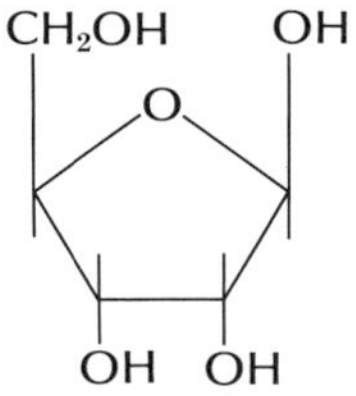

ribose

A similar cyclic compound of biochemical importance is one in which the –OH group on carbon number two has been replaced by a hydrogen. In other words, the oxygen of the –OH group can be considered to have been removed. This compound is called **deoxyribose** and can be represented as

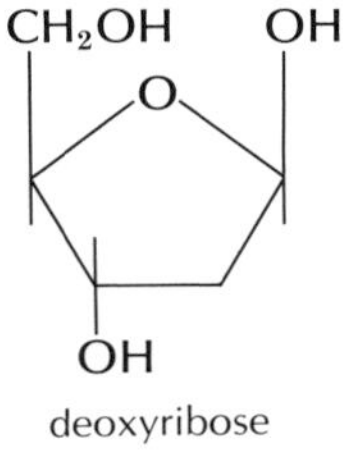

deoxyribose

As we shall see in Section 13-8, ribose and deoxyribose are involved in the structure of DNA and RNA.

Glucose in the combined form or as the free monosaccharide is undoubtedly the most abundant organic compound in nature. It is an important food source (a quick energy source) which is found in plants, fruits, and vegetables, and is present in the bloodstream of certain animals. Two glucose molecules can link together through certain –OH groups to form the compound **maltose**, a disaccharide (two monosaccharides joined):

CH$_2$OH   CH$_2$OH   CH$_2$OH   CH$_2$OH

α-glucose α-glucose α-maltose

The two α-glucose units are linked through carbon one and carbon four by what is called an α linkage. Another disaccharide of interest is **sucrose** (table sugar) which involves α-glucose linked to the β form of the five-membered ring of the monosaccharide called fructose:

CH$_2$OH   CH$_2$OH   CH$_2$OH   CH$_2$OH

α-glucose β-fructose   sucrose

Table 13-1 lists some important carbohydrates.

Many glucose units can bond together to form polymeric molecules known as **polysaccharides.** These polysaccharides are found in plants and animals. The polysaccharides involve long chains of many (hundreds and thousands) glucose units. **Starch** is a polysaccharide found in the seeds and roots of many plants. (See Figure 13-1.) It can be digested and is used as a food. Actually, there are two forms of starch that occur in plants. **Amylose** consists of long chains of α-glucose units. **Amylopectin** consists of branched chains of α-glucose units. **Glycogen** is a polysaccharide which is found in animals. Carbohydrates are stored in the body as glycogen. The structure of glycogen is similar to that of amylopectin. Another important polysaccharide is **cellulose,** which makes up the structural material of many plants. (See Figure 13-2.) Cellulose consists of long chains of β-glucose units. It cannot be digested by humans, but it can be broken down by certain microorganisms. Cellulose is obtained in large amounts from trees and the cotton plant. It is used in large amounts in the manufacture of paper, cotton cloth, cellophane, cellulose nitrate (used as an explosive), cellu-

Table 13-1 Common Carbohydrates	Source
Disaccharide	
Sucrose	Sugar cane, sugar beets, fruits, honey
Lactose	Milksugar
Maltose	Hydrolyzed starch
Cellobiose	Hydrolyzed cellulose
Monosaccharide	
Glucose	Hydrolyzed sucrose, lactose, maltose, and cellobiose
Fructose	Hydrolyzed sucrose (fruits and honey)
Galactose	Hydrolyzed lactose

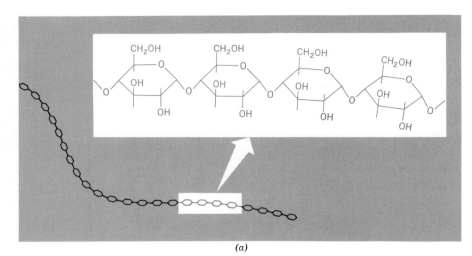

Figure 13-1
Amylose and
amylopectin starch.

(a)

A portion of an amylose molecule.

(b)

A portion of an amylopectin molecule.

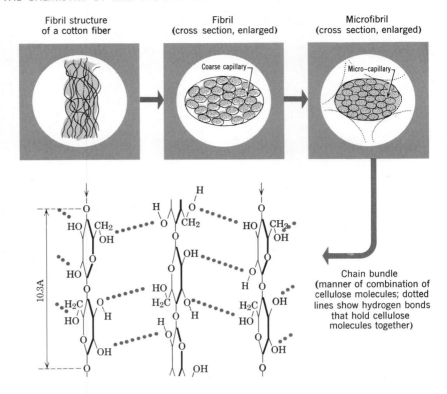

Figure 13-2
Details of a plant fiber showing how the aggregation of cellulose molecules make up the fibers. Cellulose molecules consist of long chains of carbon-one–carbon-four linked β-glucose units. From *Elements of General and Biological Chemistry*, 3rd ed., by John Holum, John Wiley & Sons, New York, 1972, as adapted from *Matthews' Textile Fibers*, 6th ed., edited by H. R. Mauersberger, John Wiley & Sons, New York, 1954, pp. 73 and 77.

Fibril structure of a cotton fiber

Fibril (cross section, enlarged)

Microfibril (cross section, enlarged)

Coarse capillary

Micro–capillary

Chain bundle (manner of combination of cellulose molecules; dotted lines show hydrogen bonds that hold cellulose molecules together)

lose acetate, and rayon (synthetic fibers used to make textiles). Over 50 million tons of cellulose products are used in the United States each year. The difference between starch and cellulose is that starch consists of α-glucose molecules bonded to form polymer molecules and cellulose consists of β-glucose molecules bonded to form polymer molecules.

13-3 LIPIDS

When the cells of living organisms are crushed and mixed with solvents, certain chemical components can be dissolved. Some cellular components are water-soluble, while others are sparingly soluble in water. Other components are quite soluble in nonpolar organic solvents. Those biochemical compounds which can be extracted from crushed cells by organic solvents are called **lipids.** Lipids represent a wide variety of compounds, as shown in Table 13-2. Only the simple lipids and steroids will be considered here.

Simple lipids are esters of glycerol and long-chain (around 20-carbon) carboxylic acids. These esters can be represented as

Table 13-2
Types of Lipids

	Example
Simple lipids	
Fatty acid esters of glycerol	Vegetable oils and animal fats
Waxes	Fruit and vegetable waxes
Steroids	Cholesterol, sex hormones
Phosopholipids	
Lecithins	
Cephalins	Lipids found in nerve tissues
Phosphatidylserines	
Sphingolipids	
Sphingomyelins	Lipids found in brain tissue
Cerebrosides	

The R, R', and R" refer to different possible carbon sequences. Such esters of glycerol are called **fats** or **oils.** The difference between fats and oils is that fats are solid, while oils are liquid at room temperature. Fats are found mainly in animals and oils in plants. Fats and oils are a food source. Certain fats are stored in our body in adipose tissue, which can serve as protective covering of certain parts of our body. The fat is also a reserve source of energy for the body. However, too much reserve fat is not advantageous.

Glycerol esters can be enzymatically broken down into glycerol and the constituent acids which are called **fatty acids:**

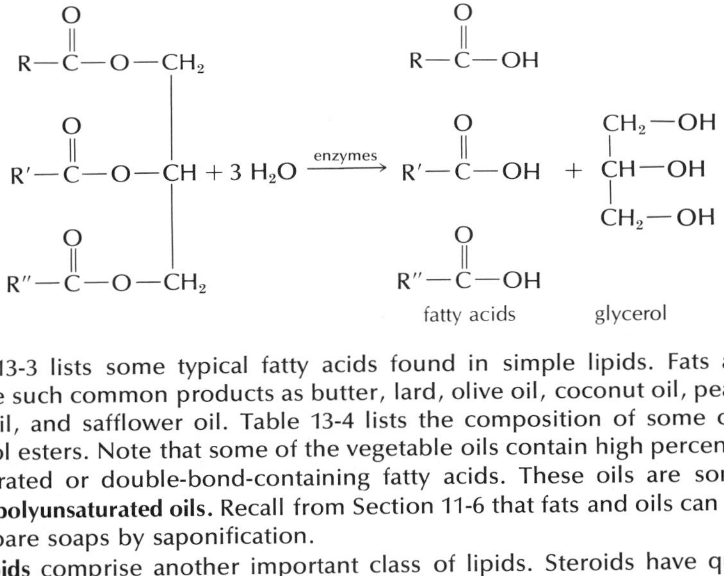

$$
\begin{array}{c}
R-\overset{\displaystyle O}{\overset{\|}{C}}-O-CH_2 \\[2ex]
R'-\overset{\displaystyle O}{\overset{\|}{C}}-O-CH + 3\ H_2O \xrightarrow{\text{enzymes}} \\[2ex]
R''-\overset{\displaystyle O}{\overset{\|}{C}}-O-CH_2
\end{array}
\qquad
\begin{array}{c}
R-\overset{\displaystyle O}{\overset{\|}{C}}-OH \\[2ex]
R'-\overset{\displaystyle O}{\overset{\|}{C}}-OH \quad + \quad
\begin{array}{c}
CH_2-OH \\
CH-OH \\
CH_2-OH
\end{array} \\[2ex]
R''-\overset{\displaystyle O}{\overset{\|}{C}}-OH \\
\text{fatty acids} \qquad \text{glycerol}
\end{array}
$$

Table 13-3 lists some typical fatty acids found in simple lipids. Fats and oils include such common products as butter, lard, olive oil, coconut oil, peanut oil, corn oil, and safflower oil. Table 13-4 lists the composition of some common glycerol esters. Note that some of the vegetable oils contain high percentages of unsaturated or double-bond-containing fatty acids. These oils are sometimes called **polyunsaturated oils.** Recall from Section 11-6 that fats and oils can be used to prepare soaps by saponification.

Steroids comprise another important class of lipids. Steroids have quite different chemical structures from simple lipids. Most steroids have the basic polycyclic ring structure shown in Figure 13-3. A wide variety of steroids are found in the body, and most have very important physiological activities. Let us consider the structures and activities of some steroids.

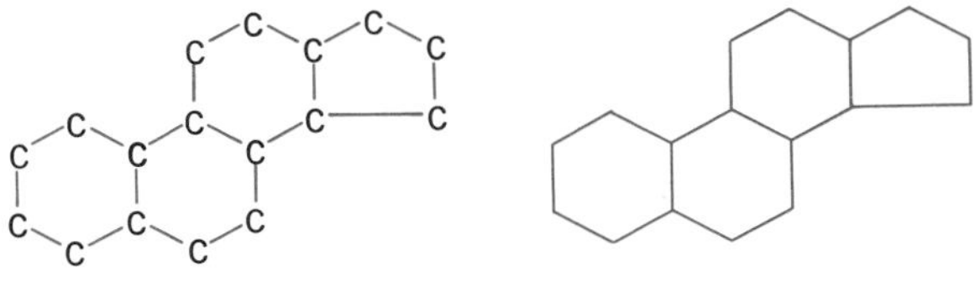

Figure 13-3
The common carbon skeletal structure of steroids.

Table 13-3
Typical Fatty Acids

Name	Formula
Myristic acid	$CH_3(CH_2)_{12}C\overset{\displaystyle O}{-}OH$
Palmitic acid	$CH_3(CH_2)_{14}C\overset{\displaystyle O}{-}OH$
Stearic acid	$CH_3(CH_2)_{16}C\overset{\displaystyle O}{-}OH$
Oleic acid	$CH_3(CH_2)_7CH{=}CH(CH_2)_7C\overset{\displaystyle O}{\underset{OH}{\diagdown}}$
Linoleic acid	$CH_3(CH_2)_4CH{=}CHCH_2CH{=}CH(CH_2)_7C\overset{\displaystyle O}{\underset{OH}{\diagdown}}$
Linolenic acid	$CH_3(CH_2)_4CH{=}CHCH_2CH{=}CHCH_2CH{=}CHCH_2CH{=}CH(CH_2)_3C\overset{\displaystyle O}{\underset{OH}{\diagdown}}$

From *Introduction to Chemistry* by T. R. Dickson, John Wiley & Sons, New York, 1971.

Table 13-4
Composition of the Fatty Acids Obtained by Analysis of Common Fats and Oils

	Average composition of fatty acids (percent)					
Fat or oil	Myristic acid	Palmitic acid	Stearic acid	Oleic acid	Linoleic acid	Other
Animal fats						
Butter	8–15	25–29	9–12	18–33	2–4	3–4 butyric
Lard	1–2	25–30	12–18	48–60	6–12	1–3 palmitoleic
Beef tallow	2–5	24–34	15–30	35–45	1–3	1–3 palmitoleic
Vegetable oils						
Olive	0–1	5–15	1–4	67–84	8–12	0–1 palmitoleic
Peanut	—	7–12	2–6	30–60	20–38	0–1 palmitoleic
Corn	1–2	7–11	3–4	25–35	50–60	0–2 palmitoleic
Cottonseed	1–2	18–25	1–2	17–38	45–55	0–2 palmitoleic
Soybean	1–2	6–10	2–4	20–30	50–58	4–8 linolenic
Linseed	—	4–7	2–4	14–30	14–25	25–58 linolenic
Safflower	—	1–5	1–5	14–21	73–78	
Marine oils						
Whale	5–10	10–20	2–5	33–40		
Fish	6–8	10–25	1–3			

From *Introduction to Chemistry* by T. R. Dickson, John Wiley & Sons, New York, 1971.

Cholesterol is the most prevalent steroid in the body:

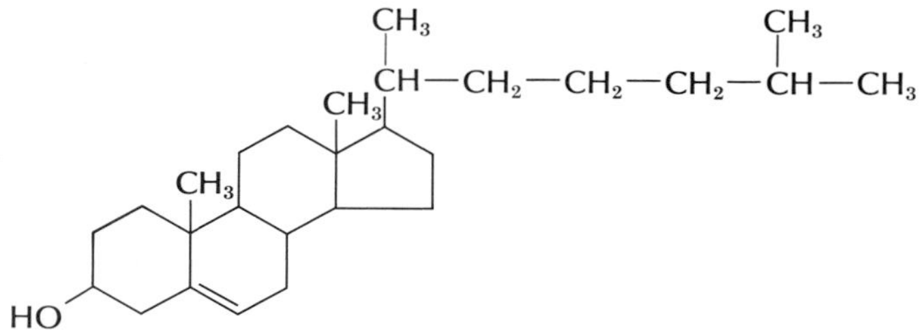

$$CH_3$$
$$CH_3$$
$$CH-CH_2-CH_2-CH_2-CH-CH_3$$
$$CH_3$$
$$CH_3$$
HO

It is found in all tissues, in the bloodstream of humans, and mainly in the brain, spinal cord, and nerve tissue of humans. The physiological activity of cholesterol is unknown, but it may be an intermediate used by the body to form other steroids.

The male and female sex hormones are steroids. A variety of hormones are involved in the development and activities of human sex glands. It is known that these hormones chemically stimulate sexual processes, but specifics of such functions are not known. The structure of the male hormones testosterone and androsterone produced in the testes are shown below.

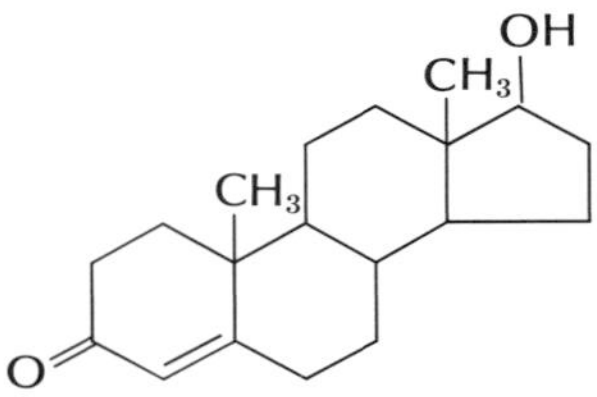

OH
CH_3
CH_3
O

testosterone

O
CH_3
CH_3
HO

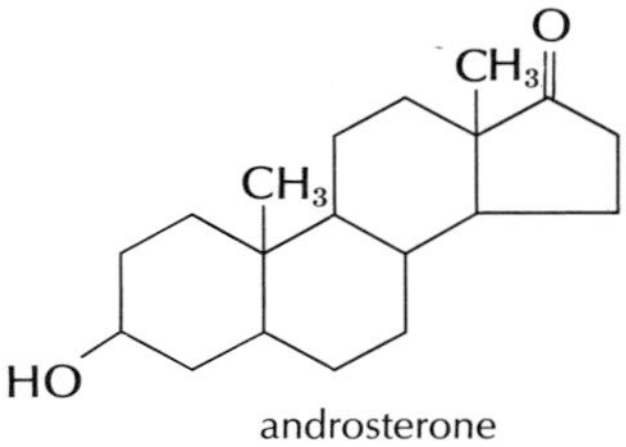

androsterone

The female hormones are involved in control of the menstrual cycle and the maintenance of the fetus during pregnancy. (See Section 15-2.) The structures of two important female sex hormones are shown below.

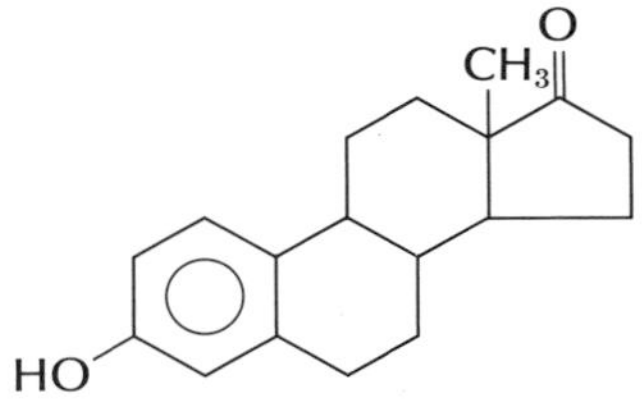

O
CH_3
HO

estrone

O CH_3
C
CH_3
CH_3
O

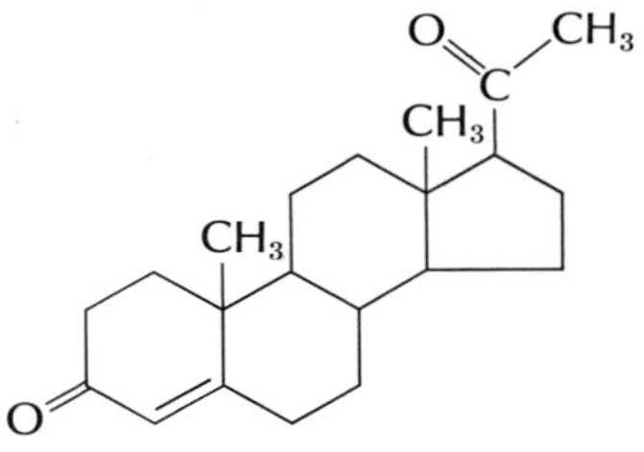

progesterone

13-4 PROTEINS

Proteins are polymeric materials found in living organisms. They serve as structural materials in the body and are fundamental to many of the processes of life. Proteins are polymers of amino acids and are produced by plants and animals. Proteins are synthesized from amino acids in the cells of our bodies. Proteins from animals and some plants are an important food since they provide amino acids that are essential to the body in the production of needed proteins. When proteins are digested, they are broken down by digestive enzymes into the constituent amino acids. The amino acids then become available to the cells for protein synthesis.

Table 12-6 lists some important amino acids. Recall that amino acids are both amines and carboxylic acids (see Section 12-13). It is possible for the carboxyl group of one amino acid to react with the amino group of another to form an amide linkage between the two amino acid units. For example,

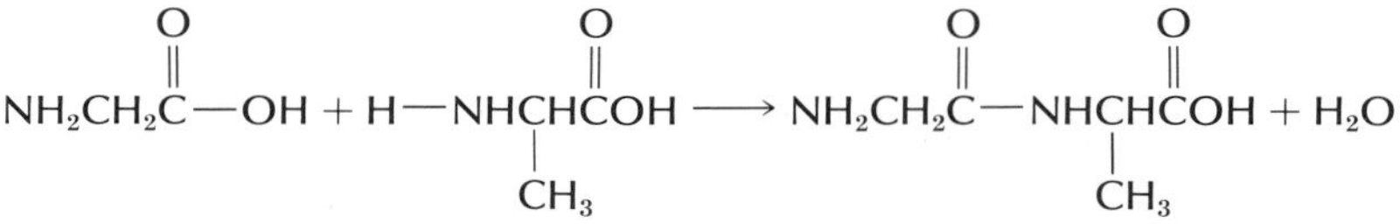

$$
\underset{\text{glycine}}{NH_2CH_2\overset{\displaystyle O}{\overset{\|}{C}}{-}OH} + \underset{\underset{\displaystyle CH_3}{|}}{\underset{\text{alanine}}{H{-}NHCH\overset{\displaystyle O}{\overset{\|}{C}}OH}} \longrightarrow \underset{\underset{\displaystyle CH_3}{|}}{\underset{\text{glycylalanine}}{NH_2CH_2\overset{\displaystyle O}{\overset{\|}{C}}{-}NHCH\overset{\displaystyle O}{\overset{\|}{C}}OH}} + H_2O
$$

Compounds involving two or more amino acids linked by amide linkages are called **peptides**. Glycylalanine shown above is a dipeptide. The amide linkages in peptides

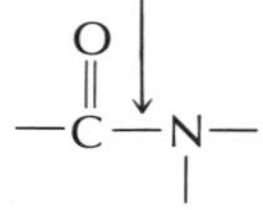

$$
-\overset{\displaystyle O}{\overset{\|}{C}}-\underset{\displaystyle |}{N}-
$$

are sometimes called **peptide bonds** or **peptide linkages**. Table 12-6 lists the official abbreviations that can be used to represent the amino acids. These abbreviations can be used to represent peptides formed from the amino acids. Thus, glycylalanine can be represented as

Gly-Ala

When using this notation, it is assumed that the amino acid on the left contributes the –OH to the formation of the peptide bond and that the amino acid on the right contributes the $-NH_2$. These abbreviations provide a more convenient way to show which amino acids comprise a peptide. Three amino acids linked by peptide bonds constitute a **tripeptide**. Using three amino acids, such as glycine, Gly, alanine, Ala, and valine, Val, it is possible to form six different tripeptides:

Gly-Ala-Val   Gly-Val-Ala
Val-Gly-Ala   Val-Ala-Gly
Ala-Val-Gly   Ala-Gly-Val

It is apparent that as the number of amino acid units increases, the number of possible peptides increases greatly.

Proteins are peptides involving hundreds or thousands of amino acid units. Thus, proteins are called **polypeptides**. Proteins serve as structural units of cells, skin, muscles, bone interior, and nerves, as enzymes, hormones, and in many other important functions in the body. Table 13-5 lists some of the functions of proteins. In spite of the diversity of uses, all proteins are polypeptides. Analysis of proteins reveals that all proteins are made up of about 20 different amino acids. Table 13-6 lists the amino acid unit composition of some proteins.

The sequences in which the amino acids are linked has been determined for a few proteins. This requires a tremendous amount of laboratory work and analysis. The determination of the amino acid sequence for insulin, for instance, took about ten years. The amino acid sequence involved in a protein is called the **primary structure** of the protein. The primary structure of beef insulin is shown in Figure 13-4. It consists of 51 amino acid units composing two chains. The chains are bonded by two disulfide linkages (–S–S–) which are characteristic of the amino acid cystine (see Table 12-6 for the formula of cystine).

The primary structure of a protein gives no indication of the three-dimensional arrangement of the protein. Actually, the protein chain can take on a shape that is determined by hydrogen bonding and other forces of attraction

Structural proteins (insoluble in water)
  Collagens—found in connective tissue
  Elastins—found in tendons and arteries
  Myosins—found in muscle tissue
  Keratins—found in hair and nails
Globular proteins (can be dispersed in water solutions)
  Albumins—found in blood
  Globulins—involved in oxygen transport in the body (hemoglobin) and in body defense against disease (gamma globulin)
Conjugated proteins (complexes of proteins linked to other molecules)
  Nucleoproteins—protein–nucleic acid complexes
  Lipoproteins—protein–lipid complexes
  Phosphoproteins—protein–phosphorus compound complexes
  Chromoproteins—protein–pigment complexes (i.e., hemoglobin)
Enzymes (many enzymes are conjugated with coenzymes)
Hormones (not all hormones are proteins)
Antibodies

Table 13-5
Functions of Proteins

between the amino acid groups along the chain. The configuration of the protein chain is called the **secondary structure** of the protein. A common secondary structure for proteins is the α helix in which the protein chain is coiled in a three-dimensional helical shape. The α helix is illustrated in Figure 13-5. Other secondary structures are known. These protein helices may actually exist in some folded or twisted form. This further modification of the form is called the

	Formula weights of proteins		
Amino acid	Human insulin, 6,000	Horse hemoglobin, 68,000	Egg albumin, 45,000
Glycine	4	48	19
Alanine	1	54	35
Valine	4	50	28
Leucine	6	75	32
Isoleucine	2	0	25
Phenylalanine	3	30	21
Tryptophan	0	5	3
Proline	1	22	14
Serine	3	35	36
Threonine	3	24	16
Tyrosine	4	11	9
Hydroxyproline	0	0	0
Aspartic acid	3	51	32
Glutamic acid	7	38	52
Lysine	1	38	20
Arginine	1	14	15
Histidine	2	36	7
Cysteine	0	4	5
Cystine	3	0	1
Methionine	0	4	16

Table 13-6
Amino Acid Composition of Proteins[a]

Data from G. H. Haggis, D. Michie, A. R. Muir, K. B. Roberts, and P. B. M. Walter, *Introduction to Molecular Biology*, John Wiley & Sons, New York, 1964, pp. 40–41.
[a] Expressed as number of amino acid residues per molecule when the formula is known.

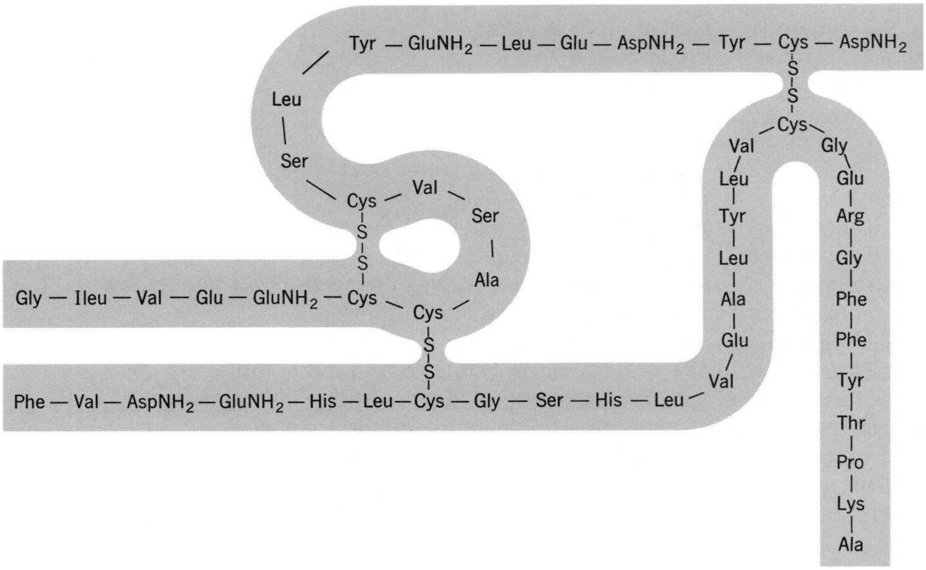

Figure 13-4
The primary structure (amino acid sequence) of beef insulin. This protein consists of 51 amino acids comprising two chains joined by disulfide linkages characteristic of the amino acid cystine. This figure shows the amino acid sequence but is not meant to convey the three-dimensional shape of the protein. Insulins from other animals have similar primary structures. From *Introduction to Chemistry* by T. R. Dickson, John Wiley & Sons, New York, 1971.

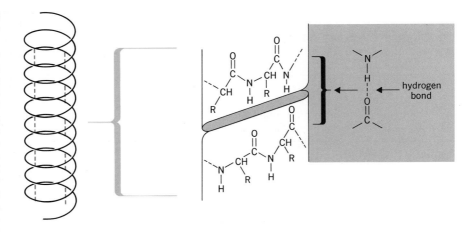

Figure 13-5
The secondary structure of a protein. Some proteins tend to coil in a form called an α helix. This coiling produces a three-dimensional tubular aspect to the protein chains. As shown, hydrogen bonds along the chain tend to hold the helix in place. Under certain drastic conditions (such as heating) the hydrogen bonds can break and the chain takes on a random form. This is called denaturation of the protein. From *Introduction to Chemistry* by T. R. Dickson, John Wiley & Sons, New York, 1971.

tertiary structure of the protein. Once this structure is known, it may be possible to determine how the protein functions in the body. The tertiary structures for a few proteins have been determined; one for the protein myoglobin is illustrated in Figure 13-6.

13-5 ENZYMES

The numerous chemical reactions which occur in the body are referred to collectively as **metabolism.** Metabolic reactions include the reactions of food digestion and the reactions in which certain food molecules are utilized by the body for energy. Other metabolic reactions are involved in the breakdown of certain body cells and the formation of new cells. Most of the thousands of metabolic reactions require specific catalysts. These catalysts are synthesized by the body. Such biological catalysts are called **enzymes.**

Enzymes are large, polymeric molecules of varying complexity. All enzymes contain protein. Some enzymes are composed entirely of protein. Many enzymes consist of a protein portion called the **apoenzyme** and a nonprotein por-

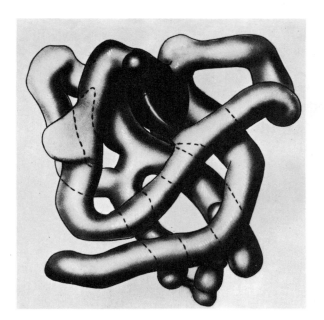

Figure 13-6
(a) Tertiary structural features of the myoglobin molecule. The sausage-like portion contains the protein chain; where it is relatively straight, the chain is coiled in an α helix. It is estimated that 70 percent of the molecule has this secondary feature. The darker, disk-like section represents a heme unit. Myoglobin stores and transports oxygen at muscles. (Courtesy of John C. Kendrew, Cambridge University.) From *Elements of General and Biological Chemistry* by John Holum, John Wiley & Sons, New York, 1968.

tion called the cofactor. Such enzymes cannot function unless the apoenzyme and **cofactor** are joined. The detailed method by which enzymes catalyze chemical reactions is not known. The chemical reactions involved in life processes need specific enzymes which must be present at the correct time and place. It is thought that an enzyme functions by interacting with one of the reactants involved in a reaction. Such a reactant is called the **substrate.** A given enzyme will act as a catalyst for only one kind of reaction. An obvious question is how can an enzyme distinguish one kind of reactant from another?

A commonly accepted theory of enzyme behavior is the **lock-and-key theory** illustrated in Figure 13-7. According to this theory the enzyme has a definite three-dimensional structure which is arranged correctly so that the substrate molecule can just fit into the structure. In this way only a specific kind of substrate can fit into a given enzyme. Once the enzyme and substrate form an aggregate, the substrate is exposed for the reaction. After the reaction occurs, the products move away from the enzyme, leaving it available to catalyze the reaction of another substrate.

Certain metabolic processes involve a long sequence of enzyme-catalyzed reactions. Each enzyme must be present at the correct time for the process to occur. If any of the required enzymes is missing, the process is disrupted. Since such metabolic processes are vital to life, the absence of an enzyme can lead to illness and death. Some hereditary diseases result from the inability of the body to produce certain enzymes. The absence of certain chemicals (vitamins and minerals) in the diet can lead to enzyme deficiencies. See Sections 14-6 and 14-8 for a discussion of vitamins and minerals. Several potent poisons, such as cyanide and mercury, inhibit the functioning of critical enzymes. A sufficient dose of cyanide ion can result in death in a matter of seconds by interfering with an enzyme involved in cell metabolism. The absence of this enzyme causes death due to lack of the ability to use oxygen.

13-6 NAMING ENZYMES

Numerous enzymes have been isolated from plants and animals. The functions of these enzymes are known, but the exact chemical structures of only a few are known. Some enzymes, such as the digestive enzymes pepsin and trypsin, have

Figure 13-7
Illustration of the lock and key theory of enzymatic activity. (a) The three-dimensional shape of the enzyme in the vicinity of the active site is arranged to accommodate the substrate molecule. The active site is the portion of the enzyme which catalyzes the reaction of the substrate. (b) The substrate is held in position by the enzyme while the reaction of the substrate occurs. (c) The products or product of the substrate reaction leave the enzyme, and the enzyme is available for another substrate.

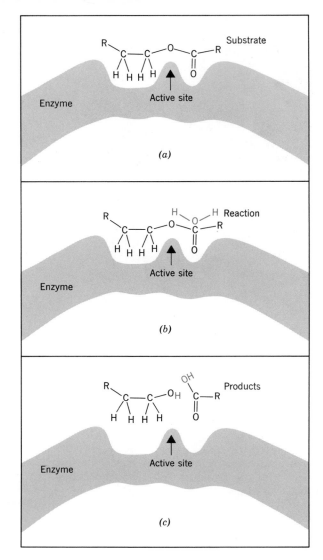

names that do not reveal their function. However, most enzymes have names that indicate the substrate upon which the enzyme functions or the kind of chemical reaction which the enzyme catalyzes. Such names have an -ase ending which indicates an enzyme. A few examples will illustrate these kinds of names: The enzyme amylase catalyzes the decomposition of amylose starch into smaller carbohydrates. The enzyme maltase catalyzes the decomposition of maltose into glucose. Peptidase enzymes catalyze the breaking of peptide bonds in proteins. Oxidases catalyze biological oxidation reactions. Table 13-7 lists some enzymes and their functions.

13-7 BODY CELLS

Plants and animals consist of a large number of microscopic cells held together by structural compounds. Cells differ in size and shape depending upon their function. However, cells have several subcellular components in common. A typical animal **cell** is illustrated in Figure 13-8. Important bodily processes occur in such cells. The cell is surrounded by a cell **membrane** which allows passage of

Name	Location	Substrate	Products	
Carbohydrases		Carbohydrates	Simple sugars	Table 13-7 Some Enzymes
A. Amylases		Starches	Maltose	
1. Ptyalin	Saliva	Starches	Maltose	
2. Amylopsin	Pancreatic juice	Starches	Maltose	
B. Lactase	Intestinal juice	Lactose	Glucose, galactose	
C. Maltase	Intestinal juice	Maltose	Glucose	
D. Sucrase	Intestinal juice	Sucrose	Glucose, fructose	
Esterases		Esters	Acids, alcohols	
A. Cholinesterase	Nerve tissues	Acetylcholine	Choline, acetic acid	
B. Lipases		Fats, oils	Fatty acids, glycerol	
1. Steapsin	Pancreatic juice	Fats, oils	Fatty acids, glycerol	
2. Gastric lipase	Gastric juice	Fats, oils	Fatty acids, glycerol	
C. Phosphatase	Body tissues	Organic phosphates	Phosphoric acid, organics	
Nucleases		Nucleic acids and related molecules	Hydrolysis products	
A. Nucleicacidase	Intestinal juice	Nucleic acid	Nucleotides	
B. Nucleotidase	Intestinal juice	Nucleotides	Nucleosides, phosphoric acid	
C. Nucleosidase	Body tissues	Nucleosides	Ribose, deoxy-ribose, bases	
Peptidases		Peptides	Smaller peptides, amino acids	
A. Aminopolypeptidase	Intestines	Polypeptides	Smaller peptides, amino acids	
B. Carboxypolypeptidase	Pancreatic juices	Polypeptides	Smaller peptides, amino acids	
C. Dipeptidase	Intestines	Dipeptides	Amino acids	
D. Prolinase	Intestines	Polypeptides containing proline	Amino acids, smaller peptides	
Proteases		Proteins	Hydrolysis products	
A. Chymotrypsin	Pancreatic juice	Proteins	Proteoses, peptones, poly-peptides	
B. Pepsin	Gastric juice	Proteins	Proteoses, peptones, poly-peptides	
C. Rennin	Gastric juice	Casein (milk protein)	Paracasein	
D. Trypsin	Pancreatic juice	Proteins	Proteoses, peptones, poly-peptides	

Figure 13-8
The components of a
typical animal cell.

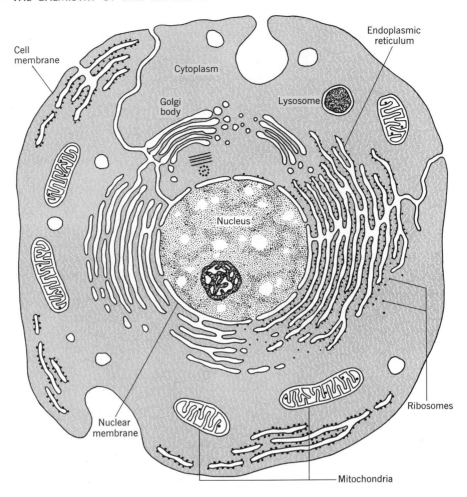

food molecules and chemical wastes in and out of the cell. The **cytoplasm** of the cell contains various specific cellular components. The cytoplasm consists of cellular fluids containing such substances as proteins and enzymes. The **nucleus** of the cell, surrounded by a nuclear membrane, contains the **chromosomes** which control the hereditary characteristics of the cell and is involved in cell division and protein synthesis. The **mitochondria** are membrane-contained components of the cell which are involved in energy production of the cell. The **endoplasmic reticulum** are areas within the cell containing the ribosomes, which are involved with protein synthesis. The **lysomes** of the cell appear to be involved with breakdown and dissolving of the cell if the cell membrane is ruptured.

13-8 NUCLEIC ACIDS: THE CHEMICALS OF HEREDITY

It has been found that a polymeric substance called **deoxyribonucleic acid (DNA)** is the fundamental constituent of genes found in the chromosomes of the cell nucleus. The **genes** are hereditary units of the cell, so an understanding of the structure of DNA is needed to interpret hereditary processes. Actually, DNA belongs to a class of polymeric substances called **nucleic acids.** Nucleic acids can be broken down into monomer units called **nucleotides.** The nucleotides can be further broken down into phosphoric acid, some heterocyclic amines (or bases), and either deoxyribose or ribose sugar. This is illustrated in Figure 13-9. Nucleic acids that contain deoxyribose are called **deoxyribonucleic acids (DNA)**, and those that contain ribose are called **ribonucleic acids (RNA)**.

The nucleotide monomer units of nucleic acids can be considered to be composed of a phosphoric acid unit, a ribose (or deoxyribose) unit, and a unit of adenine, guanine, cytosine, and thymine or uracil instead of thymine in RNA. The formation of a typical nucleotide is illustrated below.

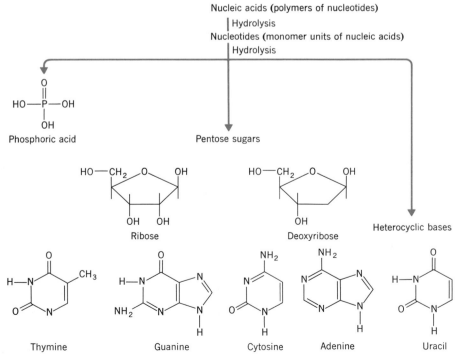

phosphoric acid   ribose   adenosine monophosphate, "AMP"

These nucleotide units are linked together to form the nucleic acid polymer as shown in Figure 13-10. Thus, the nucleic acids consist of a sequence of alternating phosphate units and pentose sugar units. Protruding from this sequence are the heterocyclic amines. A shorthand representation of nucleic acids is given in Figure 13-11. The secondary structure of DNA is important to the visualization of how it functions in the cell. One of the most significant theories of contemporary science involves the structure of DNA as proposed by F. H. C. Crick and J. D. Watson in 1953. They proposed the double helix structure of DNA in which two DNA strands intertwine in the form of two helices. The intertwining is

Nucleic acids (polymers of nucleotides)

| Hydrolysis

Nucleotides (monomer units of nucleic acids)

| Hydrolysis

Phosphoric acid

Pentose sugars

HO—CH₂, O, OH

OH OH

Ribose

HO—CH₂, O, OH

OH

Deoxyribose

Heterocyclic bases

Thymine   Guanine   Cytosine   Adenine   Uracil

Figure 13-9
Nucleic acids form nucleotides upon hydrolysis. Upon further hydrolysis the nucleotides are broken down into phosphoric acid, a sugar (ribose or deoxyribose), and heterocyclic bases.

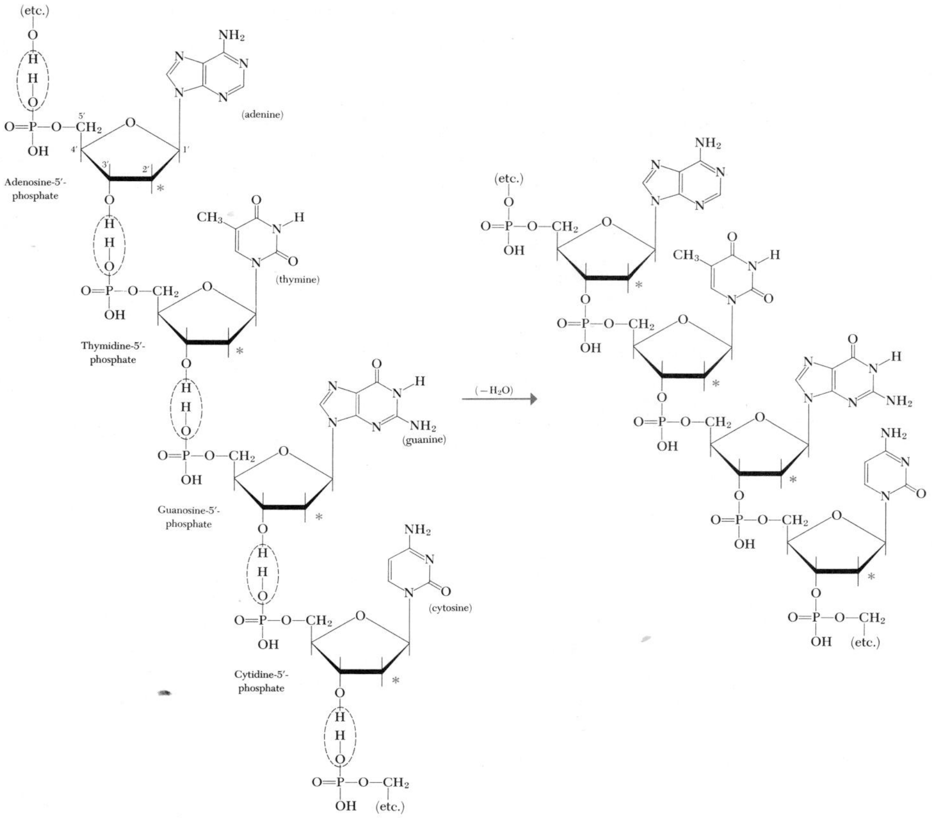

(adenine)
Adenosine-5'-phosphate
(thymine)
Thymidine-5'-phosphate
(guanine)
Guanosine-5'-phosphate
(cytosine)
Cytidine-5'-phosphate
(etc.)
$(-H_2O)$

Figure 13-10 Manner of formation of a nucleic acid chain. Shown here is a segment of a DNA chain. If the sites marked by asterisks were given –OHs, the example would be for RNA (assuming uracil replaced thymine). The sequence of the heterocyclic amines is purely arbitrary in this drawing, but one each of the four amines common to DNA has been included. A molecular weight of 2.8×10^9 for the DNA of one species (*E. coli*) has been reported. Using a value of 325 as the average formula weight of each nucleotide, this DNA would be made of 8,600,000 nucleotide units. It is probable that such a DNA molecule would make up a collection of genes rather than just one. Genetic studies indicate that the average gene size is 1500 nucleotide pairs (of a double helix). From *Elements of General and Biological Chemistry*, 3rd ed., by John Holum, John Wiley & Sons, New York, 1972.

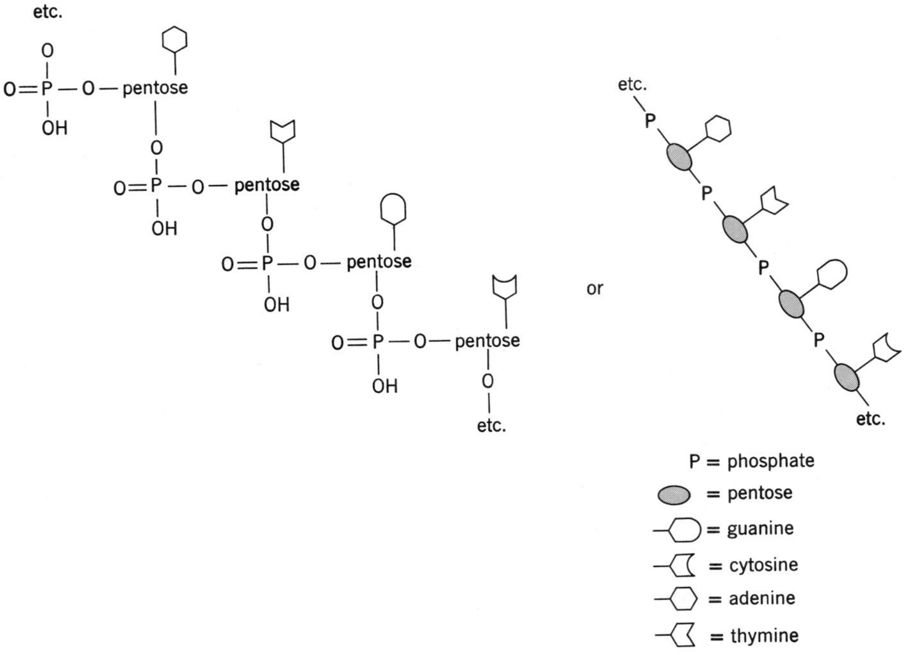

etc.
pentose
or
P = phosphate
= pentose
= guanine
= cytosine
= adenine
= thymine

Figure 13-11 A shorthand representation of a nucleic acid chain which consists of an alternating sequence of phosphate units and sugar units from which the heterocyclic amines protrude. From *Introduction to Chemistry* by T. R. Dickson, John Wiley & Sons, New York, 1971.

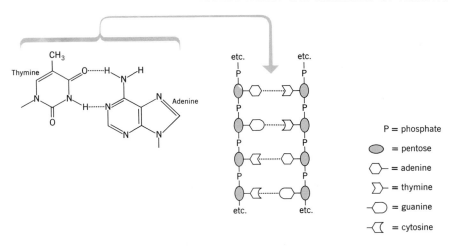

Figure 13-12
The DNA strands in the double helix form of DNA are held together by hydrogen bonding between the heterocyclic amines. From *Introduction to Chemistry* by T. R. Dickson, John Wiley & Sons, New York, 1971.

P = phosphate

⬭ = pentose

⬡─ = adenine

⧁─ = thymine

─○ = guanine

─⧀ = cytosine

accommodated by the hydrogen bonding between thymine and adenine units and guanine and cytosine units on opposite strands. This hydrogen bonding is illustrated in Figure 13-12, and a model of the double helix structure of DNA is shown in Figure 13-13. DNA in the genes serves as a starting point for the biological processes that are carried out in the cell. Apparently, genetic information is contained in the genes by specific sequences of heterocyclic bases on the DNA strands. In other words, the genetic information of a cell is contained as various

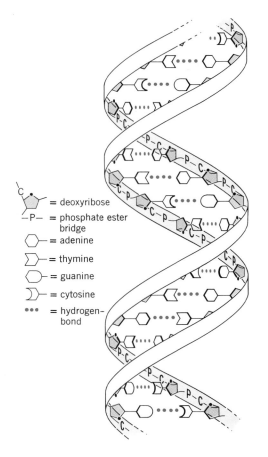

Figure 13-13
Schematic representation of a DNA double helix. From *Elements of General and Biological Chemistry*, 3rd ed., by John Holum, John Wiley & Sons, New York, 1972.

= deoxyribose

─P─ = phosphate ester bridge

⬡─ = adenine

⧁─ = thymine

○─ = guanine

⧁─ = cytosine

••• = hydrogen-bond

sequences of these heterocyclic bases which are present along the DNA strands that comprise the genes of the chromosomes. These sequences constitute the **genetic code** of the genes. The term "genetic code" refers to hereditary information which can be transmitted to new cells and to new generations.

13-9   DNA AND THE GENETIC CODE

All cells in our bodies carry the genetic code. The vast majority of cells which structure our bodies are called **somatic** (Greek: body) cells. Certain cells located in the gonads are capable of being formed into sperm or eggs. These are called **germ** (Latin: sprout, bud) cells. Human somatic cells contain 46 chromosomes, but the germ cells contain only 23 chromosomes. At sexual maturity the germ cells become active, enabling the female to ovulate and the male to produce sperm. Human sperm consists of a protein sheath surrounding the DNA structured male chromosomes. The female egg carries the female chromosomes. At conception the egg is united with the sperm, and the chromosomes pair up to form a new cell called a **zygote;** a fertilized egg. The zygote multiplies by cell replication as the development of a new individual takes place. This new being carries within its cells the genetic codes donated by the mother and the father. As development proceeds, the number of cells increases rapidly and some take on different forms and functions which result in the various specialized parts of the body. That is, some cells structure the muscles, some structure the body organs, and so on. The new individual takes on inherited characteristics of the parents. Sometimes, one characteristic will be dominant over the other. For instance, the brown eye trait of the mother may be dominant over the blue eye trait of the father—then the offspring will have brown eyes. All the cell specialization and inherited characteristics are related to the genetic code contained in the cellular genes. As a result of various factors, including selective inherited characteristics, conditions of growth, and the environment, the new individual does not turn out to be an exact physical replica of the mother or the father. Nevertheless, it is through the genetic code that humans propagate a physical pattern of themselves to new generations.

When cells divide (undergo mitosis) to form new cells, it is necessary for the genes to be replicated before the cell splits, so that the new cells will contain the genetic code. As shown in Figure 13-14, the double helix theory provides an explanation of how the **replication** of DNA can take place. Furthermore, the DNA provides a pattern on which the synthesis of RNA can take place. The RNA is involved in the synthesis of enzymes. **Enzyme synthesis** is a very important cellular process. The enzymes of the body are being continuously replaced, and new cells are being produced. The amino acids for these enzymes come from the breakdown of old protein and from protein-containing foods. A variety of amino acids called the amino acid "pool" is normally available in a cell. Specific enzymes are synthesized in a cell according to the genetic code of the genes. These enzymes catalyze specific reactions taking place in the cells. The genetic code carries inherited information, resulting in the entire organism taking on various characteristics.

13-10   PROTEIN SYNTHESIS

Enzyme protein synthesis is a complicated process, but the basic process is summarized in Figure 13-15. As shown in the figure, **RNA synthesis** occurs in the nucleus using the DNA as a pattern. There are three major types of RNA produced. The **ribosomal RNA** migrates from the nucleus and becomes incorporated in the **ribosomes**. Ribosomal RNA apparently prepares the ribosomes for enzyme synthesis. The **messenger RNA** (*m*RNA) contains the pattern (as a

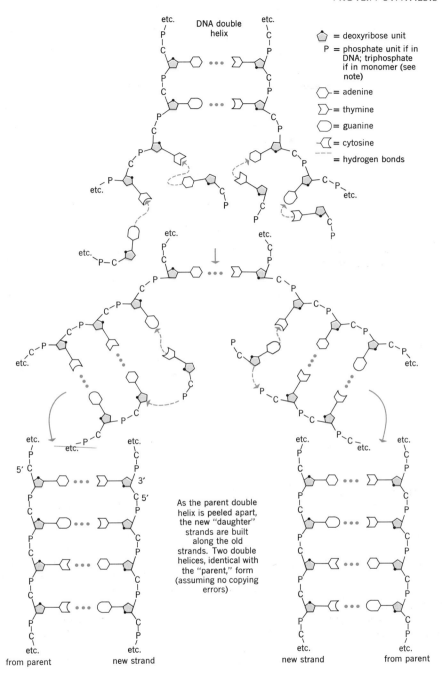

Figure 13-14
A possible mechanism for the replication of DNA. From *Elements of General and Biological Chemistry*, 3rd ed., by John Holum, John Wiley & Sons, New York, 1972.

As the parent double helix is peeled apart, the new "daughter" strands are built along the old strands. Two double helices, identical with the "parent," form (assuming no copying errors)

DNA double helix

= deoxyribose unit
P = phosphate unit if in DNA; triphosphate if in monomer (see note)
= adenine
= thymine
= guanine
= cytosine
= hydrogen bonds

from parent   new strand   new strand   from parent

sequence of bases complementary to those on DNA) upon which protein synthesis is based. The *m*RNA migrates from the nucleus and becomes attached to one or more ribosomes. The point of attachment of the ribosomes and *m*RNA is the region at which protein synthesis occurs. A variety of **transfer RNA's (*t*RNA)** migrate from the nucleus and become attached to specific amino acids from the amino acid pool. There are at least twenty kinds of *t*RNA—one for each of the amino acids. The *t*RNA's with the amino acids attached migrate to the ribosomes

Figure 13-15
The relations of DNA
to various RNAs and
protein synthesis.
From *Elements of
General and Biological
Chemistry*, 3rd ed., by
John Holum, John
Wiley & Sons, New
York, 1972.

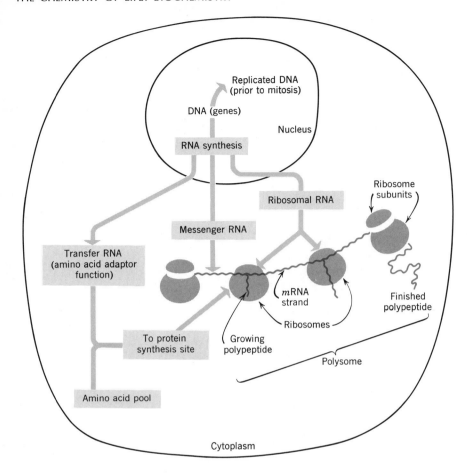

on the *m*RNA. The point of attachment of the *m*RNA and the ribosome involves a certain part of the genetic code. This part of the genetic code is catalytically activated by the ribosome. The active genetic code calls for a specific amino acid. Thus, the *t*RNA with this amino acid fits into the activated portion of the *m*RNA, and the amino acid it carries breaks off to become part of a growing chain. The ribosome then moves along the *m*RNA to activate a portion of the genetic code adjacent to the previously used portion. The *t*RNA–amino acid pair corresponding to this new code moves into the ribosome, and the amino acid becomes linked to the previous amino acid on the protein chain. The movement of the ribosome continues with a new amino acid being added to the growing protein chain each time. Finally, after hundreds or thousands of amino acids have been added, the ribosome reaches the end of the *m*RNA, and the newly formed protein breaks off. In this manner, the genetic code carried by the *m*RNA from the DNA of the genes directs the synthesis of a specific protein which then functions in a certain manner. This process is very specific for the synthesis of many enzymes. In fact, human cells contain enough DNA to contain codes for about 7 million different proteins. It is thought that only about 10 percent of the codes are active, but this represents some 700,000 proteins to regulate human growth, development, and metabolism.

Certain diseases such as galactosemia (failure to metabolize galactose) and sickle cell anemia (the production of abnormal red blood cells) are attributed to **genetic defects** in which the genetic code is not normal. The abnormal code can cause a genetic disease to be passed on to the next generation, depending upon whether or not the offspring develops the characteristic related to the defect. Even if the characteristic does not dominate in the off-spring, it is still present in the cells and can be passed on to the next generation. See Section 14-18 for a discussion of genetic defects.

13-11 BLOOD

The human body contains 5–6 liters of blood circulating through the veins and arteries. Blood is a suspension, colloidal dispersion, and solution of vital materials involved in life processes. The respiratory function of blood involves the transport of oxygen to body cells and the removal of carbon dioxide. Digested food is transported by the blood to the cells of the body, and waste chemicals are carried from the cells. The blood contains several kinds of ions called the blood electrolytes. Some of these ions help maintain the pH of the blood and are called the blood buffers.

Microscopic examination of blood reveals that it is a suspension of certain kinds of cells in a liquid. The suspended cells are called the **formed elements,** and the liquid is called blood plasma. The formed elements are the **red cells** called erythrocytes, the **white cells** called leukocytes, and the blood platelets. The plasma is a solution and colloidal dispersion of proteins, ions, lipids, and other molecules involved in the life processes. Coagulated or clotted blood has the formed elements removed, along with some of the molecules in the plasma. The removal of these components leaves a yellow liquid called **blood serum.**

13-12 RED BLOOD CELLS

The **erythrocytes** are small disk-shaped cells produced in the red bone marrow. Normally, the red cell count in blood is about 5 million cells per cubic millimeter of blood. The count is accomplished by microscopically counting a small portion of blood. The color of blood is caused by the protein called hemoglobin contained in the red cells. **Hemoglobin** is a complex protein responsible for the transport of oxygen in the blood. The behavior of hemoglobin is illustrated in Figure 13-16. Hemoglobin consists of four heme groups (iron-containing red pigments) per each globin. The globin is the globular protein part. The heme groups embedded in the globin are capable of a reversible reaction with oxygen which allows for the transportation and release of oxygen by the hemoglobin. Hemoglobin (Hb) combines with oxygen to form **oxyhemoglobin:**

$$Hb + O_2 \rightleftarrows HbO_2$$

Blood in the veins is purple-red, and blood in the arteries is bright red. The difference is caused by the different amounts of oxygen in the blood. Oxyhemoglobin is bright red in color and hemoglobin is purple-red. Oxyhemoglobin can be broken down to release oxygen to the body cells and reform hemoglobin. The oxygen is used in the metabolic oxidation processes taking place in the cells. There is a continuous decomposition and formation of red blood cells in the body. Much of the iron is reused by the body, but some is excreted. Thus, iron is an important mineral needed in the diet.

Oxygen combines with hemoglobin in the lungs, and the oxygen-containing blood flows through the arteries to various parts of the body. The oxyhemoglobin is unstable in the presence of carbonic acid, H_2CO_3, formed from the metabolic waste carbon dioxide $(CO_2 + H_2O \rightarrow H_2CO_3)$. The carbonic acid causes the oxyhemoglobin to release the oxygen. The oxygen need of a body cell is directly related to the amount of carbon dioxide released by the cell. The greater the amount of carbon dioxide given off by the cell, the greater the

Figure 13-16
Hemoglobin combines with oxygen in the lungs and is carried to the body cells through the arteries to the capillaries. Under the stimulation of carbon dioxide, oxyhemoglobin releases oxygen to be used by the cells. The hemoglobin returns to the lungs in the veins to be combined with more oxygen. The carbon dioxide released by the metabolic oxidation in the cells occurs in the blood in the three forms shown. Carbamino hemoglobin is a hemoglobin–carbon dioxide complex.

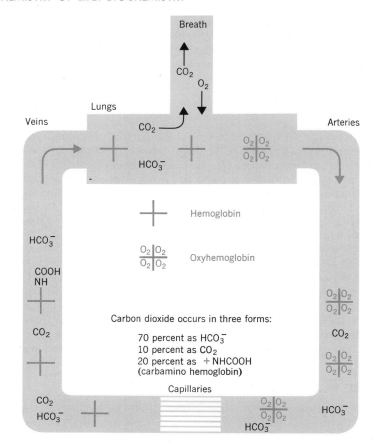

amount of oxygen which will be released to the cell. The carbon dioxide as a waste is transported by the blood to the lungs, where it can be exhaled. Carbon dioxide exists in the blood in three major forms: dissolved CO_2, the hydrogen carbonate ion HCO_3^-, and combined with hemoglobin. If the compound carbon monoxide is inhaled, it can form the compound carboxyhemoglobin ($CO + Hb \rightarrow HbCO$). Carboxyhemoglobin is more stable than oxyhemoglobin and can remain in the bloodstream for hours. The nature of such carbon monoxide poisoning is discussed in Section 8-6.

13-13 WHITE BLOOD CELLS

White blood cells or **leukocytes** are formed mainly in the yellow bone marrow. These cells are involved in the protection of the body against infection by disease-causing microorganisms. The five kinds of leukocytes are listed in Table 13-8. The usual white cell count is about 7,000 per cubic millimeter of blood. The

Table 13-8
Types of Leukocytes in Blood

Type	Average percentage	Type	Average percentage
Neutrophils	62	Eosinophils	2.3
Lymphocytes	30	Basophils	0.4
Monocytes	5.3		

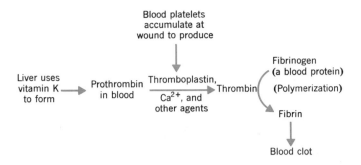

Figure 13-17
The chemistry of
blood clot formation.

monocytes and neutrophils are chemically attracted to bacteria and foreign molecules in the blood. These protective leukocytes absorb and dissolve these infection-causing agents. During an infection, the white blood cell count increases greatly. The accumulation of damaged or inactive leukocytes in infected tissue produces a yellow fluid called pus.

The blood **platelets** are formed in various parts of the body. The exact function of the platelets is not known, but they appear to be involved in blood clotting. See Figure 13-17.

13-14 OTHER BLOOD CHEMISTRY

Blood plasma is about 90 percent water and 10 percent dissolved substances. These dissolved substances are mainly proteins and other organic compounds, and the remainder consists of inorganic ions, the **electrolytes**. The electrolyte composition of blood is shown in Table 13-9. Lymph fluid which flows in the lymphatic system of the body acts as an intermediate fluid between the blood and the cells. Certain chemicals pass through the lymph fluids from the blood to the cells and vice versa.

The three important types of proteins in the blood are albumins, globulins, and fibrinogen. The albumins function as **blood buffers** and control the osmotic pressure of the blood along with the blood electrolytes. The buffering systems which maintain the pH of the blood are illustrated in Table 13-10. As illustrated in Figure 13-17, the fibrinogen is involved in blood clotting. The globulins are involved in the formation of antibodies (see below) and immunity against certain diseases.

When certain foreign bodies (microorganisms or foreign chemicals) enter the bloodstream, the blood often responds in a chemical fashion intended to protect against the effect of the foreign body. The presence of the foreign body causes the formation of a serum-soluble protein called an **antibody**. Any substance which stimulates the formation of an antibody is called an **antigen**. The antibody interacts with the antigen to prevent it from having any harmful effect. A specific antibody is produced for each antigen which enters the blood. This

Positive ions	Negative ions
Sodium ion, Na^+	Chloride ion, Cl^-
Potassium ion, K^+	Hydrogen carbonate ion, HCO_3^-
Calcium ion, Ca^{2+}	Hydrogen phosphate ion, HPO_4^{2-}
Magnesium ion, Mg^{2+}	Sulfate ion, SO_4^{2-}
	Organic acid ions
	Negatively charged proteins

Table 13-9
Electrolytes in Blood

Table 13-10 Blood Buffer Systems[a]	1. Carbon dioxide—hydrogen carbonate ion $2 H_2O + CO_2 \rightleftharpoons H_3O^+ + HCO_3^-$ 2. Dihydrogen phosphate ion—hydrogen phosphate ion $H_2O + H_2PO_4^- \rightleftharpoons H_3O^+ + HPO_4^{2-}$ 3. Albumins $H_2O + $ albumin proteins $\rightleftharpoons H_3O^+ + $ albumin ions

[a] Each of these buffer systems involves an equilibrium reaction. Any increase in H_3O^+ in the blood will cause one or more of these reactions to shift (to the left as written) to consume the excess H_3O^+. Any decrease in H_3O^+ in the blood will cause one or more of these reactions to shift (to the right as written) to replenish the H_3O^+. These buffer systems maintain the pH of blood at 7.4.

antibody—antigen reaction protects the body from certain microorganisms (bacteria) and poisons (toxins). In Section 14-9, the use of the antibody–antigen reaction to combat disease is discussed.

13-15 BLOOD TYPING

The four general blood types are A, B, AB, and O. The distribution of these blood types in the United States is shown in Table 13-11. Blood is typed according to the presence of certain antigens and antibodies. Red blood cells may or may not contain one or two substances which can coagulate to clot the blood. These substances are **agglutinogen A** and **agglutinogen B.** Blood serum contains one of two substances capable of coagulating the agglutinogens in the red blood cells. These substances are **anti-A-agglutinin** and **anti-B-agglutinin.** If a person is given a blood transfusion of the wrong type, it can result in the coagulation of the blood.

Blood is typed as follows (see Table 13-12): If red cells in a blood sample are mixed with serum containing anti-A-agglutinin and serum containing anti-B-agglutinin and no coagulation occurs, the blood is type O. If coagulation occurs when serum containing anti-A-agglutinin is added to a blood sample and when serum containing anti-B-agglutinin is added to a similar sample, the blood type is AB. If a blood sample is coagulated by anti-A-agglutinin serum but not anti-B-agglutinin serum, the blood is type A. If a blood sample is coagulated by anti-B-agglutinin serum but not anti-A-agglutinin serum, the blood type is B.

The red blood cells may contain another antigen called the **Rh factor.** A large

Table 13-11
Distribution of Blood
Types Among
Americans

Type	Percentage of population	Type	Percentage of population
O	45	B	12
A	39	AB	4

Table 13-12
Simple Blood Typing[a]

Type	Anti-A-agglutinin	Anti-B-agglutinin
O	No clumping	No clumping
AB	Clumping	Clumping
A	Clumping	No clumping
B	No clumping	Clumping

[a] Separate blood samples are tested with each antigen.

portion of the population has the Rh factor and is said to be Rh-positive. The remainder of the population does not have the factor and is said to be Rh-negative. The serum does not normally contain the Rh-antibody. A person with Rh-negative blood receiving a transfusion of Rh-positive blood will form the antibody in their serum. If another transfusion of Rh-positive blood is given, coagulation of the blood can occur. Since the Rh factor can be inherited, an unborn fetus can be affected. Suppose an Rh-positive male mates with an Rh-negative female and the fetus is Rh-positive. The Rh-positive blood of the child causes the formation of the antibody in the serum of the mother. The child is not affected, but the presence of the antibody in the mother can affect a second child. The blood of an Rh-positive fetus in a second pregnancy will be mixed with the antibody from the blood of the mother. This can cause the coagulation of the blood of the fetus and death.

13-16 HORMONES

Many of the metabolic processes in the body are carried out on a continuous basis. On the other hand, some metabolic processes occur only at certain times and for certain periods. The body has a system by which these processes are initiated when they are needed. This system is based upon chemical communication between certain glands and other parts of the body. The **endocrine glands** secrete chemicals called hormones into the bloodstream so that they are spread to all parts of the body. **Hormones** are chemicals which stimulate various life processes. As shown in Figure 13-18, there are several important endocrine glands. Hormones carry chemical information to all parts of the body through the bloodstream. However, a specific hormone will stimulate a process in a certain part of the body. For example, certain hormones stimulate growth in children. Insulin is a hormone which controls the amount of glucose in the blood. Digestive hormones stimulate the digestive process. Certain sex hormones stimulate puberty and the development of male and female sex characteristics. Other hormones stimulate and control the female menstrual cycle and act when pregnancy occurs. See Section 15-2 for a discussion of the role of hormones in the menstrual cycle. Overall hormones exercise control over many of the vital body processes. Table 13-13 lists some important hormones. Many of the body hormones are proteins or have a protein portion and a nonprotein portion. The sex hormones (see Section 13-3) are steroids.

13-17 DIGESTION

There are three basic **foods** utilized by humans. These foods are carbohydrates, proteins, and lipids or fats. All these foods are important in the diet, since they are utilized in certain metabolic processes. When food is eaten it must be digested before the chemicals are useful to the body. **Digestion** is the process in which food is degraded to simple molecules which are absorbed into the body through the intestines. Nondigestible wastes pass through and are excreted. The digestion process is summarized in Figure 13-19. As shown in the figure, digestion takes place in three areas of the digestive system. Digestive juices secreted in various parts of the system contain enzymes which are utilized in the breakdown of food.

The digestive juice secreted in the mouth is the **saliva.** Over 1 liter of saliva is produced each day. Saliva contains electrolytes, proteins, and the enzyme ptyalin, an α-amylase. The electrolytes and proteins keep the pH of the saliva nearly neutral (pH 7). Certain proteins make the saliva slippery. The ptyalin catalyzes the breakdown of starches to form partially decomposed starch and some maltose. Chewing of food is an important process in the mouth, since it

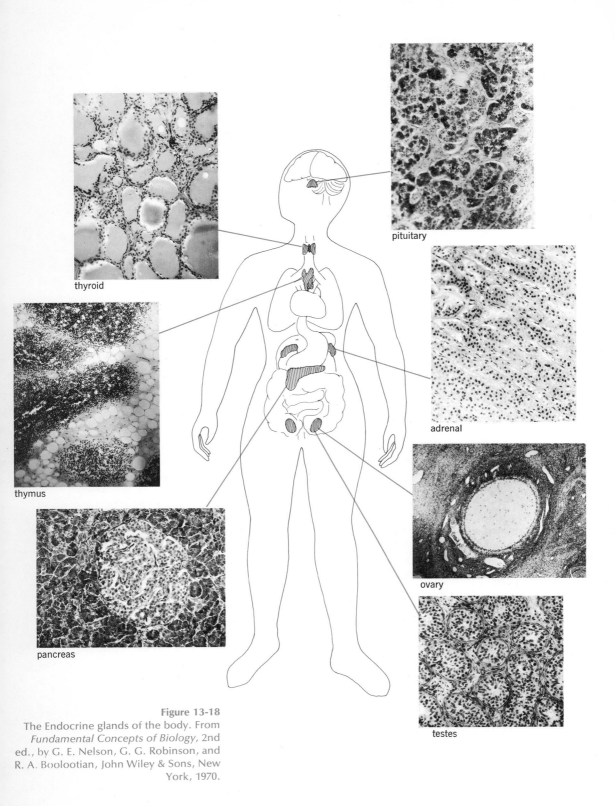

thyroid

pituitary

thymus

adrenal

pancreas

ovary

testes

Figure 13-18
The Endocrine glands of the body. From
Fundamental Concepts of Biology, 2nd
ed., by G. E. Nelson, G. G. Robinson, and
R. A. Boolootian, John Wiley & Sons, New
York, 1970.

Table 13-13
Human Hormones
and Sources

Gland or tissue	Hormone	Major function of hormone
Thyroid	1. Thyroxine	1. Stimulates rate of oxidative metabolism and regulates general growth and development
	2. Thyrocalcitonin	2. Lowers the level of calcium in the blood
Parathyroid	Parathormone	Regulates the levels of calcium and phosphorus in the blood
Pancreas (Islets of Langerhans)	1. Insulin	1. Decreases blood glucose level
	2. Glucagon	2. Elevates blood glucose level
Adrenal medulla	Epinephrine (adrenalin)	Various "emergency" effects on blood, muscle, temperature
Adrenal cortex	Cortisone and related hormones	Control carbohydrate, protein, mineral, salt, and water metabolism
Anterior pituitary	1. Thyrotropic	1. Stimulates thyroid gland functions
	2. Adenocortiocotropic	2. Stimulates development and secretion of adrenal cortex
	3. Growth hormone	3. Stimulates body weight and rate of growth of skeleton
	4. Gonadotropic (2 hormones)	4. Stimulates gonads
	5. Prolactin	5. Stimulates lactation
Posterior pituitary	1. Oxytocin	1. Causes contraction of some smooth muscles
	2. Vasopressin	2. Inhibits excretion of water from the body by way of urine
Ovary (follicle)	Estrogen	Influences development of sex organs and female characteristics
Ovary (corpus luteum)	Progesterone	Influences menstrual cycle, prepares uterus for pregnancy; maintains pregnancy
Uterus (placenta)	Estrogen and progesterone	Function in maintenance of pregnancy
Testis	Androgens (testosterone)	Responsible for development and maintenance of sex organs and secondary male characteristics
Digestive system	Several gastrointestinal	Integration of digestive processes

From G. E. Nelson, G. G. Robinson, and R. A. Boolootian, *Fundamental Concepts of Biology*, 2nd ed., John Wiley & Sons, New York, 1970.

Figure 13-19
Summary of the
digestive process.

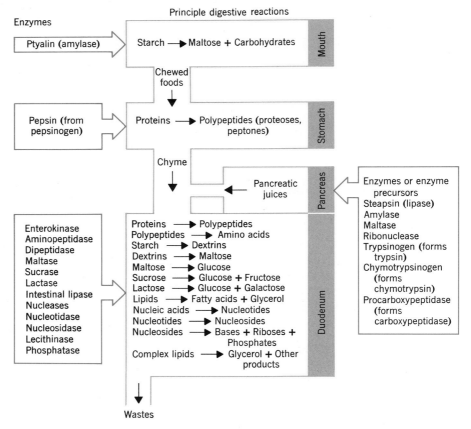

produces small food particles which are more easily digested. Some partial digestion of starches begins in the mouth, but proteins and lipids are unchanged as they pass to the stomach.

The stomach serves as a storage area and a digestive organ. **Gastric** or **stomach juices** contain electrolytes, enzymes, and hydrochloric acid. Over 2 liters of gastric juice are produced each day. When food enters the stomach the hormone gastrin is secreted to the bloodstream. Gastrin stimulates the flow of gastric juices. Gastric juice is highly acidic (pH of about 2). The gastric juice contains a substance pepsinogen which is converted to the enzyme pepsin in the acidic environment. Pepsin is a proteolytic enzyme, which means that it catalyzes the breakdown of proteins. In the presence of pepsin, proteins in food are partially degraded to shorter polypeptide chains. The only other digestion that occurs in the stomach is the breakdown of some carbohydrates in the highly acidic juices. As a result of the physical churning of the stomach and the mixing of the food with juices, the stomach converts the food to a liquefied suspension called **chyme.** The chyme passes from the stomach into the duodenum.

The first 30 centimeters of the small intestine are called the **duodenum.** The duodenum contains enzyme-rich intestinal juices and pancreatic juices. The **pancreas** is located near the duodenum and secretes juices directly to the duodenum. The juices in the duodenum are basic enough to neutralize the acidic chyme entering from the stomach. The presence of chyme in the duodenum causes the hormone secretin to be secreted. Secretin stimulates the flow of pancreatic juices into the duodenum. Intestinal juices are secreted by the walls of the intestine. As shown in Figure 13-19, these juices contain a variety of proteolytic enzymes which aid in the breakdown of proteins and polypeptides. The duodenum juices also contain a variety of enzymes which aid in the degradation of carbohydrates and starches. Lipids in the duodenum form a suspension of fat and oil globules. The formation of such a suspension is aided by certain secretions from the liver called **bile juices.** The duodenum juices contain the enzyme steapsin which is lipolytic; this means that it catalyzes the degradation of lipids. The major part of digestion occurs in the duodenum.

The digestive process breaks foods down to simple molecules which are absorbed into the body through the walls of the upper intestine. Let us consider the digestive products of the three food types. The carbohydrates most often eaten are starch, cellulose, maltose, lactose, and sucrose. Degradation of cooked starch begins in the mouth, and some carbohydrates are broken down in the stomach. The duodenum juices complete the carbohydrate breakdown to form the simple molecules glucose, fructose, and galactose. These sugars enter the body. Cellulose is not broken down by digestion and passes through the digestive system to be excreted with other wastes. The digestion of simple lipids occurs in the duodenum and produces a variety of fatty acids and glycerol which are absorbed into the body. The digestion of proteins begins in the stomach and is completed in the duodenum. A variety of amino acids are formed upon digestion of proteins. These amino acids are absorbed into the body and enter the amino acid pool.

13-18   METABOLISM OF FOOD

Digested food is used in the body in numerous ways. Basically, the food chemicals are used to build new body cells, as starting material for vital molecules and as a source of energy for life processes. Let us consider the utilization of each of the three food types.

CARBOHYDRATES   Figure 13-20 summarizes the uses of digested carbohydrates. Carbohydrates enter the blood as blood sugar. A portion of digested carbohydrates is converted to glycogen (body starch) and stored in the liver and muscles. Glycogen serves as carbohydrate storage for the body. Excess carbohydrates can be converted to lipids or proteins (using some source of nitrogen) by certain metabolic reactions. Carbohydrates are used in the cells as a source of energy in **metabolic oxidation** which takes place in the body cells.

The net result of the metabolic oxidation of glucose can be represented by the

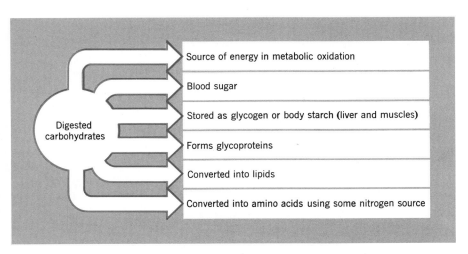

Figure 13-20
The uses of carbohydrates in the body.

equation

$$C_6H_{12}O_6 + 6\ O_2 \longrightarrow 6\ CO_2 + 6\ H_2O + energy$$

The reaction of glucose and oxygen does not occur in a single step but rather as a series of reactions which ultimately produce carbon dioxide and water. Some of the energy produced takes the form of heat, but most of the energy is captured in energy-rich compounds which are products of certain steps in the metabolic oxidation. The most important of these energy-rich compounds is **adenosine triphosphate (ATP)**, shown in Figure 13-21. ATP serves as a chemical source of energy for many metabolic reactions. This energy comes from the decomposition of ATP to ADP. (See Figure 13-21.) Rather than use the energy of the metabolic oxidation directly, the body reserves some of this energy in ATP. The ATP is distributed within the cells and provides energy for certain reactions when it is needed.

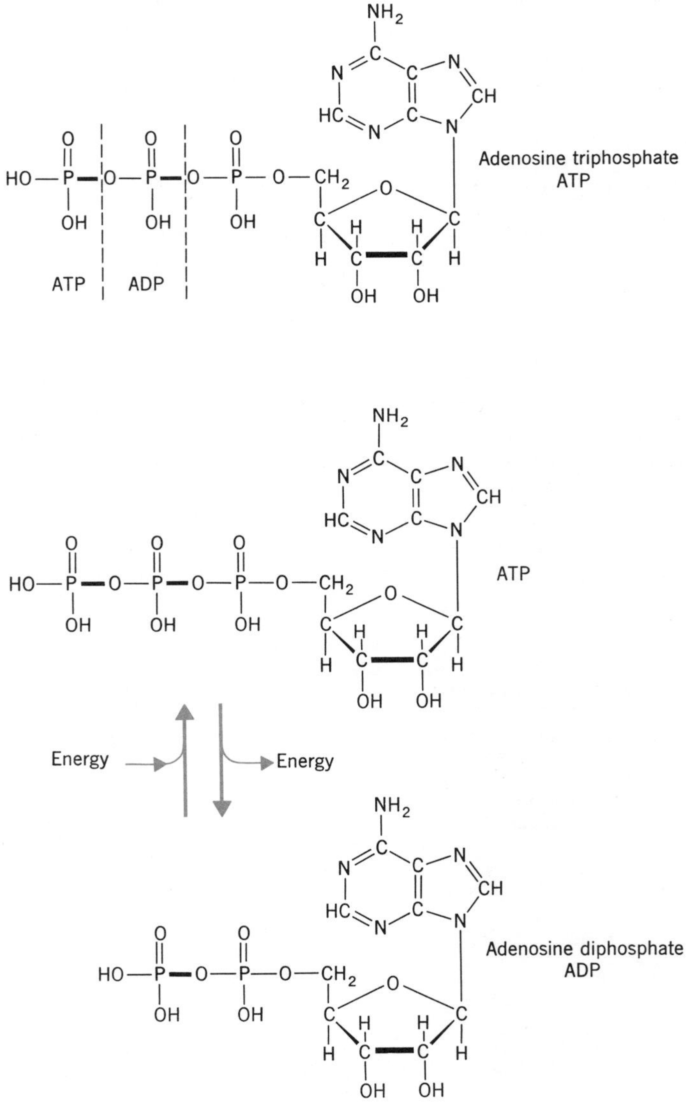

Figure 13-21
ATP forms ADP by the loss of a phosphate unit. Energy is released as ATP forms ADP. ADP forms ATP by the addition of a phosphate unit. Energy is absorbed as ADP forms ATP.

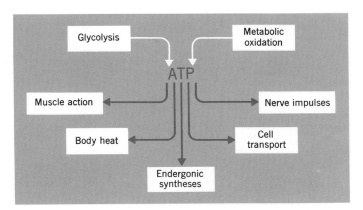

Figure 13-22
The functions of ATP
in the body.

The functions of ATP are given in Figure 13-22. ATP is an energy source utilized in the synthesis of chemicals such as nucleic acids, proteins, and fats used in new body cells. Energy from ATP is converted to mechanical work of muscle contraction. Electrical energy of nerve impulses comes from the energy of ATP. Some ATP energy is used in the transport of chemicals into and out of cells, and some is given off as body heat.

The oxidative metabolism of carbohydrates taking place in the mitochondria (see Section 13-7) of the cells begins with a sequence of reactions in which glucose is converted to lactic acid. This sequence, called **glycolysis,** results in the formation of some ATP. See Figure 13-23. Some of the lactic acid enters another sequence of reactions called the **Krebs cycle,** or **citric acid cycle,** illustrated in Figure 13-24. Note that body fats also supply starting material for the citric acid cycle. As illustrated, the citric acid sequence of reactions is cyclic in the sense that the final product in the series of reactions returns to begin the reaction sequence again. Note that carbon dioxide is given off in the cycle. This is the carbon dioxide which is one of the products of the oxidative metabolism. Note

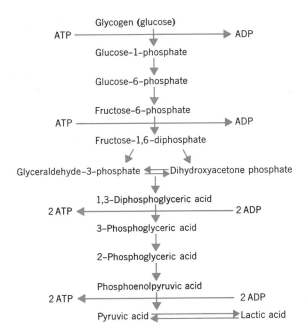

Figure 13-23
The glycolysis reaction path in which glucose is converted to lactic acid.

Figure 13-24
Citric acid, or Krebs
cycle, and the
respiratory chain.

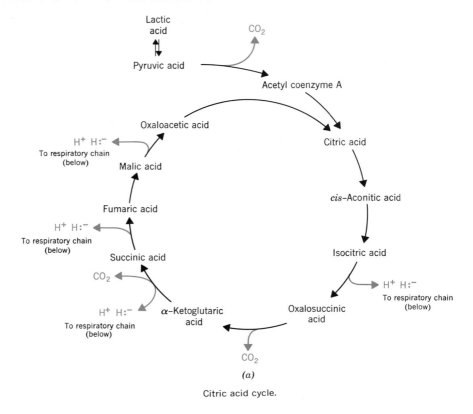

(a)

Citric acid cycle.

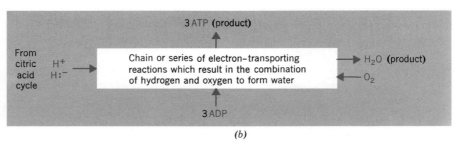

(b)

Respiratory chain.

also that at certain points in the citric acid cycle hydrogens (both nuclei and electrons of hydrogens) are transferred to a third reaction sequence called the **respiratory chain.** The respiratory chain results in the combination of the hydrogen with oxygen to form water. The oxygen is supplied to the cells by the hemoglobin in the blood. The water produced is the other product of oxidative metabolism. The respiratory chain also produces ATP. In fact, the respiratory chain is the cell's main source of ATP. (See Figure 13-24.) The coupled sequences of glycolysis, the citric acid cycle, and the respiratory chain is the complex reaction pattern through which the body carries out the oxidation of carbohydrates to carbon dioxide and water and captures much of the energy of this oxidation in ATP molecules.

LIPIDS The various uses of lipids in the body are summarized in Figure 13-25. Digested lipids are absorbed into the body through the lymph system. In the

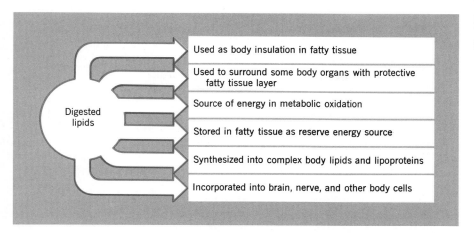

Figure 13-25
The uses of lipids in
the body.

lymph system, the digested lipids are resynthesized into glycerol esters before they enter the blood and are distributed throughout the body. In the blood, the water-insoluble lipids become associated with proteins. These lipoproteins are soluble in the blood. A certain amount of digested lipids is used to synthesize lipids which are used by the body in various cells. Furthermore, the body can synthesize certain lipids from carbohydrates in a sequence of reactions called the **lipigenesis cycle.** Nerve and brain tissue contain significant amounts of such biosynthesized lipids.

A portion of digested lipids are deposited in certain areas of the body tissue as adipose tissue. This **adipose tissue** is known as body fat. A certain amount of this fat is needed for body insulation. However, too much body fat is not advantageous and can be a strain on the heart. Eating too much fat- and carbohydrate-

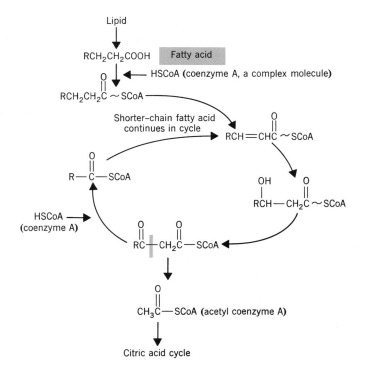

Figure 13-26
The fatty acid
cycle—fatty acids from
lipids are degraded to
acetyl coenzyme A,
which can enter the
citric acid cycle. A
given fatty acid will
continuously pass
through the cycle until
it is degraded.

containing foods can result in deposition of excess body fat. Of course, individuals vary greatly in their ability to lose or gain weight by changing their eating habits. Consequently, dieting for weight loss should be done cautiously and according to the advice of a doctor. Most of the organs of the body are surrounded by fat layers for protection against injury.

Body fat is also a form of stored energy in the body. The lipids in the body fat can be utilized for energy. Lipids can leave the adipose tissue and become oxidized in a sequence of reactions called the **fatty acid cycle.** As shown in Figure 13-26, the product of this cycle can enter the citric acid cycle. Thus, body fats can serve as energy sources much like carbohydrates. This is an example of the versatility of the metabolism of the body. That is, the body can use carbohydrates directly for energy, convert carbohydrates to lipids for storage, and use stored lipids for energy when needed.

PROTEINS   The various uses of digested proteins are summarized in Figure 13-27. The amino acids from digested proteins enter the amino acid pool of the body. From the amino acid pool, various proteins are synthesized to be incorporated into new cells for body tissue. In fact, the primary use of digested proteins is the formation of new cells. Other uses are the synthesis of certain enzymes, hormones, and plasma proteins. Of course, to synthesize specific proteins the body needs the correct amino acids. Certain amino acids which are needed are synthesized from other amino acids in the amino acid pool. However, the body is not capable of synthesizing some types of amino acids. These amino acids are called the **essential amino acids** and must be present in the diet. See Section 14-3 for a discussion of the essential amino acids.

Since amino acids are nitrogen-containing compounds, they can serve as starting material for the synthesis of some nitrogen-containing physiologically active compounds. An example of such a nitrogen-containing compound is epinephrine or adrenalin:

$$CHOHCH_2NHCH_3$$

OH

OH

Epinephrine is a hormone which is released by the adrenal glands at times of stress or emergency. This hormone acts as a heart stimulant and increases the

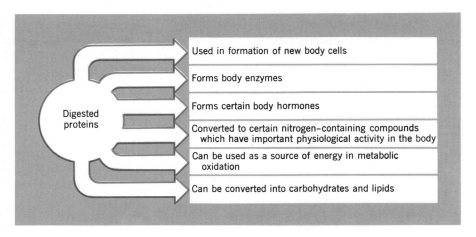

Figure 13-27
The uses of digested proteins in the body.

Digested proteins

Used in formation of new body cells

Forms body enzymes

Forms certain body hormones

Converted to certain nitrogen–containing compounds which have important physiological activity in the body

Can be used as a source of energy in metabolic oxidation

Can be converted into carbohydrates and lipids

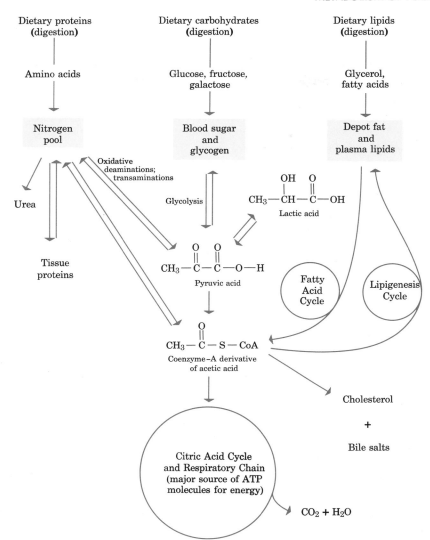

Figure 13-28
Interrelations of metabolic pathways. From *Elements of General and Biological Chemistry*, 3rd ed., by John Holum, John Wiley & Sons, New York, 1972.

rate of the heartbeat. Another function is to increase the rate of glycogen decomposition into glucose which is used in muscles to form ATP used in muscle contraction. Thus, epinephrine is capable of stimulating extra body energy when it is needed.

Proteins can also be converted to molecules which in turn can be converted to carbohydrates or lipids (see Figure 13-28). Furthermore, proteins can be converted to molecules which can enter the citric acid cycle. Thus, in this way proteins can act as a source of energy. If a diet consists of insufficient carbohydrates and fats but excess proteins, some of the protein can be converted to fats and carbohydrates. Nutrition is discussed in Section 14-2. A starving person will begin to use amino acids from body proteins for energy. This can result in the inability to generate new cells and protein deficiency. In fact, a starving or undernourished person will actually consume his own body cells for necessary amino acids and energy.

The overall relationships and functions of the proteins, carbohydrates, and lipids are summarized in Figure 13-28.

BIBLIOGRAPHY BOOKS

Asimov, I. *The Chemicals of Life*. New York: New American Library, 1962.
Holum, J. R. *Elements of General and Biological Chemistry*. 3rd ed. New York: Wiley, 1972.
Maxwell, K. F. *Chemicals and Life*. Belmont, Ca.: Dickenson, 1970.
Watson, J. D. *The Double Helix*. New York: New American Library, 1968.
Wood, J. M. *Enzymes and the Environment*. Tarrytown-on-Hudson, N.Y.: Bogden & Quigley, 1972. 36 pp.

ARTICLES AND PAMPHLETS

Bonner, J. T. "Hormones in Social Amoebae and Mammals" *Scientific American* **220**, 78 (June 1969).
Clark, B. F. C., and K. A. Marcker. "How Proteins Start." *Scientific American* **218**, 36 (January 1968).
Crick, F. H. C. "The Genetic Code." *Scientific American* **207**, 66 (Oct. 1962).
Crick, F. H. C. "The Genetic Code: III." *Scientific American* **215**, 55 (Oct. 1966).
Edelman, G. M. "The Structure and Function of Antibodies." *Scientific American* **223**, 34 (Aug. 1970).
Fraser, R. D. B. "Keratins." *Scientific American* **221**, 87 (Aug. 1969).
Goodenough, U. W., and R. P. Levine. "The Genetic Activity of Mitochondria and Chloroplasts." *Scientific American* **223**, 22 (Nov. 1970).
Gross, J. "Collagen." *Scientific American* **204**, 121 (May 1961).
Kendrew, J. C. "The Three-Dimensional Structure of a Protein Molecule." *Scientific American* **205**, 96 (Dec. 1961).
Loewenstein, W. R. "Intercellular Communication." *Scientific American* **222**, 79 (May 1970).
Mayerson, H. S. "The Lymphatic System." *Scientific American* **208**, 80 (June 1963).
Nirenberg, M. W. "The Genetic Code: II." *Scientific American* **208**, 80 (Mar. 1963).
Ptashne, M., and W. Gilbert. "Genetic Repressors." *Scientific American* **222**, 36 (June 1970).
Watson, J. D., and F. H. C. Crick. "A Structure for Deoxyribose Nucleic Acid." *Nature* **171**, 737 (1953).
Watson, J. D., and F. H. C. Crick. "Genetical Implications of the Structure of Deoxyribonucleic Acid." *Nature* **171**, 964 (1953).
Wood, J. E. "The Venous System." *Scientific American* 86 (Jan. 1968).
Yanofsky, C. "Gene Structure and Protein Structure." *Scientific American* **216**, 80 (May 1967).
Zuckerkandl, E. "The Evolution of Hemoglobin." *Scientific American* **212**, 110 (May 1965).

QUESTIONS AND PROBLEMS

1. What is biochemistry?
2. Which four elements comprise 99 percent of all atoms of living organisms?
3. What is a carbohydrate? Give an example.
4. Give an example of a disaccharide and indicate which monosaccharides make it up.
5. Describe the structures of starch and cellulose in terms of glucose units.

6. What are simple lipids?

7. What are fatty acids?

8. What does the term polyunsaturated mean?

9. What are steroids? Give an example.

10. What are proteins?

11. Using the abbreviations for the amino acids, give the six possible tripeptides which can be formed from the amino acids glycine (Gly), serine (Ser), and Leucine (Leu).

12. Describe some of the functions of proteins in the body.

13. How many different kinds of common amino acids are found in proteins?

14. What is the difference between the primary, secondary, and tertiary structures of proteins?

15. What are enzymes?

16. Describe the lock-and-key theory of enzyme function.

17. Using the names of the following enzymes, state the substrate with which each enzyme is associated:

    (a)   sucrase   (b)   lactase   (c)   dipeptidase

18. Describe the primary structures of RNA and DNA in terms of the building units which comprise these nucleic acids.

19. Where is DNA located in cells?

20. Give a description of the double helix theory of the structure of DNA.

21. What is meant by the term genetic code?

22. How is genetic information passed from parents to children?

23. Describe the DNA-controlled synthesis of enzymes including the role of ribosomes, mRNA, and tRNA.

24. What three major components make up the formed elements of blood?

25. What is the difference between blood plasma and blood serum?

26. What function do erythrocytes have in blood?

27. Describe the oxygen transporting function of hemoglobin in the blood.

28. What is the main function of leukocytes in the blood?

29. Explain how blood is typed into A, B, AB, and O groups.

30. Describe the Rh factor in blood.

31. What are hormones and where do they originate in the body?

32. Describe what happens to the following foods during digestion and indicate the part of the digestive tract involved in the digestion of each food:

    (a)   carbohydrates   (b)   lipids (fats and oils)   (c)   proteins

33. State some of the uses of the following foods in the body:

    (a)   carbohydrates   (b)   lipids   (c)   proteins

34. What are the function and source of ATP in the body?

CHEMISTRY: HEALTH, MEDICINE, AND DRUGS

14

14-1 CHEMISTRY AND BODILY FUNCTION

The human body is an intricate collection of chemicals working in concert to maintain vital functions. The chemical complexity of bodily functions has not been matched by any chemical processes devised by humans. In fact, we have just begun to understand the chemistry of a few of the numerous chemical processes which occur in a living being. Of course, we do know that the body can malfunction because of a genetic defect, dietary deficiency, invasion of bacterial or viral enemies, or some physical injury.

Throughout the centuries of recorded history man has sought to relieve sickness and disease by surgical entry into the body or by administration of foreign materials to the ailing individual. Modern medicine has developed from this tendency of man to try to repair his bodily malfunctions rather than accept them as inevitable. The ancients certainly contributed to the development of medicine, but a Swiss physician named Paracelsus (1493–1541) is considered to be one of the first to apply chemistry to medicine in a logical manner. He introduced the use of numerous drugs. Since Paracelsus, the knowledge of bodily functions and the use of drugs to combat disease has grown into the highly sophisticated practice of medicine known today. However, most drugs were discovered by trial and error using test animals and sometimes humans. Numerous medicinal drugs were discovered as natural products in plants, and more recently many drugs have been synthesized by pharmaceutical chemists.

Today, a doctor has literally thousands of medicinal chemicals which can be used to treat a variety of diseases. It is interesting that we do not know exactly how most medicines function in the body, but we know the effects. For instance, aspirin is obviously an effective pain reliever, but we do not know its exact physiological function. In fact, the effects of medicines illustrate how little we know of the intricate functions taking place in our bodies. Who would have thought that a chemical produced by the mold *Pencillium notatum* would turn out to be the wonder drug called penicillin? Of course, a fortuitous accident coupled with the prepared mind of Alexander Fleming did allow the discovery of the antibiotic properties of penicillin and established the basis for the discovery of other antibiotics. (See Section 14-9.)

The history of medicine is filled with numerous startling discoveries. The realization that certain bacteria caused disease and that our bodies combatted these diseases in a chemical manner allowed for the development of vaccines and sera (see Section 14-9). Our thought processes involve chemical reactions, and we have learned that certain simple chemicals can change our moods or in some cases actually play chemical tricks on the mind (hallucinations). Some chemicals are depressants and some are stimulants, while others, like LSD, are hallucinogens. (See Sections 14-10–14-17.) Actually, we are just beginning to unravel the intricate chemistry of the mind. It is interesting that certain mental disorders once considered to be wholly psychological can be controlled by chemicals. For instance, many manic depressives once thought to be incurably insane can now be cured with small doses of lithium chloride.

Medical evidence and clinical observation has shown us that we have to eat the correct kinds of foods to remain healthy. Nutrition is discussed in Section 14-2. Moreover, it has been found that certain organic substances (vitamins) and certain inorganic substances (minerals) must be present in tiny but sufficient amounts so that our bodies can function normally. The absence of just one of these critical substances can lead to sickness and death. Vitamins are discussed in Section 14-6 and minerals and trace elements are discussed in Section 14-8.

14-2   NUTRITION

The body requires certain kinds and amounts of **food nutrients.** Carbohydrates, fats, and proteins are nutrients which meet the energy needs of the body. Nutrients involved in the building and renewal of body cells and metabolic functions of the body include carbohydrates, fats, proteins, vitamins, and mineral elements. Let us consider the sources and needs of the three basic food nutrients—carbohydrates, fats, and proteins.

CARBOHYDRATES   Carbohydrate-containing foods are the most common source of food energy in the world. People in underdeveloped but highly populated countries (e.g., in Latin America, India, etc.) consume larger percentages of carbohydrate-containing foods in their diets than the people in developed countries. Carbohydrate-containing foods include cereal grains, sugars, legumes (beans and peas), and roots. The sources of food carbohydrates are shown in Table 14-1. Rice and wheat are the nutrient staples of the world, while corn, legumes, potatoes, and other root crops are used only in certain parts of the world. On a per capita basis, developed countries use much more sugar (sucrose) as a carbohydrate source than the underdeveloped countries. Annual sugar consumption in industrialized countries averages about 50 kilograms per capita, while underdeveloped countries average about 3 kilograms per capita.

Recall from Section 13-18 that the main function of carbohydrates is as a source of energy. A small amount (around 0.5 kilogram) of carbohydrates is

Table 14-1 Carbohydrate Sources	Carbohydrate	Source
	Glucose	Fruits, vegetables, honey, corn syrup, invert sugar
	Fructose	Fruits, vegetables, honey, corn syrup, invert sugar
	Sucrose	Sugar cane, sugar beet (refined sugar, powdered sugar, brown sugar)
	Maltose	Malted cereal grains, hydrolysis of starch
	Lactose	Milk sugar
	Starch	Grains and cereal foods, legumes, potatoes, other root vegetables, and green bananas
	Cellulose (nondigestible)	Cereal bran, fruit and vegetable peeling and fiber

	Component (percent of weight)				
Product	Water	Fat	Protein	Lactose	Minerals
Milk	87.3	3.7	3.5	4.8	0.7
"Skim" milk	90.5	0.2	3.5	4.0	0.8
Light cream	72.4	20.0	3.0	4.0	0.6
Buttermilk	90.8	0.5	3.0	5.0	0.7
Butter	15.6	81.0	1.0	—	2.4
Cheese, "American"	37.0	33.0	24.0	—	6.0
Cottage cheese	70.0	1.0	25.0	—	21.0
Creamed cheese	43.0	40.0	14.0	—	3.0
Whey	93.0	0.3	1.0	5.0	0.7

Table 14-2
Typical Composition
of Dairy Products

stored in the body as glycogen or body starch. Excess carbohydrates in the diet can be converted to body fat. Overall carbohydrates are essential in the diet, but in excess they can contribute to overweight. The minimum carbohydrate content of a typical diet should be about 100 grams per day.

FATS OR LIPIDS Fats are also a good source of energy in the diet and, since some fats are stored in fatty tissue deposits in the body, they serve as a reserve energy source. Most of our fat intake comes from meat, dairy products, and vegetable oils (cooking oils, shortenings, and margarines). The fat content of meats and foods containing vegetable oils is quite variable. However, the fat content of dairy products is known with more certainty, as shown in Table 14-2. Per capita fat consumption in developed countries is much higher than in underdeveloped countries.

The amount of fat needed in the diet has not been established. It is certainly essential as a source of energy and to provide fatty deposit insulation for the body. However, excess fat consumption is one factor in overweight and possibly heart disease. Certain clinical studies have indicated that a diet high in polyunsaturated fats (see Section 13-3) and low in saturated fats is good for persons who are prone to heart attacks. These reports and subsequent public demand have stimulated the food industry to produce cooking oils and margarines high in polyunsaturated fats. Regular margarine might contain 6–15 percent polyunsaturated acids, while the polyunsaturated variety will contain 22–48 percent polyunsaturated acids. Authorities differ on whether it is necessary to beneficial for the general population to increase their consumption of polyunsaturated fats.

PROTEINS Proteins (Greek: *proteios* = of the first quality) are fundamental food nutrients. Protein-containing foods supply the nitrogen and some phosphorus and sulfur which are vital to the body. The amino acids from digested proteins are needed to structure the various proteins of the body (see Section 13-8). Common protein-containing foods include meat, fish, poultry, dairy products, and eggs. However, legumes, nuts, and cereal grains also can supply protein.

Recall from Section 13-4 that about twenty amino acids are common to animal and plant proteins. The body can synthesize some of these amino acids that are needed to form body proteins. However, the amino acids that the body cannot synthesize must be present in the diet. These amino acids are called the **essential amino acids.** The eight essential amino acids are

isoleucine	lysine	phenylalanine	tryptophan
leucine	methionine	threonine	valine

These amino acids must be present in the amino acid pool (see Section 13-8), and they must be present at the same time so that the body can build proteins. Moreover, these amino acids must be present in the body in proper proportions so that specific proteins can be formed.

14-3 PROTEINS IN FOODS

Protein-containing foods differ in the content of essential amino acids. The total amino acid content is not important, but the total content of essential amino acids and their relative amounts is important. In other words, a person could eat large amounts of protein-containing foods and still suffer from protein deficiency if the food does not have the correct balance of essential amino acids. If the correct amino acid balance is not present, the excess amino acids cannot be used in protein synthesis but are used by the body as a source of energy. To compare foods it is desirable to have a way to express the amino acid content of the food in terms of the essential amino acid distribution needed by the body. The **biological value** of a food is the ratio of the nitrogen retained in the body for growth and maintenance to the nitrogen absorbed in digestion. This expresses the fraction of the protein in the food which is used by the body. Another factor is how much of the protein of a food is digested. This is termed the **digestibility**. The **net protein utilization (NPU)** of a food is the product of the biological value and digestibility. The NPU expresses the fraction of the amount of protein we eat that becomes available as the required amino acid mixture. Thus, the NPU values of foods can be used to compare the foods as usable protein sources. Table 14-3 shows the total protein content and NPU values for some typical foods. Note that certain foods have very high NPU values and are thus excellent protein sources. This does not mean that foods low in NPU values should not be eaten. In fact, as shown in Figure 14-1, if protein foods are selected so that their essential amino acid contents are complementary, they can be eaten together as an effective protein source.

The daily amount of protein required by a person has been determined. Of course, individuals differ in their protein needs, and foods differ in their protein quality. Medical researchers monitor the nitrogen content of food eaten by test subjects and, on the same day, measure the nitrogen loss by the body in urine and feces. If the nitrogen in the food balances the nitrogen loss, a nitrogen balance exists. Then the subjects are placed on protein-free diets and the nitrogen loss is measured. The total nitrogen loss equals the minimum amount of protein which must be eaten. The minimum protein needs can be adjusted for individual differences. Such research reveals that the daily protein requirement is about 0.3 gram per pound of body weight. This requirement refers to usable protein and should be adjusted for the NPU values of the food eaten. This adjustment can be

Table 14-3
Protein Content and NPU Values for Typical Foods

Food	Protein content (percent of weight)	NPU
Soybean flour	42	61
Cheeses	22–36	70
Meat and poultry	20–32	67
Nuts and seeds	18–32	43–58
Dried legumes	20–25	38–61
Fish	18–24	80
Grains	8–14	51–70
Eggs	13	94
Milk	3.5	82

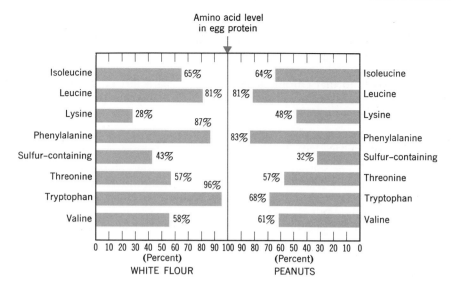

Amino acid level in egg protein

WHITE FLOUR	PEANUTS
Isoleucine 65%	64% Isoleucine
Leucine 81%	81% Leucine
Lysine 28%	48% Lysine
Phenylalanine 87%	83% Phenylalanine
Sulfur–containing 43%	32% Sulfur–containing
Threonine 57%	57% Threonine
Tryptophan 96%	68% Tryptophan
Valine 58%	61% Valine

0 10 20 30 40 50 60 70 80 90 100 90 80 70 60 50 40 30 20 10 0
(Percent)   (Percent)
WHITE FLOUR   PEANUTS

Figure 14-1
White flour and peanuts when eaten together are nearly as good a protein source as egg protein. Egg protein is one of the best sources of essential amino acids. White flour and peanuts compliment one anothers' essential amino acid deficiencies. However, lysine- and sulfur-containing amino acids correspond to only about 75 percent of egg protein, making flour and peanuts only about 75 percent as good as egg protein.

made by dividing the requirements by the average NPU value of foods and multiplying by 100:

$$\left(\begin{array}{c} \text{basic} \\ \text{protein requirement,} \\ \text{grams per pound} \end{array}\right)\left(\frac{100}{\text{NPU}}\right) = \text{actual protein requirement}$$

For instance, a diet consisting of the protein foods meat, dairy products, and eggs is typical in the United States. Such a diet has an average NPU of about 70 percent. Thus, the actual protein requirement is

$$\left(\frac{0.3 \text{ gram}}{\text{pound}}\right)\left(\frac{100}{70}\right) = \frac{0.4 \text{ gram}}{\text{pound}}$$

The daily protein intake by an individual can be calculated by multiplying the actual requirement by the body weight. Using your body weight, calculate your own protein requirement. Of course, if your diet consists of other sources of protein, you could calculate the actual requirement using the NPU values of the foods you eat.

14-4 MALNUTRITION

Malnutrition is apparently not a major problem in the United States, but a 1968 survey by the U.S. Public Health Department was revealing. The survey of the nutritional level of 12,000 randomly selected persons in low-income areas of the country showed that some 17 percent existed on diets that did not meet acceptable nutritional standards. Many of the people were found to be suffering from vitamin and protein deficiencies. In contrast, it is estimated that 15 percent of Americans are overweight.

Malnutrition is prevalent in the underdeveloped countries of the world. It is said that two-thirds of the population of the world goes to sleep hungry each night. One of the major health problems in the world is an infant protein deficiency disease called **Kwashiorkor.** It is common in Africa, Latin America, and parts of the Orient. The disease usually arises when an infant is weaned from the mother's breast and put on a high-carbohydrate, low-protein diet. The infant fails to grow, muscles develop poorly, the skin and hair change texture and color, diarrhea occurs, and the abdomen bloats. Kwashiorkor often leads to death and, if the child recovers, he usually suffers permanent brain damage. The disease has been successfully prevented in some regions by the distribution of high-protein food supplements.

Nutritionists hold that chronic malnutrition is characterized by physical and mental inactivity. Furthermore, a pregnant woman suffering from malnutrition can produce a baby having permanent brain damage.

14-5 CALORIE COUNTING

Perhaps you or someone you know has been on a low-calorie diet which involves keeping a tally of the number of Calories of food consumed each day. Calorie counting is related to the energy content of food and the energy requirements of the body.

The basic foods can be utilized as energy sources in metabolism:

$$fats + O_2 \longrightarrow CO_2 + H_2O + energy$$
$$carbohydrates + O_2 \longrightarrow CO_2 + H_2O + energy$$
$$proteins + O_2 \longrightarrow CO_2 + H_2O + urea + energy$$

The ability of these substances to provide energy is called the **food value,** or **fuel value.** Fats provide about 9 kilocalories or 9 Calories per gram. The **dietetic Calorie** is 1,000 calories. To distinguish between the calorie and the dietetic Calorie, we will use a capital C when we speak of the dietetic Calorie. Carbohydrates provide about 4 Calories per gram, as do the proteins. Actually, much of the protein is utilized in building body protein rather than as a source of energy. The fuel values for some typical foods are shown in Table 14-4.

The body uses the energy from food for endergonic metabolic reactions, body heat, and physical work. The energy needs of a person depend on many factors. These needs include energy for basic metabolic processes to maintain life and energy for activity or external work. The energy required by these factors and other minor factors can be determined by certain clinical methods. As might be expected, the number of Calories of energy needed by a person depends upon the age, height, weight, sex, and kinds of activities. Table 14-5 gives a list of some typical daily Calorie requirements. Calorie intake can be counted if the percentage of protein, carbohydrate, and fat of food is known. The number of grams of each can be found from the grams of food and multiplied by the energy provided per gram. Alternately, if the fuel value of the food is known, we just

Table 14-4
Some Food Values

Food	Component (percent of weight)			Food value (Calories per gram)
	Protein	Fat	Carbohydrate	
Meat (round steak)	20.5	16.6	—	1.8
Bacon	9.9	65	—	6.5
Milk	3.5	3.7	4.8	0.7
Eggs	13.4	10.5	—	1.6
Cheese, "cheddar"	27.7	36.8	4.1	4.6
Butter	1.0	85	—	7.9
Margarine	1.2	83	—	7.8
Peanuts	25.8	38.6	21.9	5.6
Beans (navy)	22.5	1.8	55.2	3.5

	Daily Calories		
Weight (pounds)	18–35 years	35–55 years	55–75 years
Women			
99	1,700	1,500	1,300
110	1,850	1,650	1,400
121	2,000	1,750	1,550
128	2,100	1,900	1,600
132	2,150	1,950	1,650
143	2,300	2,050	1,800
154	2,400	2,150	1,850
165	2,550	2,300	1,950
Men			
110	2,200	1,950	1,650
121	2,400	2,150	1,850
132	2,550	2,300	1,950
143	2,700	2,400	2,050
154	2,900	2,600	2,200
165	3,100	2,800	2,400
176	3,250	2,950	2,500
187	3,300	3,100	2,600

Table 14-5
Calorie Requirements for People of Average Activity

Based on data of the Food and Nutrition Board of the National Academy of Sciences, National Research Council.

multiply by the number of grams of food. For instance, the number of Calories in 10 grams of typical cheddar cheese is found as follows:

10 grams cheese	28 percent protein	2.8 grams protein
	37 percent fat	3.7 grams fat
	4 percent carbohydrate	0.4 gram carbohydrate

$$2.8 \text{ grams protein} \left(\frac{4 \text{ Calories}}{\text{gram protein}} \right) = 11.2 \text{ Calories}$$

$$3.7 \text{ grams fat} \left(\frac{9 \text{ Calories}}{\text{gram fat}} \right) = 33.3 \text{ Calories}$$

$$0.4 \text{ gram carbohydrates} \left(\frac{4 \text{ Calories}}{\text{gram carbohydrates}} \right) = 1.6 \text{ Calories}$$

$$\text{Total Calories} = 46 \quad \text{Calories}$$

Or, if the fuel value 4.6 Calories per gram of cheese is known (see Table 14-4), the number of Calories in 10 grams of cheese is

$$10 \text{ grams cheese} \left(\frac{4.6 \text{ Calories}}{\text{gram cheese}} \right) = 46 \text{ Calories}$$

In this way it is possible to determine the total daily Calorie intake. Extensive tables of the Calorie contents of typical foods are available.

If food containing more Calories than needed is eaten, much of the food representing the excess Calories can be deposited as body fat. If the Calorie intake is below the requirement, reserve energy (fat deposits) is used up. As a result of this, low-calorie foods and diets are used in an attempt to lose weight.

However, doctors and nutritionists point out that a diet chosen on a basis of Calorie content only may not contain the proper balance of nutrients. Foods are classified into four groups which have complimentary nutrient content. These groups are

1. Meat, fish, poultry, eggs, and dried peas and beans
2. Milk and dairy products
3. Vegetables and fruits
4. Breads and cereals

Daily selection of foods within each of these four groups in reasonable amounts should provide proper nutrient balance. In a nonmeat diet proper protein-containing foods must be carefully selected for essential amino acid content.

14-6 VITAMINS

Organic compounds, other than proteins, carbohydrates, and fats, which are needed in the diet for specific metabolic reactions, are called **vitamins.** Humans cannot synthesize vitamins, but they occur naturally in many foods. The correct balance of such foods must be eaten to supply the vitamins. **Vitamin deficiency** can cause poor health and certain diseases. The discovery of vitamins and the determination of their functions was a very important development in the proper nutrition of humans. Vitamins were first known by letter names, but, as the chemical natures of some were determined, other names came into use. The known vitamins, which are often classed as either fat-soluble or water-soluble, are listed below.

Water-soluble	Fat-soluble
Ascorbic acid (vitamin C)	Vitamin A
Thiamine (vitamin B_1)	Vitamin D
Riboflavine (vitamin B_2)	Vitamin E (α-tocopherol)
Niacin (nicotinic acid)	Vitamin K (menadione)
Cyanocobalamin (vitamin B_{12})	
Vitamin B_6 (pyridoxine)	
Pantothenic acid	
Biotin	
Folic acid (folacin)	

Much scientific effort and research was involved in recognition of specific vitamins and the determination of their identities and their functions. Several important diseases have been found to be results of vitamin deficiencies. Consider, as an example, the case of the vitamin niacin and the disease pellagra. Pellagra, a disease characterized by dermatitis, a sore mouth, diarrhea, delerium, and mental depression, has been known for centuries. It was prevalent among people who consumed corn or maize as a staple of their diet. No cure was known around 1900 and the disease was common to the Southern United States. In 1914, Dr. Joseph Goldberger headed a research team to investigate the causes of the disease. He found that the disease was not caused by infection, but seemed to be dietary. He was able to induce the diesease in some volunteers by feeding them a diet similar to the poor people of the South. The search for a dietary factor which would cure pellagra was carried out for years, and in 1920 an antipellagra factor was found in vitamin B. Vitamin B was used as a cure for another vitamin-deficiency disease called beriberi. Further investigation of vitamin B revealed that it was composed of several vitamins. Vitamin B_1, or thiamine, is the antiberiberi factor. Vitamin B_2, or riboflavine; vitamin B_6 or pyridoxine; and niacin, or nicotinic acid, were the other factors. It was found that niacin,

was the antipellagra factor. These B vitamins were found to be easily synthesized, making artificial supplements available. To prevent dietary deficiencies of some B vitamins in the United States, a program of white flour enrichment has been established for the past 30 years. Enrichment refers to the addition of thiamine, riboflavine, niacin, and iron to white flour. Today, around 80 percent of white flour and bread are enriched, as are certain noodle products.

The biochemical function of all vitamins is not known. However, it is known that many vitamins become part of enzymes. Many coenzymes (see Section 13-5) are vitamins or are synthesized by the body using vitamins as vital parts. It appears that the body is not capable of synthesizing vitamins, but they are vital to the formation of necessary compounds. Niacin, for example, becomes incorporated as part of the coenzyme nicotinamide adenine dinucleotide (NAD) and nicotinamide adenine dinucleotide diphosphate (NADP). These coenzymes are involved in cellular respiration reactions. The deficiency of the vitamin results in impaired cellular respiration. The function and deficiency effects of some common vitamins is shown in Table 14-6.

The recommended daily allowance of some vitamins has been determined by clinical studies. The **U.S. Recommended Daily Allowances (RDA)** are established by the Food and Nutrition Board of the National Academy of Sciences. The RDA replaces the outdated Minimum Daily Requirement (MDR) and is generally twice the MDR. For a few vitamins, the RDA is not known. The requirements vary with age, height, weight, and sex, but an average adult allowance is often used. Vitamins A, D, and E are measured in terms of international units (IU). The units are based upon clinical tests with animals. For example, the RDA for vitamin E is 45 IU. The daily allowance of other vitamins is stated in milligrams (10^{-3} gram) and sometimes micrograms (10^{-6} gram). For example, the RDA for vitamin C is 90 milligrams.

14-7 VITAMIN DEFICIENCIES AND VITAMIN THERAPY

The discovery of vitamins has had a profound effect on the well-being of the population. Unfortunately, there are millions of people in the world who suffer from vitamin deficiencies because of an inadequate diet or lack of knowledge. It is estimated that around 1 million people in India are blind as a result of vitamin A deficiencies. The blindness could probably have been prevented with small doses of vitamin A in their diets.

Reserves of fat-soluble vitamins are stored in body fats. The water-soluble vitamins are not stored and must be ingested daily. Massive doses of some fat-soluble vitamins can cause a disease known as **hypervitaminosis.** However, it is interesting that massive doses of certain vitamins appear to be useful in treating certain diseases. For example, some schizophrenics are treated by massive doses of certain B vitamins and some arthritics are treated with massive doses of vitamin A.

The Nobel prize-winning chemist, Linus Pauling, started a medical controversy in 1970 when he argued that massive vitamin C (ascorbic acid) doses could

Table 14-6
Common Vitamins

Vitamin	General function	Deficiency symptoms	Common food sources
Vitamin A	Needed for healthy bone, teeth, skin, and mucous membrane; helps vision in dim light	Night blindness; easily infected	Fish, liver, eggs, butter, cheese, carrots
Vitamin D	Needed for absorption of calcium and phosphorus; involved in calcium deposit in bones and teeth	Ricketts: softened bones, enlarged joints, deformation of the spine and chest	Exposure to sunshine allows body to synthesize this vitamin
Vitamin K	Needed for proper blood clotting	Hemorrhage and internal bleeding	Available in many foods; can be synthesized in the intestines
Vitamin E	Protects vitamin A and polyunsaturated acids from decomposition; general antioxidant	No deficiency symptoms	Wheat germ oil, cotton seed oil
Ascorbic acid (vitamin C)	Involved in production of collagen proteins; prevents tooth loosening in gums; increases resistance to infection	Scurvy: bleeding under the skin and bruising; loose teeth and bleeding gums	Citrus fruits, tomatoes, cabbage
Thiamine (vitamin B_1)	Encourages healthy nerves and good digestion; involved in glycolysis	Tiredness; depression; loss of appetite; sore muscles; beriberi: polyneuritis (inflammation of nerves), swelling due to fluids in tissue; heart failure	Grain germs, legumes, nuts, milk
Niacin	Needed for respiration enzymes; good digestion; healthy nerves and skin	Pellagra: dermatitis, sore mouth, mental depression, diarrhea	Red meat, liver, coffee

Table 14-6 (Continued)

Vitamin	General function	Deficiency symptoms	Common food sources
Riboflavine (vitamin B_2)	Glucose and protein metabolism; healthy skin and good vision	Cheilosis: lip cracking and bleeding, patchy skin, eyes burning and itching	Red meat, liver, milk, egg whites, green vegetables, fish
Vitamin B_{12}	Coenzyme for several metabolic reactions	Pernicious anemia	Meat, liver, eggs, fish

prevent or combat the common cold. He suggested that individuals have a varying need for the vitamin, and he recommended daily doses of 0.25–10 grams. Many doctors criticized Pauling's idea and claimed that massive doses of vitamin C would not help prevent a cold and could cause other complications. Clinical studies carried out since 1970 have not clarified the issue. The results of the studies have not definitely confirmed or refuted the use of vitamin C therapy for colds. Since vitamin C can easily be purchased, many individuals have experimented on themselves. Some claim that vitamin C definitely prevents colds or at least diminishes the severity of a cold. Others claim that vitamin C does not help them at all. Perhaps further clinical studies will provide an answer to the controversy, or perhaps the effect of vitamin C varies greatly from individual to individual.

14-8 MINERALS AND TRACE ELEMENTS

Around 96 percent by mass of the body consists of chemical combinations of the elements carbon, hydrogen, oxygen, and nitrogen. The remaining 4 percent consists of the mineral elements and trace elements. The **minerals** are inorganic ions which must be part of a healthy diet. Table 14-7 lists these minerals. Among other uses, the minerals make up the body electrolytes. A brief description of the function of these minerals is given below.

CALCIUM The calcium ion, Ca^{2+}, is an important component of bones and teeth. Bones and teeth are essentially a protein substructure with a deposit of the mineral salt hydroxyapatite, $Ca_{10}(PO_4)_6(OH)_2$. Ninety percent of body calcium is in the bones and teeth. Calcium ion is also important in the body fluids. Calcium is thought to be involved in the regulation of ion transport across cell membranes. Calcium appears to be needed in the activation of certain enzymes and in blood clotting. The recommended daily intake of calcium for adults is 800 milligrams. Milk is one of the best sources of calcium.

Table 14-7
Mineral Elements in the Body

Element	Percentage of body weight	Element	Percentage of body weight
Calcium	1.5–2.2	Chlorine	0.15
Phosphorus	0.8–1.2	Sodium	0.15
Potassium	0.35	Magnesium	0.05
Sulfur	0.25	Iron	0.004

PHOSPHORUS About 85 percent of the phosphorus in the body is found combined with calcium in the bones and teeth. The dihydrogen phosphate ($H_2PO_4^-$) and hydrogen phosphate ions (HPO_4^{2-}) are found in body fluids as a buffer system. Phosphorus is incorporated into many important body compounds such as ATP, phospholipids, DNA, and RNA. Sufficient phosphorus is usually available in a normal diet.

MAGNESIUM Much of the magnesium of the body is found as magnesium ion in the bones. The function here is not known. Magnesium ion is found in relatively high concentrations in cell fluids. Magnesium appears to be necessary for the function of many important enzymes involved in respiration and protein and carbohydrate metabolism. Incidentally, magnesium is also an important part of the chlorophyll molecules involved in photosynthesis. The recommended daily intake of magnesium is 300–350 milligrams. Nuts, cereal grains, green leafy vegetables, and seafoods are good sources of magnesium.

SODIUM The sodium ion is the positive ion of highest concentration in the extracellular fluids (interstitial fluid and blood plasma). The amount of sodium ion has an influence on the osmotic pressure of these extracellular fluids. Sodium ion can penetrate cell membranes, and nerve transmission and muscle use involve a temporary exchange of extracellular sodium ion with cellular potassium ion. Large amounts of sodium are usually available in our diets, especially if we use liberal amounts of salt. However, recent research has indicated that excessive use of salt can contribute to high blood pressure.

CHLORIDE The chloride ion is the negative ion of highest concentration in extracellular fluids. It also is present in cellular fluids. Along with sodium ion, it influences the osmotic pressure of extracellular fluids. Chloride is an activator of amylase enzymes and is needed in the formation of gastric hydrochloric acid. Sufficient chloride is available in the diet.

POTASSIUM The potassium ion is the major positive ion in cellular fluids, where it influences the osmotic pressure within the cells. Cellular potassium ion is temporarily exchanged with extracellular sodium ion during nerve transmission and muscle contraction. Potassium ion is necessary for carbohydrate and protein metabolism and protein synthesis. Usually, potassium is abundant in the diet.

SULFUR Most sulfur enters the body in the form of the sulfur-containing amino acids, methionine, cystine, and cysteine. Much of the sulfur is incorporated into the body protein in these amino acid forms. Some sulfur exists as sulfate ion, SO_4^{2-}, in the body fluids. Several important sulfur-containing compounds are involved in metabolism.

IRON Iron exists in the body as iron (II) ion, Fe^{2+}, and iron (III) ion, Fe^{3+}. However, these ions are not found as free ions but in combination with other chemicals. Most of the iron of the body is contained in the hemoglobin and myoglobin, where it is directly involved in oxygen transport. Iron is stored in the cells of the liver, spleen, and bones in the form of an iron–protein combination. This serves as a reserve of iron for the body. However, iron is still needed in the diet. The recommended daily intake for men is 10 milligrams and for women 15 milligrams. Meat, liver, egg yolk, and green leafy vegetables are good iron sources.

TRACE ELEMENTS Trace elements are elements which are required in the diet in very small amounts. The utilization of the important trace elements is in-

Element	Approximate amount in body (milligrams per kilogram)	Location or function in the body
Chromium	0.08	Related to function of insulin in glucose metabolism
Cobalt	0.04	Required for function of several enzymes, part of vitamin B_{12}
Copper	1.4	Required for function of respiratory enzymes and some other enzymes
Iodine	0.4	Located in thyroid gland, need for hormone thyroxine
Manganese	0.3	Required for function of several digestive enzymes
Molybdenum	0.07	Required for function of several enzymes
Zinc	23	Required for function of many enzymes
Fluorine		Found in bones and teeth; thought to be essential, but function unknown
Selenium		Essential for liver function
Silicon		May be essential in humans
Tin		May be essential in humans

Table 14-8
Trace Elements in the Body

dicated in Table 14-8. Usually, the trace elements are supplied in our normal diet. However, a deficiency can cause sickness and disease. To be sure of a sufficient supply of iodide ion, iodized salt (salt with a trace of potassium iodide) is available. The use of iodized salt in the United States has drastically decreased the incidence of goiter, which is an enlargement of the thyroid gland. Very recently, there is evidence of an increase in the incidence of goiter. This seems to be due to the fact that much of the population is using greater amounts of presalted prepared foods which do not contain iodized salt.

14-9 CHEMOTHERAPY

Throughout history, humans have discovered that certain chemicals isolated from plants or compounded from other chemicals could combat certain human diseases. Today, a vast number of natural and synthetic chemicals are used in modern medicine. The treatment of diseases by use of chemicals is called **chemotherapy.** Table 14-9 lists some chemicals used in chemotherapy. Let us consider some of these chemotherapeutic agents.

As shown in Table 14-10, certain diseases are caused by microorganisms such as bacteria, protozoa, and viruses. These microorganisms can infect the body, producing various kinds of side effects. The treatment of virus-caused infections has not been as successful as the treatment of other infections. Nevertheless, the use of chemotherapy against infectious diseases has effectively controlled such infections. Forty years ago, infectious diseases were a major cause of death in the United States. Today, they are a minor cause of death.

Microorganisms often invade the body through the blood stream. In the blood, the white blood cells protect against foreign organisms. When the white cells are not effective or are overwhelmed, an infection occurs. Chemotherapy involves finding agents which will counteract or kill the microorganisms without

Table 14-9
Some Chemicals Used
in Chemotherapy

General class, with examples	Use
Vitamins Vitamins C, D, B_1	Treatment of vitamin-deficiency diseases
Heart drugs Nitroglycerin (dilates arteries) Digitalis (effect rate of heartbeat)	Treatment of heart ailments
Diuretics — regulators of amounts of body fluids Chlorothiazide	Treatment of high blood pressure and heart disease
Alkaloids — nitrogen-containing compounds with a wide variety of physiological activities Ephedrine (asthma and hayfever) Novacaine (dental anesthetic) Quinine (malaria)	Treatment of a variety of ailments and as local analgesics (pain relievers)
Analgesics — pain relievers Aspirin (most widely used analgesic, also an effective antipyretic-fever reducer) Morphine (addictive, but excellent analgesic) Codeine (addictive, but good for minor pain and coughs)	Used to alleviate minor or major pain
General anesthetics Ethyl ether (the first anesthetic discovered; it is seldom used today) Sodium pentothal (used for minor operations) Nitrous oxide (N_2O, sometimes called laughing gas; sometimes used in dental surgery) Halothanes (various halogen–alkane compounds; most often used for major surgery)	Induce sleep and alleviate pain for surgical operations
Antibiotics and sulfa drugs Penicillin (most widely used antibiotic) Sulfapyridine (treatment of pneumonia)	Treatment of certain bacterial and viral infections
Hormones Insulin (treatment of diabetes) Cortisone (treatment of arthritis and other disorders) Sex hormones (used to stimulate sexual activity and as birth control agents)	Treatment of hormone deficiency and to control body processes

Table 14-9
(Continued)

General class, with examples	Use

Vaccines and immune sera
  Scarlet fever, polio, cholera, diphtheria, typhoid, and tetanus vaccines
  Smallpox, measles, and botulism immune sera
	Provides immunity against certain infectious diseases

Psychiatric drugs
  Simple chemical agents
    Lithium chloride (controls or cures certain manic depressives)
  Stimulants or antidepressants
    Imipramine
    Amphetamine
  Depressants and tranquilizers
    Chlorpromazine
    Reserpine
    Meprobamate
  Barbiturates
    Seconal
    Amytal
    Phenolbarbital
  Hallucinogenics
    LSD (lysergic acid diethylamide—general hallucinogenic)
    Mescaline (color hallucinogenic; alkaloid from certain cactus species)
	Used in psychotherapy and to treat psychological disorders

Anticancer agents
  Radioactive phosphorus (treatment of bone cancer)
  Radioactive iodine (treatment of thyroid cancer)
  Methotrexate (treatment of acute leukemia)
  5-Fluorouracil (treatment of breast cancer)
  Mechlorethamine (treatment of Hodgkin's disease)
  Actinomycin D (treatment of certain tumors)
	Counteract, control, and cure certain cancers

killing or injuring the patient. Some chemicals which are effective are called bacteriostatic agents or bacteriocidal agents. **Bacteriocidal** agents kill the microorganisms, and **bacteriostatic** agents control the spread of microorganisms.

By far the most useful chemicals in chemotherapy are the antibiotics. **Antibiotics** are chemicals synthesized by certain microorganisms such as molds and which are bacteriostatic or bacteriocidal. Table 14-11 lists some common antibiotics. The most famous and most often used antibiotic is penicillin, discovered by Alexander Fleming in 1929. Since the discovery of penicillin, many other antibiotics have been found. Many of these were found to be produced by microorganisms in soils. Soil samples have been collected from various parts of the

Table 14-10
Some Diseases
Caused by
Microorganisms

Bacteria- and fungus-caused	Virus-caused	Protozoa-caused	Rickettsiae-caused
Anthrax	Colds	Amebic dysentary	Epidemic typhus
Botulism	Influenza	Malaria	Rocky mountain spotted
Cholera	Measles	Sleeping sickness	fever
Diphtheria	Mumps		Q fever
Gonorrhea	Poliomyelitis		
Lockjaw	Smallpox		
Meningitis	Yellow fever		
Pneumonia			
Scarlet fever			
Syphilis			
Tuberculosis			
Typhoid fever			

Table 14-11
Some Antibiotics

Penicillin G	Tyrothricin
Penicillin V	Terramycin
Streptomycin	Tetracycline
Dihydrostreptomycin	Aureomycin
Bacitracin	Erythromycin
Chloromycetin	

world and investigated for these microorganisms. This soil research has been quite fruitful. For instance, the antibiotic chloromycetin was isolated from Venezuelan soil. The search for new antibiotics is continuing today. Some of the antibiotics are quite effective for treatment of certain bacterial infections. A few of the antibiotics can be used to combat some specific viral infections.

One of the ways the body protects against some infectious diseases is through the production of antibodies which counteract the invading microorganisms which act as antigens. The antigen–antibody phenomena in blood is discussed in Section 13-14. When a person contracts certain infectious diseases for the first time, antibodies are produced in the blood. These antibodies protect against further infection of this type. The person is said to have acquired **immunity** against the disease.

It is possible to stimulate immunity against certain infectious diseases in two ways. One method involves administering **vaccines** containing dead or inhibited disease-causing microorganisms. These vaccines cause the body to produce the appropriate antibodies which result in immunity. The other method involves the administration of **immune sera.** Serum containing the desired antibodies is obtained from immunized animals and injected into a person. The antibodies remain in the blood and protect against a specific infectious disease. Today, it is standard preventative medical practice to be treated with specific vaccines and immune sera.

Aspirin is undoubtedly the most widely used chemotherapeutic agent. Aspirin is acetylsalicylic acid,

Aspirin is an effective analgesic (pain reliever) and antipyretic (fever reducer). It is used in great amounts in most parts of the world. Over 40 billion aspirin tablets, or 15 million kilograms of aspirin, are manufactured in the United States each year. A typical aspirin tablet contains 5 grains or 0.32 gram of aspirin mixed with a binder such as clay or starch. In addition to relieving minor pain and reducing fever, aspirin is used for arthritis treatment. Prolonged use of aspirin can lead to stomach and intestinal disorders. Aspirin is toxic in large doses. A number of deaths are caused by overdoses of aspirin each year.

14-10   DRUGS

The term "**drug**" refers to a foreign substance which, when taken into the body, produces a biological change. Modern medicine is based upon the use of a wide variety of medicinal drugs. However, some of these chemicals are used excessively and illicitly. Furthermore, the search for medically useful compounds has resulted in the discovery of addictive and hallucinogenic drugs. Of course, all drugs are medically useful but potentially dangerous when used incorrectly.

The United States consumes large amounts of medicinal drugs. In 1968, Dr. Paul D. Stolley of Johns Hopkins University compiled data on the drug purchases made in a typical city of 112,000 people. In that year, pharmacies in the city filled about 200,000 prescriptions consisting of around 9 million doses. This amounted to over 80 doses per person annually. This count did not include drugs used and dispensed by hospitals. Around 17 percent of the drugs used in the homes were of the psychotropic or mood-changing type such as stimulants, depressants, and tranquillizers. As a result of his study, Stolley stated the opinion that such psychotropic drugs were overprescribed. He also stated that antibiotics were excessively prescribed for minor illnesses. It is estimated that 1,113,811,000 prescriptions were filled in the United States in 1971. That amounts to over five prescriptions for each man, woman, and child.

Another aspect of drug use in the United States is the high cost. In 1972, Senator Gaylord Nelson of Wisconsin revealed some comparative drug costs. He claimed that one pharmaceutical firm sells 100 tablets of Darvon to American pharmacists for about $7, but charges British pharmacists only about $2. Another firm charges American pharmacists about $40 for 1,000 Serpasil tablets, but sells the same drug to the U.S. Defense Department under the generic (chemical) name of reserpine for 60¢ per 1,000 tablets. Still another firm charges about $20 for 100 tablets of Terramycin in the United States and only about $3.68 in New Zealand. Senator Nelson revealed several additional cases of variable pricing of drugs and is proposing a bill called "The Public Health Price Protection Act."

14-11   ALCOHOL

A socially acceptable drug worthy of discussion is ethyl alcohol. Ethyl alcohol, commonly called alcohol, is the drug found in alcoholic beverages. The alcohol in such beverages is prepared by the fermentation process discussed in Section 12-8. The main effect of alcohol consumption is the depression of the central nervous system. The drinker tends to relax and become less inhibited. Around 0.2 percent alcohol in the blood usually produces drunkenness which takes the form of loss of self-control and physical balance. Frequent consumption of alcohol over long periods can result in **alcoholism.** The alcoholic becomes both physically and psychologically dependent on alcohol. (See below for a discussion of dependence.) In the United States there are thought to be 5–9 million alcoholics of which around 80 percent are males between 30 and 55 years old. Alcoholism is one of the most common diseases in this country and often results in the physical, mental, and social decay of the afflicted person.

Large amounts of alcoholic beverages are consumed in the United States each

Table 14-12
The Alcohol Content
of Some Beverages

Beverage	Percent by volume ethyl alcohol	Proof[a]
Whisky, bourbon, gin, rum, scotch, vodka	40–50	80–100
Liqueurs (many), brandy, cognac	40–50	80–100
Fortified wines: sherry, port, madeira, specialty wines, etc.	20	40
Table wines: burgundy, chablis, claret, champagne, etc.	10–15	20–30
Beer, ale, hard cider	2–8	4–16

[a] United States proof is double the percentage alcohol.

year. Enough alcohol is consumed annually to supply every living American with over 25 gallons of beer and 4 gallons of wines and liquors. The alcohol content of some common beverages is listed in Table 14-12.

14-12 DRUG DEPENDENCE

In addition to the wide-scale use of legal drugs, it is common knowledge that there is a significant amount of illegal drug use in the United States. Several terms are commonly used in discussing drug abuse. Some drugs are called **soft drugs,** since their use is often not as dramatic and devastating as the narcotics or **hard drugs.** The World Health Organization has established definitions for the terms "drug habit" and "drug addiction." These are the following:

> **Drug habit**—a condition resulting from the repeated consumption of a drug. The characteristics of a habit are:
>
> 1. A desire—but not a compulsion—to continue taking the drug
> 2. Little or no tendency to increase dosage
> 3. Some degree of psychological dependence on the drug, but no physical dependence and no withdrawal symptoms
> 4. Any detrimental effects are primarily on the individual user
>
> **Drug addiction**—a state of periodic or chronic intoxication produced by the repeated use of a drug. The characteristics of addiction are:
>
> 1. An overpowering desire or need to continue taking the drug and to obtain it by any means
> 2. A tendency to increase the dosage
> 3. A psychological and generally a physical dependence with possible **withdrawal symptoms**
> 4. Detrimental effects on the individual user and society

Psychological dependence is a mental need for the drug, while **physical dependence** is an actual biological need of the body for the drug. Since the difference between a drug habit and addiction is not always clear, the more general terms **"drug dependence," "physical dependence,"** and **"psychological dependence"** are used in describing drugs. Certain drugs produce a drug tolerance in the user. **Tolerance** results in the need for an increased dosage to obtain the same effect.

With these terms in mind, let us consider some common drugs.

14-13 DEPRESSANTS

BARBITURATES The barbiturates are a group of depressant (hypnotic or sleep-producing) drugs sometimes called "downers." The names and formulas of some common barbiturates are given in Table 14-13.

Barbiturates produce a sedated or relaxed feeling which results from the depression of the central nervous system. A user can experience loss of muscle coordination and mental confusion. Excessive use of barbiturates can result in physical and psychological dependence and a degree of tolerance. Withdrawal from prolonged use can result in anxiety, convulsions, delirium, and even psychotic-like experiences. It is not uncommon for a heavy user of barbiturates to die of an overdose. Tolerance and loss of time perception causes the user to take large doses. The depressing effect slows down the absorption of the drug into the body. The sedative effect produces sleep, with much of the large dose remaining in the stomach. Often, such a dose causes death.

Barbiturates usually are contained in colored capsules and are taken orally. They are among the most abused drugs in the United States. It is estimated that there are over 1 million barbiturate-dependent persons in this country. There is evidence that a large portion of the barbiturates sold illegally in the United States are manufactured by legitimate pharmaceutical companies. This does not mean that these companies are a direct supply of these drugs, but some of their drugs end up on the illegitimate market. Of course, there is also a large supply of illegally manufactured barbiturates, some of which are of questionable purity.

14-14 STIMULANTS

AMPHETAMINES The amphetamines are drugs which enhance mental and physical activity. They are termed stimulants or "uppers." The formulas of the common amphetamines, benzidrine, dexedrine, and methedrine, are shown in Table 14-14. These stimulants are produced in the form of colored tablets of varying shapes. Amphetamines are manufactured by pharmaceutical companies for legitimate medical use. However, as is the case with barbiturates, large amounts of these legitimate drugs end up on the illegitimate mar-

**Table 14-13
Some Common
Barbiturates**

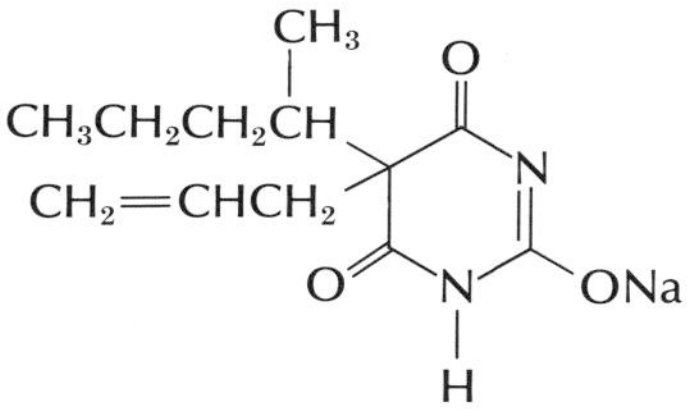

secobarbital
(Seconal)

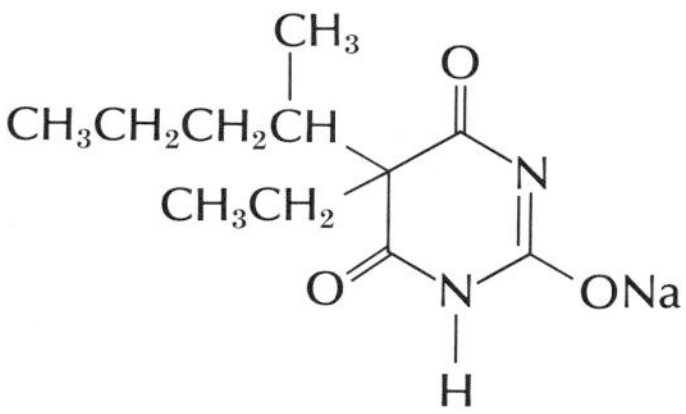

sodium pentobarbital
(Nembutal)

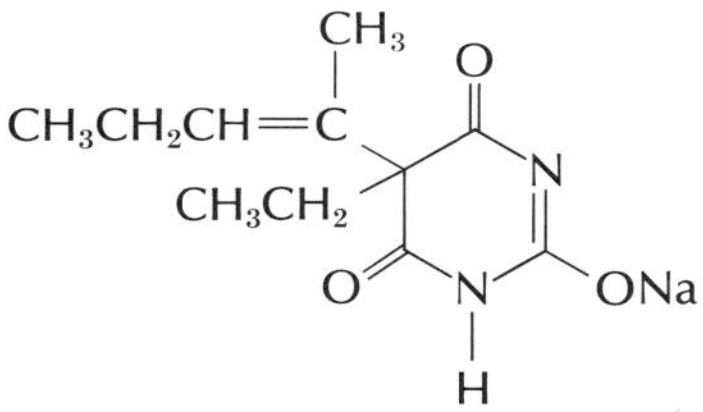

vinbarbital

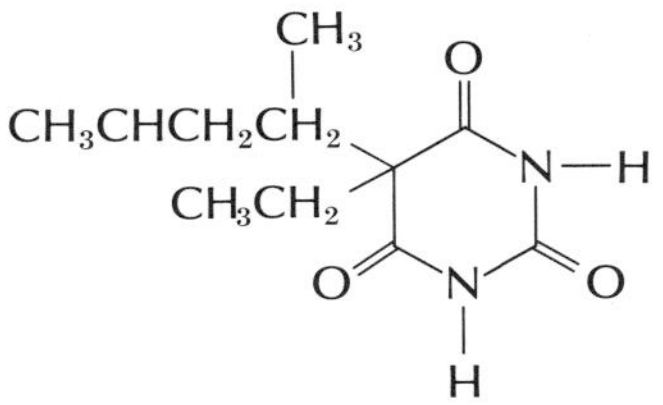

amobarbital
(Amytal)

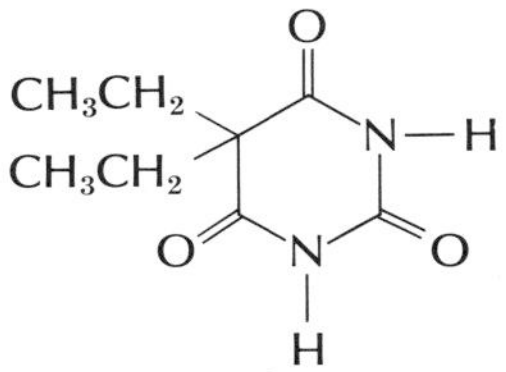

barbital

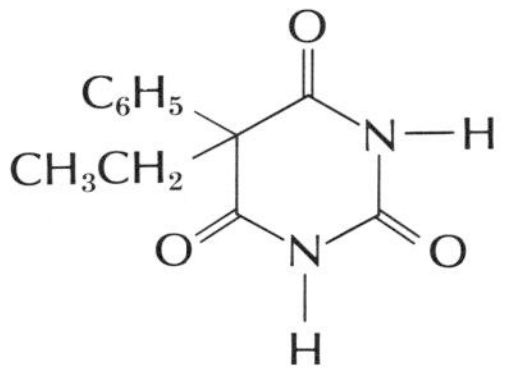

phenobarbital

Table 14-14
Some Amphetamines

benzedrine
(amphetamine)

dexedrine
(dextroamphetamine)

methedrine
(methamphetamine)

ket. Of course, there are also large amounts of illegally manufactured stimulants available.

Amphetamines are stimulants of the central nervous system. They increase the heart rate and the blood pressure, relieve drowsiness, and produce prolonged wakefulness. A typical dose will increase physical and psychological activity, accompanied by insomnia and loss of appetite. A user can develop a strong psychological dependence, and prolonged use can result in a pronounced tolerance. The elevated or "high" feeling of these drugs is often followed by increased tension and anxiety. Some users resort to the use of barbiturates to induce sleep after an amphetamine "high." In some cases, users have been known to take amphetamines and barbiturates at the same time. Needless to say, this practice can be quite damaging to the individual. Hallucinations can result from the use of amphetamines, especially if taken intravenously. Methedrine, or "speed," is the stimulant most often used in this manner for purposes of producing mental fantasies and delusions. Heavy use often results in psychotic episodes, brain damage, and insanity.

Some athletes have been known to use stimulants to increase their physical activity. Several deaths have been attributed to this practice. In many international sporting events, such as the Olympic games, winning athletes must submit urine samples which are tested for amphetamine content.

COCAINE Cocaine is a naturally occurring stimulant found in the leaves of the coca plant, which is native to certain South American countries. It can be isolated from the leaves in the form of a white powder. Presently, cocaine is not used medically in the United States, and the only cocaine available is illegally imported. Users sniff the powder into the nose, where it is absorbed into the bloodstream, or inject a solution intravenously.

Cocaine is a very strong stimulant of the central nervous system. A cocaine dose can cause intense excitement, increase in physical activity, and loss of inhibition. Heavy use can result in hallucinations and paranoid delusions. Cocaine users have been known to become physically violent. A high degree of psychological dependence can develop in users, but physical dependence is not known. Tolerance usually does not occur, and withdrawal illness is slight. The effect of cocaine is so intense that chronic users will often voluntarily abandon use. However, the psychological dependence often brings them back to heavy use. Long-term use can often lead to mental deterioration, paranoid psychological behavior, and deep depression when not using.

14-15 OPIATES

MORPHINE, CODEINE, HEROIN Opium, a chemical mixture obtained from the juices of the opium poppy, has been known for centuries as an effective pain killer and an addictive drug. The two major drugs found in opium are morphine

Table 14-15
The Opiates

Morphine   Codeine   Heroin

and codeine (see Table 14-15). Morphine is an excellent analgesic and relieves all levels of pain. Morphine is used medicinally in controlled dosages. It is highly addictive, and morphine dependence is much like heroin dependence, which is described below. Codeine, or methylmorphine, is also used medicinally as an analgesic and cough reliever. It has a weaker effect than morphine, but is less addictive.

The most notorious of the opiates is heroin which is diacetylmorphine. It is synthesized from morphine and does not occur in opium. Heroin is the most powerful analgesic opiate and can produce an euphoric or well-being state. Some users describe the sensation of a dose of heroin as somewhat like a sexual orgasm. The opiates depress the central nervous system, and induce drowsiness and sleep. They dull the senses and do not produce hallucinations.

Heroin is used in the form of a white powder. It is illegal in the United States, but large quantities are smuggled across the borders. It is a widely abused drug and is common to most large cities. Estimates of the number of heroin addicts in the United States vary from 250,000 to 600,000. Heroin can be sniffed through the nose, but it is normally taken by injection under the skin (subcutaneously) or intravenously ("main line"). Heroin is highly addictive, producing strong physical and psychological dependence with a pronounced tolerance. A chronic user develops a strong desire for larger or more frequent doses. Withdrawal illness often begins within a few hours of the last dose. If the user does not receive a dose, he experiences a profound withdrawal effect. Withdrawal symptoms include muscle cramps, insomnia, restlessness, profuse sweating, nausea, diarrhea, pain, and increased respiratory rate. The addiction can remain with the user over long periods of time, and most, if not all, users have a very difficult time giving up heroin use.

The strong dependency on heroin often causes the user to resort to criminal or antisocial acts to obtain money for the drug. A heroin habit may cost an addict around $100 per day. Moreover, the user's constant quest for the drug often results in his neglect of personal health and nutrition. Overall chronic use can result in physical and psychological degradation. An addict who uses a syringe or needle can easily infect himself or kill himself by injecting an air bubble into his vein. Much of the heroin purchased on the illegal market is of variable purity, since it is diluted ("cut") with other powdery chemicals. An addict can easily die of an overdose (OD) if he injects too much heroin. This can happen if he obtains a dose of high purity when he is used to diluted heroin. However, most heroin addicts probably die of physical ailments caused by neglect of personal health.

Some methods of cure are effective for certain addicts and useless to others. Generally, the success rates of most methods are low. One method which is being tried involves the substitution of a synthetic drug, methadone, for the heroin. Methadone has a morphine-like effect and is addictive, but the withdrawal illness is less intense. An addict receives methadone doses under medical supervision. This allows him to slowly break the habit without having to suffer the heroin withdrawal illness ("cold turkey"). If the addict feels the desire for heroin returning he can obtain more methadone. Some criticize this treatment method as the substitution of one drug for another. Others claim that methadone use can aid the cure of an addict or at least relieve him from his need for an illegal drug.

A common approach taken in the United States to control heroin is to attempt to intercept smuggled heroin and break up smuggling operations. Critics point out that this approach may decrease the heroin supply, but it drastically increases the street price of the drug. Higher prices force the addicts to commit more crimes to obtain enough money.

In Great Britain, heroin addiction is treated as a medical problem, not a criminal problem. Addicts are registered and provided with minimum heroin doses by the government. The idea is to decriminalize heroin addiction, keep contact with the addicts, and take the profit out of heroin sales. The British claim that, although the system has some problems, heroin addiction in their country is minor compared to the United States.

14-16 HALLUCINOGENS

LSD, MESCALINE The most common of the hallucination-producing drugs are LSD (lysergic acid diethylamide) and mescaline (see Table 14-16). Except for some psychiatric research and legal use in religious services (mescaline is used in a rite of the First American Church), hallucinogens originate from illicit sources. Small doses of hallucinogens can result in euphoria and perceptual and visual distortion. Sometimes, the user claims that he sees sound and hears color. Mescaline is known as a color hallucinogen, producing vivid color perception. Hallucinogen use may produce distortions of time and space and vivid hallucinations and illusions. No physical dependence appears with use, but some psychological dependence is known. Tolerance is slight if at all.

LSD use produces an experience known to users as a "trip." Such experiences are reported to vary from highly euphoric with pleasant hallucinations to depression with horrible and frightening hallucinations ("a bad trip"). The experiences vary from individual to individual and from one dose to another. In some cases, users experience a "flashback" or the repeat of a portion of a "trip" at some time after they have stopped using the drug. The effects of LSD on the mind and body are not completely understood. However, there are many cases of chronic users who develop LSD-induced psychological disorders or aggravation of already existing disorders. Sometimes, chronic users have to be restrained during a "bad trip." Furthermore, there are many recorded instances of LSD users committing suicide or exerting physical violence on others.

Table 14-16
Some Hallucinogens

LSD
(lysergic acid diethylamide)

Mescaline

14-17   THE MARIJUANA CONTROVERSY

MARIJUANA OR CANNABIS   Marijuana (pot, hashish) is prepared from the crushed leaves, stems, seeds, and flowers of the cannabis (hemp) plant. This drug has been known for centuries. Use of cannabis was largely limited to Eastern and Middle Eastern countries until the latter part of this century. It is usually smoked or ingested in food and drink. The active principle of the cannabis plant has been found to be **tetrahydrocannabinol (THC):**

THC can be synthesized, and is sometimes used in this form. Some fake THC has been known to be sold on the illicit market. This material usually consists of a mixture of other drugs and little if any of the expensive synthetic THC.

The effectiveness of marijuana is directly related to the THC content. Typical marijuana available in the United States has a THC content of about 1 percent. Domestic and Northern Mexico marijuana has a THC content ranging from 0 to about 1.5 percent. Marijuana from more tropical areas has a higher THC content, up to about 6 percent. Stronger marijuana can have a more pronounced effect on the user. Hashish is a more purified form of cannabis material and may contain up to 20 percent THC. The use and sale of marijuana is illegal in the United States, but it is widely available on the illicit market.

Marijuana is a very mild hallucinogen and its effect is described as similar but not the same as alcohol. The effect of marijuana seems to vary with the individual user and the quality (THC content) of the cannabis. The effects include paresthesia (tingling or creeping sensation on the skin), a happy carefree feeling, and depersonalization; weakness and a relaxed feeling; changes in sensory perceptions; subjective slowing of time; a slight impairment of thinking, memory, and judgment; silliness; and sleepiness. Other reported effects are somewhat uncontrollable laughter, flight of ideas, a desire to talk and be with others, and an increase in appetite. Chronic use has been reported to cause hallucinations (usually from hashish use), a state of apprehension, anxieties, and depression. Since marijuana has fairly recently come into wide use, no research on the long-term effects has been made. However, much research on the use and abuse of marijuana is currently being carried out. It is known that marijuana is not physically addictive, nor does a tolerance develop. However, it does appear to produce a mild psychological dependence with some users, and some cases of strong psychological dependence with chronic users are known.

Marijuana use is controversial. It is estimated that around 20 million Americans have smoked it. Not all of these are habitual users, but there appears to be a large number of people who smoke marijuana for enjoyment and social recreation. A few important points concerning marijuana use can be stated to help clarify the controversy of its use. There is no direct evidence that marijuana use can produce psychological disorders. Hallucinations seldom if ever occur with users of domestic-strength marijuana. There is no basis to the claim that marijuana use will lead to the use of harder drugs. It is true that many hard drug users once used marijuana, but many also used cigarettes and alcohol. However, those who use alcohol and want a better "high" generally drink more alcohol. Those who use marijuana and want a better "high" are statistically likely to experiment with other drugs. Of course, sociological group pressure and the availability of drugs encourages such experimentation. There is no direct relationship between marijuana use and violent behavior or crime. Of course, in the United States it is a crime to use marijuana. Marijuana use does not appear to encourage sexual activity, although some users claim that it does. THC has a much lower toxicity when compared to alcohol.

The legalization of marijuana use in the United States is a much discussed subject. Some say that marijuana will soon be legalized, and others claim it will never be accepted by the general public. Those who favor legalization point out that marijuana smoking is pleasurable and there is no evidence that it is any more dangerous than alcohol or tobacco use. In addition, they point out that a large number of marijuana users already exist. The situation is somewhat like Prohibition in the 1920s. That is, a large number of normally law-abiding citizens are committing legally serious crimes when they use marijuana. They would like marijuana to be as available as alcohol with similar controls. A less permissive view states that marijuana should not be legalized, but its use should not be considered as a serious crime. That is, marijuana use should be decriminalized and its use should be made a personal choice.

Those opposed to marijuana use claim that it is a dangerous drug and can lead the user to harder drugs. Some say the only way to control it is to have strict laws on sale and use. There have been a few cases in which first offenders were sentenced to 20 years in prison for marijuana possession. Of course, this is the extreme, and often first offenders are not jailed. They also point out that controls of legalized marijuana would be much more difficult than our alcohol controls. Not only would a large bureaucratic control agency be required, but marijuana can easily be stolen from growing fields or cultivated by individuals, which could encourage the development of a black market.

The more moderate view is that we have enough trouble with tobacco and alcohol use and the addition of one more socially acceptable drug seems unnecessary. More extensive research into marijuana use and abuse should be accomplished before a decision is made. It appears that such research is being carried out by professional researchers and by individuals who use marijuana illegally.

14-18 MEDICAL GENETICS

Since the discovery of the function of DNA in cells and an understanding of the chemistry of heredity developed, medical genetics has taken on new aspects. It has been known for some time that many congenital diseases result from genetic defects passed on from the genes of the parents. However, it is now possible to trace some of these inborn defects to the genetic code of the genes. As a result, a new field of **gene therapy** is developing which is concerned with alleviating inborn errors of metabolism with or without gene alteration.

Certain genetic abnormalities result from the alteration of the genetic code of a single gene. This gene does not function properly, resulting in absence or alteration of some aspect of metabolism. Over 100 known diseases are attributed to this kind of **genetic defect.** These include hemophelia, sickle cell anemia, muscular dystrophy, albinism, cystic fibrosis, and Tay–Sachs disease. Some experts believe that there may be 100,000 or more of these diseases.

The exact cause of all of these diseases is not known, but some are. As an example, let us consider sickle cell anemia, a disease of the blood common to Blacks and afflicting approximately 300,000 Americans. The characteristics of sickle cell anemia are that when red blood cells of an individual having the disease are subjected to conditions of low oxygen concentration, the blood cells become distorted from the normal circular state to a sickle-like shape, and many of the cells rupture. The sickled conditions, which often arises after exercise, can cause severe pain, bleeding, sore joints, sores that heal slowly, infections, and death. An individual with the disease usually does not live to adulthood. Studies of the blood of parents of those having sickle cell anemia revealed that the parents' blood contained two kinds of hemoglobin. One kind was normal hemoglobin and the other was sickle cell hemoglobin. These parents were not suffering from the disease, but were carriers and could produce children having sickle cell anemia. The disease is hereditary.

Extensive investigations of sickle cell hemoglobin and normal hemoglobin have shown that the two kinds of hemoglobin differ by one amino acid unit in a chain of about 150 amino acid unit protein. Sickle cell hemoglobin contains a valine unit instead of the glutamic acid unit found in normal hemoglobin. This minor difference makes the sickle cell hemoglobin less soluble in the blood under certain conditions. When this condition occurs, the sickling of the blood cells takes place and the individual suffers. Tracing back to the genetic code contained on the genes, it has been found that the abnormal hemoglobin probably results from a difference of only one nucleotide unit in a DNA strand of a gene involved in some step of hemoglobin formation.

The cause of sickle cell anemia is known, but a cure is not yet possible. It is possible to test the blood of various individuals to determine whether they have the disease or whether they are carriers of the disease that could be transmitted to future generations.

Still other genetic abnormalities, like mongolism, involve problems with chromosome numbers and incorrect chromosome arrangements. These diseases are very difficult to deal with at this time. However, treatment of single gene defects or defects of a few genes is possible. One treatment approach, which is used in certain cases, is to supply the chemical missing in the metabolism. A good example of this is the treatment of diabetes in which the individual is supplied with insulin, the enzyme which the body fails to synthesize because of the disease. Other genetic diseases are treated by similar chemical methods and, as an understanding of still other diseases is accomplished, chemical therapy may also be possible for some of them.

Another approach to genetic therapy is to correct the genetic defect by transferring genetic information to the cells. This would involve actually altering or supplementing the genetic code to correct the error. The techniques needed to accomplish this kind of genetic transfer have not been developed, but the process is not inconceivable.

The ability to detect genetic defects in parents, and other techniques such as amniocentesis, have given rise to an increase in genetic counseling. **Amniocentesis** is a technique in which an analysis of the amniotic fluids around the fetus can reveal certain genetic defects in the fetus. **Genetic counseling** is a service rendered to potential parents who are known to be carriers of genetic defects. The intent of the counseling is to inform parents of medical diagnostics and to educate them on the consequences of genetic defects. Critics of genetic counseling claim that it might have a negative impact on the relationship of a married couple and they may prefer not to know the facts. Others claim that it is the right of potential parents to know of possible genetic defects in any children they may potentially bear.

Another aspect of medical genetics is **genetic engineering.** Genetic engineering refers to gross alterations of genetic composition or altering the normal process of reproduction. Many genetic engineering techniques have been attempted with animals and plants. A few potential and tested techniques will be mentioned in this discussion. Note that this discussion is not intended to condone any of these techniques, but rather to suggest their feasibilities.

In vitro fertilization in which an egg is removed from the ovaries, fertilized, and then implanted in the uterus at the proper stage is being attempted. Several variations of this technique could be possible. A woman who produces no eggs could use the egg of another woman, have it fertilized with her husband's semen, and then have it implanted in her uterus for growth. A woman without a uterus could have one of her eggs fertilized and then implanted in the uterus of another woman who volunteers her services. Artificial wombs are conceivable in which a fetus could be developed external to the mother.

It may soon be possible to determine the sex at early fetus stages, to look at the developing fetus with tiny optical probes, and even to obtain a tissue sample or operate on a fetus.

Cloning is the production of genetically identical copies of an individual organism. Cloning has been accomplished with some plants and animals. The cloning of a frog has been carried out and can be used to illustrate the technique. The nucleus of a frog egg cell was replaced by the nucleus of a somatic cell from the intestine of an adult frog. With proper treatment and nourishment the egg behaved as though it were fertilized and developed eventually into an adult frog. The cloned frog was an exact genetic copy of its single parent from which the nucleus of the somatic cell was taken. Some experts state that cloning of mammals will be accomplished within 5 years. The cloning of humans is technically possible. Cloning certainly represents a profound deviation from normal reproduction.

Genetic therapy and genetic engineering give rise to some complex ethical and moral issues. Considering the rapidity at which many technical changes occur, many of these issues might have to be faced in the near future.

BIBLIOGRAPHY BOOKS

Blachly, P. H., ed. *Drug Abuse: Data and Debate.* Springfield, Ill.: Charles C. Thomas, 1970.

Brown, B. J., ed. *Drugs and the Young.* New York: Time Education Program, undated.

Bureau of Narcotics and Dangerous Drugs. *Drugs of Abuse.* Washington, D.C.: U.S. Government Printing Office, 1970.

Bureau of Narcotics and Dangerous Drugs. *Fact Sheets.* Washington, D.C.: U.S. Government Printing Office, 1970.

King, A. R. *Basic Information on Alcohol.* Washington, D.C.: Narcotics Education, 1964.

Pauling, L. *Vitamin C and the Common Cold.* San Francisco: W. H. Freeman, 1970.

Wagner, A. F., and Folkers, K. *Vitamins and Coenzymes.* New York: Wiley-Interscience, 1964.

ARTICLES AND PAMPHLETS

Beecher, H. K. "Anesthesia." *Chemistry* 90 (June 1959).

Collier, H. O. "Aspirin." *Scientific American* 209 (5), 97 (1963).

Faltermayer, E. K. "What We Know About Marijuana—So Far." *Fortune* 96–98, 128–132 (Mar. 1971).

Gates, M. "Analgesic Drugs." *Scientific American* 215 (6), 131 (1966).

Greenberry, L. A. "Alcohol in the Body." *Scientific American* 186 (6), 86 (1953).

Hollister, H. E. "Marijuana in Man: Three Years Later." *Science* 21 (Apr. 2, 1971).

"How Penicillin Kills Bacteria." *Chemistry* 41 (7), 44 (1968).

Numes, F. "LSD—An Historical Reevaluation." *Journal of Chemical Education* **45,** 688–691 (1968).

Solmssen, U. V. "The Chemist and New Drugs." *Chemistry* 40 (4), 22 (1967).

Tunney, J. V., and M. E. Levine. "Genetic Engineering." *Saturday Review of Science* (Aug. 5, 1972).

Young, D. P. "Science Alone Is Not Enough." *Chemistry* (Nov. 1971).

QUESTIONS AND PROBLEMS

1. What are the three basic food nutrients which meet the energy demands and other metabolic needs of the body?

2. Give three examples of carbohydrate-containing foods.

3. What are the major sources of lipids in our diets?

4. What are the sources of proteins in our diets?

5. What are essential amino acids?

6. Why is it not possible to eat just any protein-containing food and expect the body to function correctly?

7. What is meant by the net protein utilization (NPU) of a food?

8. How is it possible to obtain a good protein balance in your diet by eating nonmeat protein sources of low NPU values?

9. Daily bodily protein requirements are 0.3 gram per pound of body weight. Why is it necessary to eat more protein-containing foods than indicated by the requirement?

10. Describe the Kwashiorkor disease.

11. What is meant by daily calorie requirements and calorie counting?

12. What are vitamins? Give three examples.

13. How do many vitamins function in the body?

14. What are nutritional minerals? Give three examples.

15. What are nutritional trace elements?

16. Define the term "chemotherapy."

17. What are antibiotics? Give two examples.

18. Which chemotherapeutic agent is the most widely used in the world?

19. What are the two physiological effects of aspirin?

20. What is a drug?

21. Is alcohol a drug? Why?

22. Define the following terms:
    (a) drug habit
    (b) drug addiction
    (c) physical dependence
    (d) psychological dependence
    (e) tolerance

23. Describe the effects of barbiturates.

24. Describe the effects of amphetamines.

25. Why is cocaine often classified as a hard drug or narcotic?

26. Describe the effects of heroin.

27. What do you feel should be done about heroin addiction in the United States?

28. Describe the effects of hallucinogens.

29. What is marijuana? What is the active ingredient in marijuana?

30. Describe the effects of marijuana.

31. Rank order the following choices as they correspond to your feelings about marijuana use:

    _____ Marijuana use should be a personal decision and should be legalized.

    _____ Strict laws and law enforcement should be exercised against marijuana use.

    _____ Marijuana use should not be legalized.

32. What is a genetic defect?

33. Describe the cause and effect of sickle cell anemia.

34. Do you feel that you would like to have genetic counseling? Why?

35. Do you feel that genetic counseling should be available to married couples if they want it?

36. Rank order the following choices as they correspond to your feelings (1 = most reprehensible to 3 = least reprehensible):

    _____ Genetic therapy by transfer of genetic information

    _____ In vitro fertilization

    _____ Cloning

37. Do you feel that research into genetic engineering should be controlled or limited in any way? Explain.

38. Considering the fact that a portion of the barbiturates and stimulants sold on the illicit market come from legitimate pharmaceutical manufacturers, do you think that manufacturers should be required to keep more strict controls over the sales and distributions of these drugs? Why?

15-1 POPULATION

Modern humans have become aware of their space-time environment. We are able to look back at the ancient and recent past. Combining knowledge of the past with knowledge of the present we are able to make predictions about the future. Such predictions may or may not be true, but they can influence our behavior. One such prediction involves population growth. World population in ancient times and even the recent past is not known with much certainty, but approximations can be made. Current populations in certain countries are known with a fair amount of precision, while the populations of some underdeveloped countries can only be approximated. From the population figures it is apparent that the population of the earth is increasing as shown in Figure 15-1. The population of the earth apparently grew slowly, reaching one-half billion by 1650. Then 250 years later, in 1850, the population reached 1 billion. It took 80 more years to double to 2 billion, and is expected to reach 4 billion by 1975, only 45 years later. This population data reveals that the number of people on earth is increasing, and the rate of increase is increasing. That is, it is taking shorter and shorter time periods to double the population. The reason for this is simply that the birth rate is exceeding the death rate. People are having more children than the number needed to replace them when they die. See Figure 15-2.

The approximate rate of population increase in the world is 2 percent. At this rate, the population of the world will double in about 35 years, increasing from about 3.5 billion to 7 billion. The prediction of a dramatic increase in the population of the world indicates that much effort will have to be expended to increase food production (see Section 15-5). However, when considering population increase problems it is necessary to look at smaller geographical areas such as continents or subcontinents. The predicted population of an area depends upon the current population, the rate of increase, and the amount of migration. The predicted growths in several areas of the world are shown in Table 15-1. From the table it can be seen that developed countries have smaller growth rates and will require longer periods of time to double the population. Certain underdeveloped regions have high growth rates and population doubling is expected to occur over shorter time periods. Since many of these countries have economies and agriculture which marginally support current populations, increases in the number of people may negate any possible improvement.

Figure 15-1
Estimated world population growth since 1 AD.

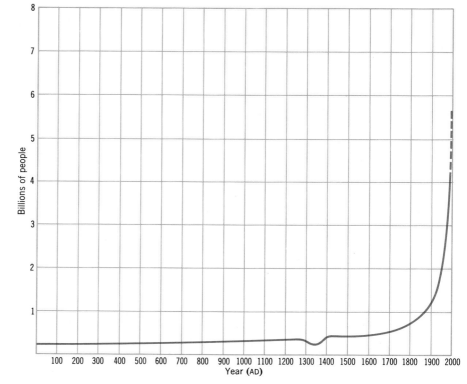

Figure 15-2
How the population growth rate increases. (a) An increase in the rate of growth of the population results simply from couples producing more than two children before they die. Of course, this simple view assumes that the children propagate as their parents did. In localized regions of the world, population growth rates are also affected by migration, a decrease or increase in the death rate, the relative number of child-bearing women, and any birth control practices. (b) Zero population growth is simply a situation in which each set of parents produces two children to replace them. Of course, actual zero population growth is complicated by the fact that not all people are involved in procreation and inward or outward migration can affect the population growth.

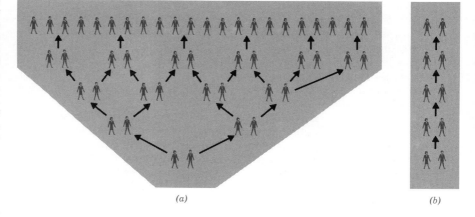

(a)   (b)

You may wonder what role chemistry has in the population dilemma. Actually, certain kinds of chemistry are quite important. Population control by chemical means has been developed (see Sections 15-2 and 15-3). Section 15-6 discusses the use of chemical fertilizers in agriculture, and Section 15-8 describes pesticide use in agriculture.

15-2 CHEMICAL CONTRACEPTION—"THE PILL"

The development of an effective birth control pill was based upon knowledge of the chemistry of the female menstrual cycle. Normally, a woman follows a monthly menstrual cycle in which ovulation occurs. The **birth control pill**

Table 15-1
Predicted Populations
and Doubling Times
for Various Regions of
the World

Region	1975	1990	2000	Approximate doubling time (years)
World	3.7×10^9	5.1×10^9	6.2×10^9	35
Asia (China, Japan, South Asia)	2.2×10^9	2.9×10^9	3.5×10^9	35
Europe	4.7×10^8	5.0×10^8	5.3×10^8	88
Africa	3.9×10^8	5.9×10^8	7.7×10^8	26
Latin America	3.3×10^8	5.0×10^8	6.4×10^8	27
Russia	2.6×10^8	3.2×10^8	3.6×10^8	64
North America	2.4×10^8	3.1×10^8	3.5×10^8	63
Oceania	2.0×10^7	2.7×10^7	3.2×10^7	35

Population figures based on average United Nations predictions.

prevents ovulation, which eliminates the possibility of pregnancy. The function of "the pill" can be understood by considering the **menstrual cycle** which is illustrated in Figure 15-3. Each month the pituitary gland secretes an egg-stimulating hormone. This hormone causes an egg to begin to develop in the ovary. Near the middle of the month the pituitary secretes another hormone which causes the developed egg to be released from the ovary. This is the ovulation step. The two pituitary hormones also cause the ovary to secrete two hormones. As shown in Figure 15-3, the ovary secretes the hormone **estrogen** in increasing amounts, reaching a peak at ovulation. The hormone **progesterone** is secreted by the ovary after ovulation. These two ovary hormones stimulate the growth of the uterus lining (endometrium) to prepare for the implantation of a fertilized egg. The presence of progesterone and estrogen in the blood act to suppress the pituitary gland. This suppression diminishes secretions by the pituitary gland of the stimulating hormone, which in turn diminishes the secretion of estrogen and

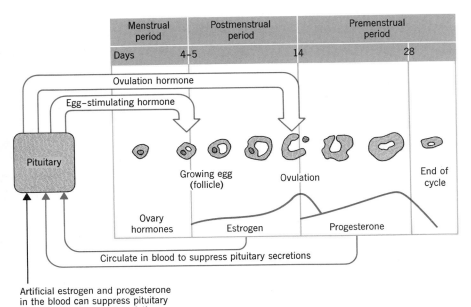

Figure 15-3
The female menstrual cycle.

progesterone from the ovary. As soon as the level of these hormones falls, menstruation occurs, which discharges the endometrium and the unfertilized egg. The drop in the estrogen and progesterone no longer suppress the pituitary and a new cycle can begin. Of course, if an egg becomes fertilized after ovulation, it becomes implanted in the uterus lining and the normal cycle is upset as pregnancy takes place.

As can be seen from the above discussion, the use of estrogen and progesterone can suppress the pituitary and prevent ovulation. Early attempts at such chemical contraception involved a series of injections of these hormones. This was very inconvenient, so an effort was made to find a way to accomplish chemical contraception by use of an orally administered drug. Several progesterone-like compounds called **progestins** were synthesized and found to be effective. The structures of some of these compounds are shown in Figure 15-4. Most oral contraceptives in use today include one of these progestins with a small amount of estrogen. Such pills when used properly have proved to be very effective contraceptives. There appear to be some mild side effects which result from the use

Figure 15-4
Some synthetic progestins used with estrogens in oral contraceptives. A typical pill contains 0.05 milligram of an estrogen and 0.5 milligram or more of a progestin.

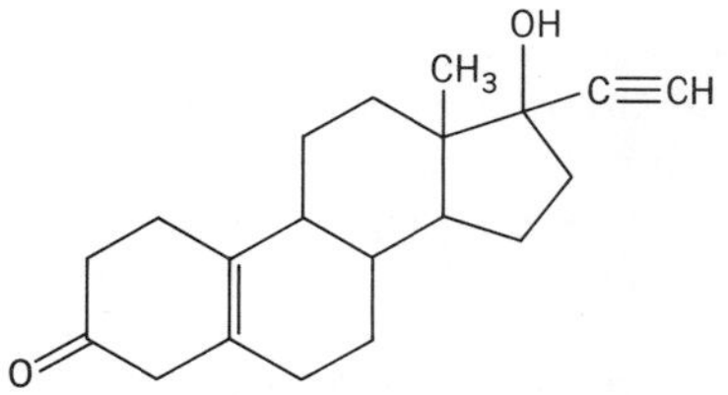

Norethynodrel

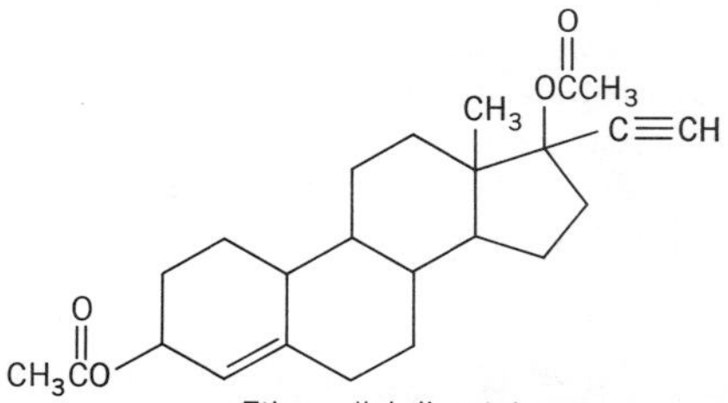

Ethynodial diacetate

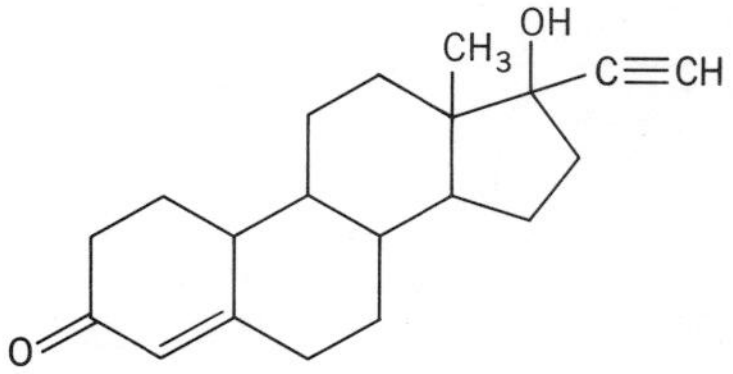

Norethindrone

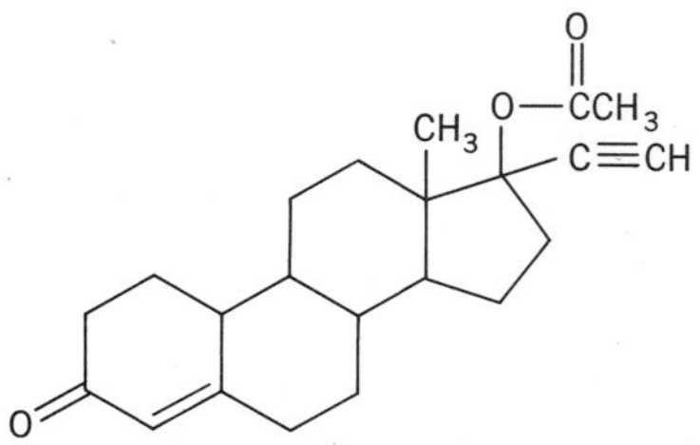

Norethindrone acetate

of oral contraceptives. Some women complain of nausea and tender breasts. Some doctors have stated that the formation of blood clots in the veins (thrombophlebitis) is more prevalent among women using oral contraceptives. However, extensive research has not revealed a direct relationship between an increased incidence of thrombophlebitis and the use of oral contraceptives. Oral contraceptives have only been in wide use for a short period of time. They do contain powerful hormones, and it is too soon to determine the long-term effects of the continuous use of such drugs. Some women do not use oral contraceptives because of the unpleasant side effects they experience or the unknown effects of using the pill over a long time period. Of course, other women do not use the pill for personal, moral, or ethical reasons.

15-3   OTHER CHEMICAL CONTRACEPTIVES

Other chemical contraceptives are being developed and may some day be available. A time-release pill which could be taken once a month may be possible. The implantation of a small plastic container of hormones in the uterus is being tested. The idea is to have the hormones slowly released in the uterus to suppress ovulation. This approach would eliminate pill-taking and could be effective for up to a year. A male contraceptive drug which suppresses sperm production is also being investigated.

Another approach involves the use of prostaglandins. **Prostaglandins** are hormone-like compounds found throughout the body. They are among the most potent biological substances known, and they appear to somehow exercise control over specific bodily processes. Several of the 14 currently known prostaglandins have been synthesized, and some are being tested in hospitals and clinics. Prostaglandins can induce abortion in early pregnancy. It may be possible to develop an oral contraceptive involving a prostaglandin which would interfere with the implantation of a fertilized egg. Other possible uses of prostaglandins are inducing labor at term, preventing and alleviating stomach ulcers, lowering blood pressure, and relieving asthma and nasal congestion. Although not in wide use, prostaglandins have the potential of becoming some of the most important chemotherapeutic agents.

15-4   AGRICULTURAL ENDEAVORS OF HUMANS

It is estimated that humans first began to domesticate plants and animals 10,000 years ago. This was the beginning of the agricultural revolution in which the behavior of humans changed from nomadic hunting to farming. The development of agriculture required the formation of villages which could be supported by the surrounding lands. Successful farming provided support of larger populations and freed people to develop nonagricultural activities and crafts. Great civilizations evolved based on agriculture and stimulated by advances such as the wheel and metallurgy. Ultimately, machines and power sources were developed, leading to the Industrial Revolution.

The advanced industrial systems of the developed countries of the world still require an agricultural base. Modern agricultural practices in such systems allow a large population to be supported by a small percentage of the population. For instance, in the United States about 5 percent of the population is involved directly in agricultural endeavors which support the rest of the population. Of course, the underdeveloped countries of the world use much greater percentages of their populations in agricultural work. The support of large populations by a small fraction of the population is one of the fundamental factors involved in the increased urbanization of people. **Urbanization** refers to the concentration of people in areas of high population density. There is a world-wide trend of people leaving rural areas to concentrate in cities. According to the trend, it is estimated that one-half of the population of the world will be living in cities of 100,000 or more by 1990. Currently, the United States is nearly 80 percent urbanized and could reach 90 percent or higher by 2000.

Modern agriculture is based upon mechanization and effective use of hybrid seeds, irrigation, pesticides, and fertilizers. Meat, dairy, and poultry products involve animal breeding, effective feeding, and the growth of large animal populations in small land areas. Agriculture in the United States consists mainly of large corporation-owned farms (agribusiness) which rely upon extensive farm-to-market road systems as well as the technological factors mentioned above. Such agriculture represents a large expenditure of energy and materials. Agriculture in developed countries depends upon the use of fossil fuels as does the industry. This kind of agriculture is said to be **fossil-fuel-subsidized.**

Food crops are cultivated on 13 million square kilometers of the earth's surface. This represents about 10 percent of the total land surface of the earth. Most good agricultural land is currently in use in the various countries of the world, and there is a relatively small amount of additional land which can be easily used. The protection of land now used for agriculture is vital. Wind and water erosion results in the loss of large amounts of top soil each year. The geological production of soil is so slow that this erosion represents a loss in agriculturally useful land. Many underdeveloped countries cannot afford to lose such land. Erosion problems can be minimized by proper irrigation and farming methods.

Around 70 percent of the world's agricultural lands are devoted to **cereal grains** (mainly wheat and rice). This means that cereal grains are the main energy source for humans. As shown in Figure 15-5, such grains account for 52 percent of the world's food energy. Meat and dairy products contribute 11 percent, while tubers (potatoes and yams) contribute 10 percent. Fruits and vegetables account for another 10 percent, and animal fats and vegetable oils 9 percent. Sugar from cane and beets represents 7 percent. Interestingly, fish only contribute 1 percent of the food energy.

Figure 15-5
Sources of the world's food energy.

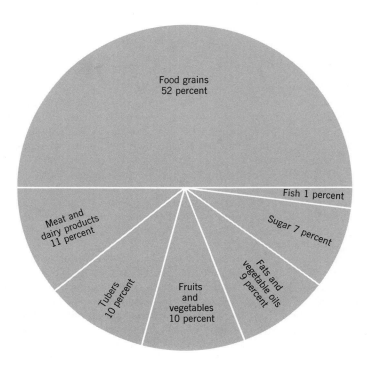

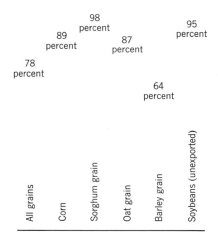

Figure 15-6
The portions of grains fed to livestock in the United States. Substantial amounts of wheat and rye grains and fish meal are also used as feed.

Of the 3.5 billion people in the world, over 2 billion suffer from hunger and malnutrition. The others, residing mainly in North America, Europe, Russia, and Australia, are generally well-nourished on a high-protein diet. The people in underdeveloped countries rely mainly on grains as food. A typical person in an underdeveloped country eats about 160 kilograms of grain each year. A typical American consumes an equivalent of 730 kilograms of grain each year. Of this, only 68 kilograms is eaten as grain products (bread and cereal) and the remainder represents the amount needed to provide meat and dairy products. On a comparative basis, an American makes over four times as much demand on an agricultural system as a person in an underdeveloped country. Of course, since grains do not have the correct amino acid balance, high-carbohydrate diets can lead to protein deficiency.

Let us look more closely at the use of crops as feeds for animals. In some underdeveloped countries, livestock are fed grasses and inedible material and produce high-protein meats and dairy foods. In contrast, advanced nations like the United States use high-protein feeds for livestock. Over 50 percent of the agricultural land in the United States is used for feed crops. Nearly 80 percent of the grain grown in this country is used for animal feed, as shown in Figure 15-6. The ability of an animal to convert proteins in feeds to protein in meats or dairy products is called the **protein conversion efficiency.** In the United States a beef cattle must be fed 21.4 kilograms of feed protein to produce 1 kilogram of meat protein. The protein conversion efficiency is said to be 21.4 to 1. The efficiencies for other kinds of livestock are shown in Figure 15-7. Considering these protein conversion efficiencies, it can be seen that a significant amount of protein is

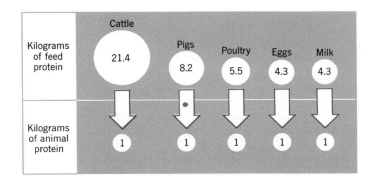

Figure 15-7
Protein conversion efficiencies for some animal products. Protein conversion efficiency is the number of kilograms of feed protein needed to produce 1 kilogram of animal protein.

wasted in producing meats and dairy products. One advantage is that the animal proteins are more tasty and have good essential amino acid balance. One disadvantage is that the amount of protein wasted in the United States each year is just about enough to make up the protein deficiencies of the undernourished people in the world. Of course, any attempt to make up the world's protein deficiency in this manner would necessitate low-meat or nonmeat diets. In some South American countries, meat sales are banned a certain number of days each month. The reason for this is to encourage the population to eat nonmeat diets to conserve valuable protein.

15-5   THE GREEN REVOLUTION

As the population of the world increases, the food demands increase. If the population doubles by the year 2000, the food production (mainly grains) will have to double. To overcome the existing nourishment problem, the food production will have to triple by the year 2000. Many underdeveloped countries have been barely able to keep food production up with population increases. However, in the last two decades scientists and agricultural experts have been able to engineer a dramatic increase in the food production of the world. This phenomenon has become known as the "**Green Revolution.**" Some consider the Green Revolution to be one of the greatest technological advances in history in the sense that it has affected more people over a shorter period of time than any other advance. This agricultural revolution included the introduction of modern agricultural techniques such as fertilization, irrigation, cultivation, and pesticide use to the underdeveloped countries. However, the most important aspect of this technological advance has been the breeding of grains which allow drastic increases in agricultural productivity. Some countries have been able to more than double their grain yields using seeds of the new grains. These hybrid grains resulted from an extensive effort by plant scientists in cross-breeding and selective breeding of plants. The first breeding efforts were carried out on wheat, resulting in a variety with short, stiff straw to support more grains. This new wheat responds well to fertilizers and is adaptable to various climates and soil conditions. Wheat of this type is now cultivated in numerous countries. After success with wheat, rice plants were bred to have similar characteristics. Two varieties called IR8 rice and IR5 rice were developed, which allowed for a doubling of rice yields. Like the new wheat, these rices have short, stiff straw, respond well to fertilizers, and are adaptable to a range of growing conditions. Furthermore, they mature in about 120 days compared with about 165 days for normal rice. Early maturation allows for the planting of a second crop in the growing season.

Once the new grains were developed, seeds were quickly distributed to various countries, resulting in dramatic increases in food production in a short period. Currently, the new species are being bred to make them more productive in the specific countries in which they are used. This also promises to help to increase productivity. Plant scientists are continually trying to breed even better grains. Wheat which can be grown in dry lands is being developed along with rice which can withstand flooding.

The new grains, which are the foundation of the Green Revolution, have been responsible for the avoidance of large-scale famine in recent years. However, these grains do not promise a final solution to food shortages, especially with pressure of increased populations. Some experts predict large-scale famine in underdeveloped countries if natural disasters or population pressures interfere with increased food productivity.

The successes of the Green Revolution lie in the use of a genetically uniform seed line. Plant scientists point out that this practice has certain inherent dangers. The problem is that the use of hybrid seeds narrows the genetic plant pool and makes key crops vulnerable to diseases. These problems can be illustrated with an example. In 1970, 70 percent of all corn in the United States was grown with only five different seed lines; and around 25 percent of the total crop was planted with one seed line. The use of such hybrid corn seeds has doubled to quadrupled the corn yield in the United States over a period of a few decades. In 1970, an epidemic of the southern corn blight infected the corn crop. The hybrid corn lines being used were more susceptible to the blight. Around one-fifth of the American corn crop was destroyed by the blight. American corn surpluses from previous years were available and no human starvation occurred, but considering that rice accounts for about one-quarter of the caloric intake of the population of the world, a disaster could occur if genetically uniform rice crops were destroyed by some disease.

Another aspect of the use of genetically uniform seeds is that crops from these seeds are replacing crops grown from local and regional grain seeds. The problem with this is that continued plant breeding requires a large genetic base for experimentation. Local and regional grain plants could become extinct as they are replaced by hybrid seeds. This loss of genetic variety could never be replaced and would allow no alternatives to disease-susceptible hybrid seeds.

Many plant scientists suggest that it is imperative that we establish world-wide plant preserves of local and regional varieties. These preserves would supply a genetic pool for future use. Further, plant breeding directed toward increasing genetic variety can be carried out to increase the genetic pool. In any case, as the southern corn blight of 1970 revealed, dependence on a limited line of hybrid seeds has to be approached with caution.

15-6 CHEMICAL FERTILIZERS

Plant growth requires chemical nutrients. Some soils are rich in such nutrients, while other soils are deficient. It has been found that crop yields can be significantly increased by the use of chemical fertilizers. Growing plants need sixteen **chemical nutrients.** Nine of these are needed in large amounts and seven in small or trace amounts. The carbon, hydrogen, and oxygen needs of plants come from air and water. The remaining nutrients are absorbed into the plants from the soil. The primary soil nutrients are nitrogen, phosphorus, and potassium. These are the soil nutrients needed in the greatest amounts. The secondary soil nutrients are calcium, magnesium, and sulfur. The trace nutrients include boron, chlorine, copper, iron, manganese, molybdenum, and zinc. The nutrient elements must be present in the soil in the correct ionic forms for plant use. The absence of any of these elements can inhibit or prevent plant growth. **Chemical fertilizers** supply the primary nutrients in the correct forms. Some fertilizers also supply the secondary nutrients. Lime and dolomite supply calcium and magnesium, and elemental sulfur and sulfates are sources of sulfur. Soils have to be tested for trace nutrient content. An absence or low level of a trace nutrient can be alleviated by use of fertilizers containing the trace nutrient.

Growing crops extract nutrients from the soil. It is estimated that 1 ton of wheat takes 18 kilograms of nitrogen, 4 kilograms of phosphorus, and 3.6 kilograms of potassium from the soil. Furthermore, if the stocks and leaves are not returned to the soil, even more nutrients are lost. In addition, this 1 ton of wheat will require about 100 tons of water in the growth period. Livestock which are fed crop feed represent even a greater drain on soil nutrients, since there is a loss in the conversion of plant foods to meat foods. It is estimated that 1 ton of beef cattle requires enough crops to remove 25 kilograms of nitrogen, 7 kilograms of phosphorus, and 6 kilograms of potassium from the soil. Since food crops remove substantial amounts of nutrients from the soil, the nutrients must be replaced by use of fertilizers.

Rich soil, high in nutrients, does not require as much fertilization as nutrient-deficient soil. In any case, fertilizer use can double agricultural yields on rich soil and triple or quadruple yields on poor soil. The use of chemical fertilizers accounts for about 25 percent of the world's food supplies. Consequently, artificial fertilizers are very necessary to modern agricultural practices. Around 60 million tons of chemical fertilizers are used in the world each year. The advanced countries use about 78 percent of this. These figures do not include China. It is difficult to determine the chemical fertilizer use in China, but it is known that China has a very efficient method of recycling human and animal wastes. Based on population predictions, it is expected that chemical fertilizer use in the world will have to triple by the year 2000. The manufacture of these fertilizers will require a large industrial effort.

It is not known when humans first began using fertilizers. Centuries ago in Europe and other parts of the world, animal droppings, blood, and carcasses were commonly used as fertilizers. Animal bones were used as phosphorus fertilizers, and wood ash was used as a source of potassium. In medieval Europe, nitrogen-fixing legumes were used along with crop rotation. Of course, in these times, the chemical nature of fertilizers was not known. They just were found to be useful. As the chemistry of plant nutrition was discovered, the organic fertilizers were replaced by inorganic chemical fertilizers. Early in the nineteenth century, potassium chloride deposits in Europe began to be used as a source of potassium fertilizer. In 1840, phosphorus was recognized as the important element in bones; and it was soon discovered that bone phosphorus could be converted to a useful phosphate form by treating bones with sulfuric acid. Later, it was found that phosphate could be leached from phosphate-containing rocks. Around 1900, Chile salt peter or sodium nitrate (mined in Chile) became an important source of nitrogen, both for fertilizer use and for use in explosives. In 1910, the Haber ammonia manufacturing method was developed, and some years later ammonia became the main source of fixed nitrogen for agriculture.

15-7 THE MANUFACTURE OF FERTILIZERS

Today, the chemical preparation of fertilizers is an enormous endeavor. The chemical formation of common fertilizers is shown in Table 15-2. Elemental nitrogen for the Haber process is obtained from air. The elemental hydrogen is obtained from coal, petroleum, or natural gas. The nitrogen and hydrogen are used to make ammonia by the Haber method. Ammonia can be injected directly into the soil as a fertilizer. However, it is usually used to make solid fertilizers which are more easily transported and used. Nitric acid is manufactured from ammonia and used to make ammonium nitrate, a solid fertilizer. Ammonia can be reacted with carbon dioxide to make urea. The combination of ammonia with sulfuric acid makes ammonium sulfate, a common solid fertilizer.

Phosphate fertilizers are manufactured from phosphate-containing rocks

---

Table 15-2 Chemical Fertilizers—Chemical Processing of Natural Products to Form Fertilizers

nitrogen and **hydrogen** produce **ammonia**: NH_3
  (air)   (oil)
ammonia and **oxygen** and **water** produce **nitric acid**: HNO_3
                (air)
nitric acid and **ammonia** produce **ammonium nitrate**: NH_4NO_3
ammonia and **carbon dioxide** produce **urea**: $(NH_2)_2CO$
sulfuric acid and **ammonia** produce **ammonium sulfate**: $(NH_4)_2SO_4$
phosphoric acid and **ammonia** produce **ammonium phosphate**: $(NH_4)_3PO_4$
phosphate rock pulverized to powdered **calcium phosphate**: $Ca_3(PO_4)_2$
sulfuric acid and **calcium phosphate** produce **superphosphate**: $CaSO_4 + Ca(H_2PO_4)_2$
phosphoric acid and **calcium phosphate** produce **triple superphosphate**: $Ca(H_2PO_4)_2$
potassium-containing ores processed to form **potassium chloride**: KCl

which are mined in various parts of the world. Such rocks contain calcium phosphate, $Ca_3(PO_4)_2$, which can be used directly as fertilizer. As shown in Table 15-2, the phosphate rock can be treated with sulfuric acid to make superphosphate fertilizer or with phosphoric acid to make triple superphosphate fertilizer. Phosphoric acid can be made from phosphate rock by reaction with sulfuric acid. Phosphoric acid can be mixed with ammonia to make ammonium phosphate, a nitrogen- and phosphorus-containing fertilizer.

Potassium fertilizers are isolated from mineral deposits of potassium chloride, KCl. Such deposits are found in numerous countries. An extremely large deposit is found in Canada. The mineral deposits are mined, and purified potassium chloride (muriate of potash) is used in fertilizer preparations.

Of the three primary nutrients, phosphorus is the least abundant. Nitrogen is quite plentiful in the form of ammonia made by fixation of atmospheric nitrogen. Large reserves of potassium are known, such as the Canadian deposits. The phosphorus reserves are limited. Only certain phosphate rock deposits are known, and large undiscovered deposits are not expected to be found. Each year, large amounts of phosphorus are washed into the oceans and deposited as calcium phosphate. Large calcium phosphate deposits are thought to exist on ocean floors. The need for phosphate fertilizers may require the exploration and mining of these deposits.

The use of fertilizers has to be carried out with care and expertise. Excessive fertilization can injure plants and cause water pollution problems. Agricultural run-off carries great quantities of phosphates and nitrates into environmental waters. Farm workers need to learn proper methods of fertilizer application. Scientists are developing slow or timed-release fertilizers which make certain amounts of nutrients available when needed. Furthermore, an attempt is being made to package fertilizers and seeds so that excessive use can be avoided.

Certain plants have been developed (by selective breeding) which respond to fertilizer use. This has resulted in increased agricultural productivity. However, only 5–10 percent of the dry weight of a plant comes from mineral nutrients. The remainder of the plant comes from photosynthesis. There is a limit to increased productivity by fertilizer use. Research is being carried out to attempt to increase the efficiency of the photosynthesis process in plants. If plants with more efficient photosynthesis processes can be developed, increased agricultural productivity can result.

15-8 PESTICIDES

Not too many years ago, malaria was one of the most prevalent diseases in the world. Today, in most regions of the world, it has been eradicated or controlled by use of DDT. The World Health Organization claims that DDT has decreased the malaria death rate in India from 750,000 a year to 1,500. Each year a significant portion of the food crops of the world are lost due to disease, insects, and weeds.

Humans have found it necessary to control certain pests that carry disease and interfere with agricultural productivity. Pest control has been accomplished by use of **chemical pesticides. Insecticides** are used to control insects, **fungicides** are used to combat fungi, **herbicides** are used to eliminate weeds, and **rodenticides** are used to control rats and mice.

There are over 1 million species of insects on earth, of which some 10,000 species cause problems to man. However, only around 100 insect species are a major problem, causing around 85 percent of insect-related damage. Needless to say, most pesticide use is directed toward these problem insects. Since insecticides are not generally specific, other insects are affected by pesticides used against the target insect.

Around 400 kinds of chemical pesticides are currently available for agricultural use. Table 15-3 gives the names and structures of some of the most common

	Table 15-3	Pesticide	Formula	Uses
Common Pesticides				

Insecticides

DDT

(1,1,1-trichloro-2,2-bis-*p*-chloro-phenyl) ethane

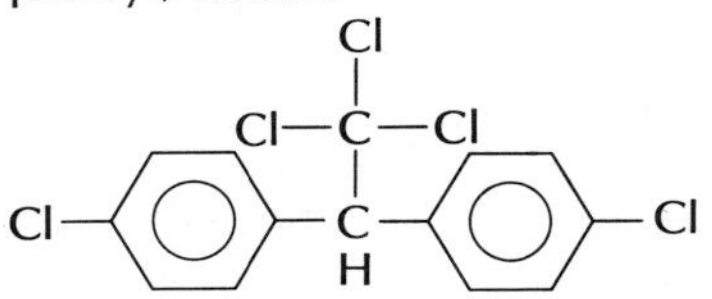

General use: insecticide
Fruit, vegetable, and cotton crops; household, livestock, and timber

Dieldrin

1,2,3,4,10,10-hexachloro-6,7-epoxy-1,4,4a,5,6,7,8,8a-octahydro-1,4-*endo-exo*-5,8-dimethano-naphthalene

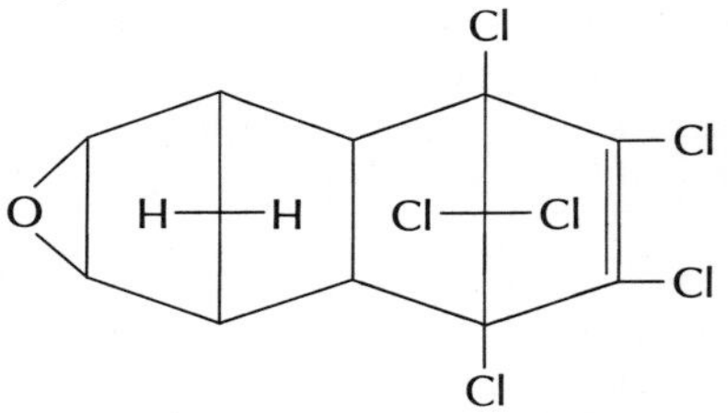

General use: insecticide
Soil insects, vegetable and fruit crops; mothproofing, public health pests

Endrin

1,2,3,4,10,10-hexachloro-6,7-epoxy-1,4,4a,5,6,7,8,8a-octahydro-1,4-*endo-endo*-5,8-dimethano-naphthalene

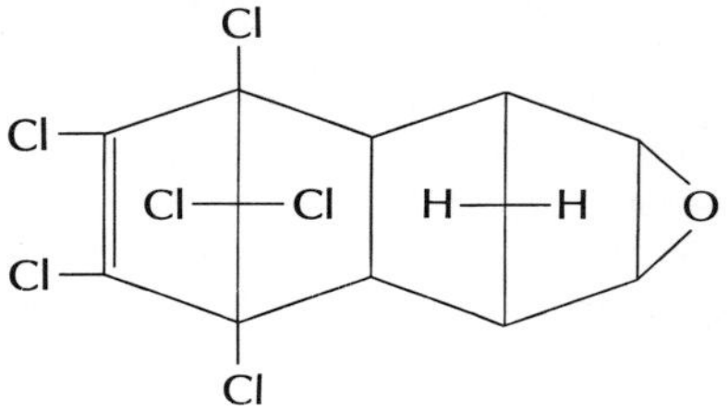

Cotton insects, cutworms, armyworms, aphids, corn borer, cabbage looper, grasshoppers, and many other insects; also used as a rodenticide

Lindane

1,2,3,4,5,6-hexachlorocyclohexane

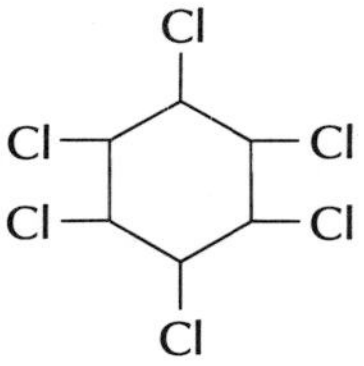

General use: insecticide
Fruit, bean, pea, tomato, and other vegetable crops; dairy, livestock, household; also used for seed treatment

Carbaryl

1-naphthyl-*N*-methylcarbamate

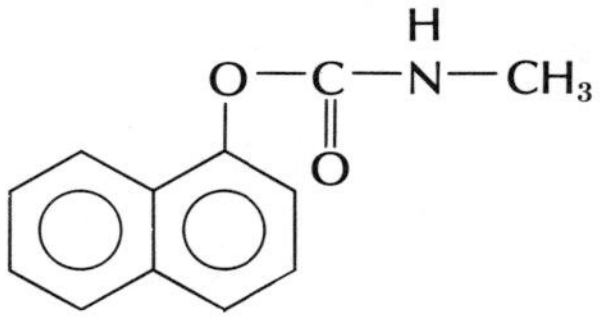

Fruit, vegetable, cotton, and other crops; used on poultry

Table 15-3
(Continued)

Pesticide	Formula	Uses
Parathion	O,O-diethyl-O-p-nitrophenyl phosphorothioate 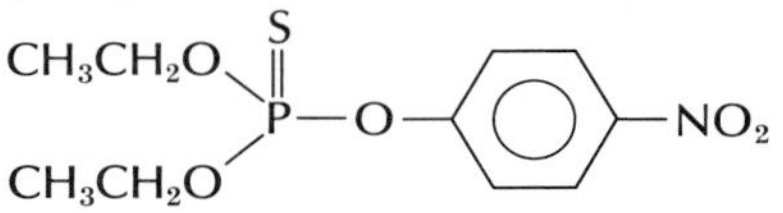	General use: insecticide Aphids, mites, beetles, leaf-hoppers, thrips, and cotton insects; soil insects

Parathion formula:

$$CH_3CH_2O \diagdown \underset{\underset{CH_3CH_2O \diagup}{}}{\overset{\overset{S}{\|}}{P}}-O-\langle\text{ring}\rangle-NO_2$$

Herbicides

Pesticide	Formula	Uses
2,4-D	2,4-dichlorophenoxyacetic acid 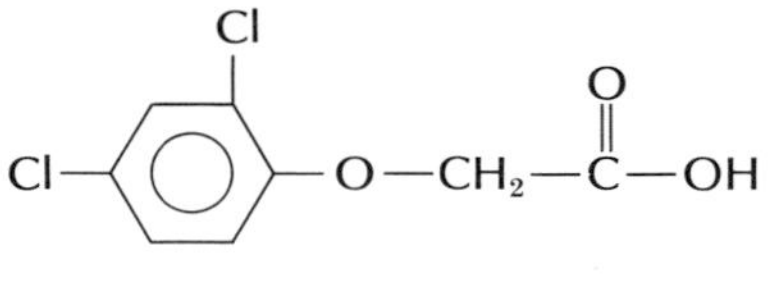	Weed control in cereal grains, corn, pastures, and lawns; aquatic weed control

2,4-D formula: $Cl-\langle\text{ring, Cl}\rangle-O-CH_2-\overset{\overset{O}{\|}}{C}-OH$

| 2,4,5-T | 2,4,5-trichlorophenoxyacetic acid 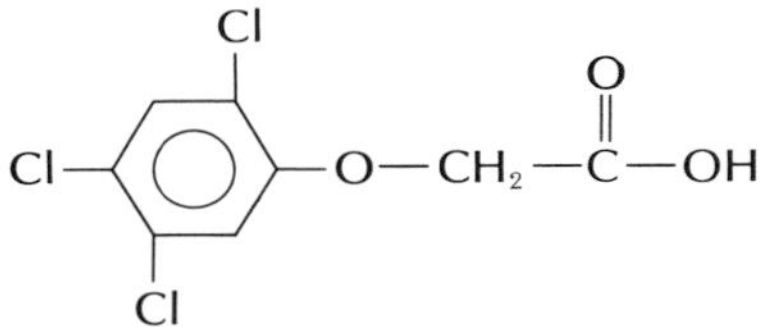 | General defoliant; brush control; aquatic weed control |

2,4,5-T formula: $Cl-\langle\text{ring, Cl, Cl}\rangle-O-CH_2-\overset{\overset{O}{\|}}{C}-OH$

Fungicides

Pesticide	Formula	Uses
Ethylmercury chloride	$CH_3CH_2-Hg-Cl$	Seed fungicide; used to coat seeds
Captan	N-trichloromethylthio-4-cyclo-hexene-1,2-dicarboximide 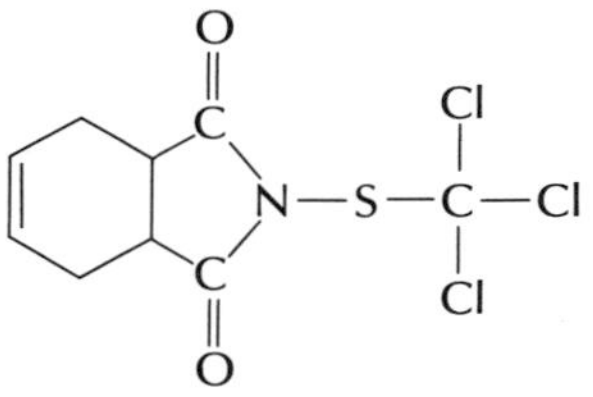	Used on fruits, vegetables, and flowers

Captan formula

Most widely used pesticides

Insecticides		Herbicides	
Carbaryl	Azinphosmethyl	Atrazine	Tribluralin
Malathion	Chlordane	Amiben	Picloran
Aldrin	Heptachlor	2,4-D	Paraquat
Diazinon	Disulfoton	2,4,5-T	Dicamba
Toxaphene	Phorate	Nitralin	Propachlor
DDT	Kelthane		
Methyl parathion	Bux Ten	**Fungicides**	
Parathion	Endrin	Dithiocarbamates	Pentachlorophenol (PCP)
		Captan	Dodine
		Mercury fungicides	

chemical insecticides, fungicides, and herbicides. Some plants have been found to be sources of insecticides. For instance, the insecticide pyrethrum is extracted from the chrysanthemum plant. It is very effective, but too expensive for wide-scale use. Among the most common synthetic pesticides are the **chlorinated hydrocarbons** including DDT, dieldrin, aldrin, heptachlor, chlordane, endrin, and lindane. Other classes of common pesticides are the **carbamates** (i.e., Sevin) and the **organophosphorus** types (i.e., parathion).

15-9   DDT

Since the discovery of the insecticidal properties of DDT in 1939, around 1.4 million tons of it have been manufactured in the United States alone. DDT was the first of the widely used synthetic pesticides. At first, it was used in an essentially unregulated fashion for agricultural, public health, and domestic pest control. After some years of use it was realized that DDT killed insects indiscriminately and was toxic to fish, birds, animals, and people when it accumulated in high enough concentrations. Furthermore, it was found that numerous pests were able to develop resistance to DDT, making insect control more difficult. In 1972, the wide-scale use of DDT in the United States was banned by the U.S. Environmental Protection Agency. This ban was based on the fact that DDT was no longer needed in this country because other alternate pesticides were available. Critics of the ban claim that some of the alternative pesticides are potentially more dangerous to use than DDT.

DDT gave birth to the large pesticide industry that exists today, which manufactures hundreds of different kinds of synthetic pesticides. Around 8 million tons of chlorinated hydrocarbon pesticides, other than DDT, have been manufactured in the United States during the few decades of the pesticide era. Since many insects develop resistance or immunity to insecticides, new insecticides are continuously being sought in numerous research laboratories in this country. The immunity problem can be illustrated by the case of a common Californian mosquito. After 25 years of continuous spraying and abatement control, this mosquito has developed an immunity to all common mosquito insecticides. The problem is compounded by the fact that this species of mosquito is a potential carrier of a Venezuelan sleeping sickness virus which can be transmitted to humans. If this virus were to enter California, it could result in a sleeping sickness epidemic. This example illustrates that once pesticides are used to control certain insects, new insecticides have to be used or developed to counteract the tendency of the insect to develop immunities by the selective effect of the pesticides. That is, some of the insects that survive spraying have acquired an immunity through a genetic quirk or mutation. These insects pass the immunity on to the next generation, and so on.

The chlorinated hydrocarbons, along with some other insecticides, have been instrumental in a significant increase in agricultural productivity. Furthermore, these insecticides have been used to eradicate or control about 30 disease-carrying insects. The most dramatic instance, as mentioned above, is the use of DDT on mosquitos, which has drastically decreased the incidence of malaria. It can safely be said that insecticides have saved the lives of millions of people. Some experts point out that DDT use has not eradicated malaria in certain tropical and subtropical regions, but only helped control it. This means that continuous applications will be needed to maintain the control. If this is not done, malaria epidemics will break out. To compound the problem, the human population in the control areas will gradually lose any built-up resistance to malaria and, thus, will be more susceptible to malaria.

Most insecticides are **biodegradable** and break down to harmless products. Chlorinated hydrocarbons are resistant to degradation and break down slowly. Thus, they are termed **persistent** or **"hard"** pesticides. Carbamate and organophosphorus pesticides degrade more quickly in the environment. They are **"soft"** pesticides, but they are more toxic to humans and, thus, more dangerous to the applicator. It is estimated that about 200 persons died from insecticides in the United States in 1970 and thousands suffered from insecticide poisoning. It is also estimated that millions of fish and animals have died for insecticide contamination, usually resulting from excessive use or accidental spills.

15-10 PESTICIDES IN THE ENVIRONMENT

Let us now consider the problem caused by the persistence of chlorinated hydrocarbon insecticides. These insecticides degrade slowly over a period of months or years. Furthermore, additional pesticides are rapidly introduced into our environment and then distributed by natural processes. As a result, the entire earth is contaminated. Of course, larger amounts of pesticides are found in agricultural areas. Nevertheless, pesticides are present in the fatty tissue of humans, animals, and fish, and mixed in the air, water, and soil. DDT residues are found in many parts of the earth. Disturbing amounts of DDT have been found in certain fish, cows' milk, and even human milk. See Table 15-4. DDT has even been detected in animals found in the antarctic, illustrating how widely pesticides have been distributed. DDT appears to interfere with calcium metabolism in birds and has been associated with the near extinction of certain species.

When pesticides are used on crops, they can be dispersed into the air and carried by wind currents to areas that are far removed from the treated areas. Pesticide residues can be washed into water supplies by irrigation waters. Once in the rivers and lakes, the pesticides can become widely distributed. Some of the pesticides in water are picked up by algae or plankton and are introduced into the food chain. These pesticides ultimately are concentrated in fish. When people, animals, or birds eat these fish, some of the pesticide residues can become absorbed in the fatty tissue, where it can remain for long periods of time. The concentration of pesticides in living organisms involved in the food

Table 15-4
DDT Levels Found in Humans and Foods

Country	DDT levels in human body fat samples (parts per million)
Arctic areas	3
Canada	4.9
England	3.9
France	5.2
Germany	2.3
Hungary	12.4
India	26
Israel	19
United States	12

Food	Average DDT content in foods[a] (parts per million)
Meat, poultry, fish	0.281
Dairy products	0.112
Grains, cereals	0.008
Legumes	0.026
Leafy vegetables	0.036
Fruits	0.027

[a] United States, 1968.

Figure 15-8
The concentration of
pesticide residues in
the food chain.
Pesticide residues
tend to accumulate in
animals at the top of
the food chain. Man
as a meat and plant
eater is at the top of a
food chain.

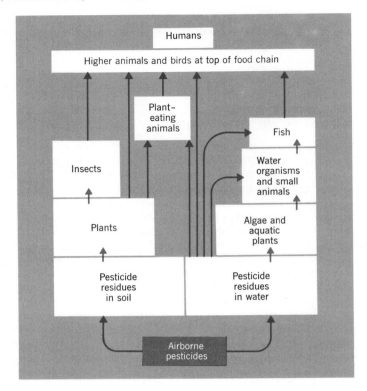

chain is illustrated in Figure 15-8. The long-term effect of pesticide residues in the fatty tissue of humans is not known. Such residues do appear to be detrimental to certain species of birds. It is estimated that 4 million tons of DDT are in the environment, and this could increase with increased use of DDT in underdeveloped countries.

The large monoculture practice of modern agriculture (acres of corn or wheat with no other plants) necessitated the use of pesticides. As discussed in the next section, there are some possible alternatives to persistent insecticides. However, until low pollution alternatives can be developed, it is necessary to ensure effective use and control excessive use of insecticides. Education of agricultural workers in proper application methods and strict controls over insecticide use are necessary. Often insecticides are applied in greatly excessive amounts and are used several times a season just as a precaution. More reasonable application methods and timing could be used. Furthermore, a greater effort could be expended to eliminate cultural conditions conducive to insect breeding. Such an approach involves better sanitation, cleaning up of crop residue, rotation of crops, and proper soil tilling.

15-11 BIOLOGICAL CONTROL

The major problems with chemical pesticides currently in use are that they are not selective to specific insects, they are toxic to humans and animals, they degrade slowly, and they are generally synthetic and, thus, can contaminate the environment. More natural methods of insect control are being developed. These methods are referred to as **biological control.** Much research and testing is being devoted to biological control. Several of these methods are discussed below.

It is possible to breed plants which have a natural resistance to certain pests.

Wheat, corn, and alfalfa strains have been developed which resist attack by common pests. This method is very promising. Unfortunately, breeding involves great research effort and long time periods. It can take years or decades to breed numerous varieties of insect-resistant plants.

Much research has been devoted to isolating the chemicals which act as sex attractants for specific insects. These chemicals, called **pheromones,** are usually secreted by female insects and serve to stimulate and attract the male insects. Incidentally, it is known that many animals and even mammals also secrete pheromones as sex attractants. The method of control involves the production of synthetic pheromones of specific insect pests. These synthetic chemicals can be used to attract insects to traps containing toxic chemicals, or they can be sprayed about to confuse males and prevent them from mating. Large-scale use of pheromones has not been accomplished, but controlled tests indicate that the method is promising.

Another chemical method involves the use of **juvenile hormones** of specific insects. Insect hormones control many metabolic reactions of insects. The juvenile hormones are involved in the development of insects in the larval stages. When the insect no longer generates these hormones, development into the adult stage occurs. The idea is to isolate and determine the structure of these juvenile hormones for specific insects. Then synthetic forms of the hormone could be manufactured. These hormones would be quite selective toward specific insects, and on contact with the larvae they would prevent the development of the insect. Some juvenile hormones have been identified, synthesized, and tested, but none are commercially available.

A method which has been successful in a few cases is the use of **sterility.** Large numbers of insects of a given species are bred and exposed to radiation to induce sterility. Then these sterile insects are released, they mate with other insects, and no offspring are produced. The sterility method was successful in the partial control of the screw worm (a cattle and livestock pest) in the United States. Sterility cannot be used against insects which infest large geographical areas, since it is not possible to release enough sterile insects. A similar insect-control technique being investigated is the breeding of insects with genetic defects (other than sterility) which, when passed on to the offspring, prevent them from developing properly, thus preventing further propagation.

Another possible biological control technique involves the use of insect diseases and predators. Many insects suffer from diseases caused by fungi, bacteria, or viruses. Usually, these microbes are quite harmless to plants and animals. If such natural diseases of specific pest insects can be discovered, the microbes can be cultured and released into the environment to attack the insects. Infectious diseases of insects are currently being researched. The use of natural predators against insects is more practical and is being used. Common predators are the lady bug and praying mantis. Large populations of the predators could be bred and released into the agricultural environment.

Most biological control methods are appealing. Some have been shown to be useful, others are not proven. It appears that combinations of biological control and chemical pesticides could be useful. For instance, an attempt, using a combination of methods, is being made to control the boll weevil, which causes hundreds of millions of dollars of damage to cotton crops in the United States. Tests of boll weevil control are currently being carried out. In these tests, chemical pesticides are being used in combination with pheromone traps and the releasing of large numbers of sterile males.

It appears that the wide use of biological control of insects will not materialize in the near future. The extensive research involved requires time and expense. Some of the methods may prove to be uneconomical. Agriculture will have to rely upon chemical insecticides. Consequently, it is important to concentrate on the control and proper use of these insecticides. Considering that chemical pesticides will continue to be used for many years, it is important to ponder the words of Rachel Carson, as written in 1962 in her book *The Silent Spring:*

> It is not my contention that chemical insecticides must never be used. I do contend that we have put poisons and biological potent chemicals indiscriminately into the hands of persons largely or wholly ignorant of their potentials for harm . . . . I contend furthermore, that we have allowed these chemicals to be used with little or no advance investigation of their effect on soil, water, wildlife, and man himself. Future generations are unlikely to condone our lack of prudent concern for the integrity of the natural world that supports all life.

15-12 FOOD ADDITIVES

There have been many reported cases of persons collapsing in Chinese restaurants with an ailment characterized by loss of breath, fainting, and chest pains resembling a heart attack. However, the ailments were not heart attacks but rather severe cases of "Chinese Restaurant Syndrome," brought on by the use of liberal amounts of the food additive monosodium glutamate (MSG) in the Chinese food. MSG is an excellent flavor enhancer when used in small amounts, but in large amounts it can cause this dramatic ailment in certain individuals.

MSG is one of the numerous food additives found in commercially pre-prepared foods. Perhaps for breakfast you had some butylated hydroxyanisole (BHA) or some butylated hydroxytoluene (BHT) or even some calcium propionate. These are common synthetic food additives found in prepared cereals and bread.

For centuries, chemicals have been added to foods. Such chemical additives were used to enhance the appearance or taste of food, to adulterate food with low-cost fillers, and to disguise spoiled or decayed food. In the United States certain government controls are exercised over the use of food additives. The use of food additives has increased with the increased consumption of instant and prepared food items. The control of food additives is under the jurisdiction of the **U.S. Food and Drug Administration (FDA),** a federal agency. The Pure Food and Drug Act of 1906 established the criteria for food additives. The act declared that food additives must be safe and serve a definite purpose. Interestingly, it was not until the passage of the Color Additive Act of 1960 that Congress declared that food processors must prove the safety of an additive to the FDA. Up to that time, the FDA had to prove that an additive was not safe. The FDA has established a list of food additives which are generally recognized as safe; the **GRAS list.** This list consists of over 650 additives. The **Delaney Amendment** to the Food and Drug Act requires the banning of an additive known to cause cancer in man or animals no matter what the intake level. Since certain additives are continuously being tested, deletions and additions to the GRAS list sometimes occur. Deletions from the GRAS list often precipitate much controversy. A good example is the case of cyclamates, a noncaloric sweetener. Based on test data obtained from injecting rats with large doses, cyclamates, the artificial sweetener, was banned from use in food and beverages on January 1, 1970. The decision was based on the Delaney Amendment. After the FDA announcement, a furor arose. The cyclamate manufacturers complained that the tests were not valid. Others felt that the Delaney Amendment is too stringent and should be interpreted more loosely. Some even claimed that sugar manufacturers were behind the ban. The FDA responded by saying that the decision would be reconsidered. However, the cyclamate ban is still in effect.

Food additives are chemicals added to prevent or retard spoilage, or to enhance the flavor or texture or nutritional quality. Some unintentional additives

Antioxidants (preservatives)	BHA (butylated hydroxyanisole) BHT (butylated hydroxytoluene) Vitamin C (ascorbic acid) Propyl gallate Citric acid Phosphoric acid	Table 15-5 Common Food Additives
Other preservatives	Calcium or sodium propionate Lactic acid Sodium acetate Sorbic acid and sodium or potassium sorbate Benzoic acid and sodium benzoate EDTA (ethylene diamine tetraacetic acid) Disodium calcium EDTA Sulfur dioxide or sodium bisulfite	
Sweeteners	Sucrose or table sugar Saccharin Sucaryl or calcium cyclamate (this sweetener was removed from the FDA GRAS list in 1970)	
Flavorings	A variety of natural and artificial flavorings including esters, essential oils, and spices	
Flavor enhancers and potentiators	MSG (monosodium glutamate) Disodium inosinate Disodium guanylate Ethyl maltol	
Food colors	Natural and artificial coloring agents—ten synthetic food colors are approved (2 blues, 4 reds, 1 green, 2 yellows, and 1 violet)	
Stabilizers and thickeners (gums from plants)	Seaweed gums (agar, algin, and carrageenan) Tree gums (arabic, larch, and tragacanth) Wood products (carboxymethylcellulose, methylcellulose, and hydroxypropylmethylcellulose)	
Emulsifiers and surfactants	Lecithin Monoglycerides and diglycerides Polysorbate 80 and Polysorbate 60	
Anticaking agents	Sodium silico aluminate Cornstarch Calcium phosphate	
Moisturizing and softening agents	Various polyhydric alcohols Glycerine or glycerol Mannitol Sorbitol Propylene glycol	
Sequestrants	EDTA (ethylene diamine tetraacetic acid) Calcium disodium EDTA Citric acid and sodium or potassium citrate Disodium phosphate Tetrasodium pyrophosphate	
Acidulants	Citric acid Phosphoric acid Tartaric acid Other fruit acids	
Alkalis	Ammonium bicarbonate Sodium carbonate Calcium carbonate	

come from pesticide use on food crops and from packaging or processing material. Table 15-5 lists some common food additives. Let us consider the function of some of these additives.

Food containing fats or oils can become rancid upon exposure to air, moisture, and/or heat. Rancidity refers to reactions in which the simple lipids decompose to form foul-smelling fatty acids such as butyric acid. Bacterial decay of protein or carbohydrate-containing food can occur, resulting in discoloration and foul odors. To prevent or retard rancidity or bacterial decay, food additives called **antioxidants** are used. Certain foods of high moisture content are subject to the growth of bacteria and molds. **Preservatives** are added to these foods to retard such microbe growth. The World Health Organization estimates that 20 percent of the world's food supply is lost by spoilage each year. Much of this loss could be prevented by proper use of preservatives and antioxidants.

Some foods need sweetening. The most common **sweetener** is sucrose. On the average in the United States, about 44 kilograms of sucrose are used by each person annually. Saccharin is a widely used low-calorie sweetener. **Flavorings** and spices are widely used food additives. They add unique flavors and odors to foods. Many flavoring agents are isolated from natural products (natural flavors) and some are synthetic (artificial flavors). Additives like monosodium glutamate (MSG) act as **flavor enhancers** by affecting the taste buds on the tongue. Other additives which intensify food flavors are called **potentiators.**

Coloring agents are added to foods to make them more attractive or give them a familiar color. For instance, butter from cows fed certain kinds of feed is white. A yellow color is added to make it look like normal butter. Some natural coloring agents are used, but about 90 percent of coloring agents are synthetic. Ten specific artificial food coloring agents are approved by the FDA.

Stabilizers and **thickeners** are additives used to thicken, give bulk to, or bind certain food products. These additives are vegetable gums or gums obtained from trees or seaweed. **Emulsifiers** and **surface-active agents** are additives used to encourage oil and water mixtures and to produce or prevent foaming.

Anticaking agents keep certain foods dry and prevent caking or clumping. Polyhydric alcohols are used to help retain moisture in foods and as softening agents. **Sequestrants** or **chelating agents** are added to some foods to react with traces of metals which might cause the decay or discoloration of the food. **Acidulants** are fruit acid additives which are used to control the acidity of foods. They are often used to produce a sour taste. **Alkalis** are additives used to counteract the acidity of certain foods. Food additives are necessary components of modern prepared foods which have to have reasonably long shelf-lives. The market for such foods is very large, amounting to some $50 billion annually. To avoid the consumption of food additives, a consumer has to give up the use of most instant, heat-and-serve, and ready-made prepared foods.

15-13 STANDARDS OF FOOD LABELING AND SAFETY

The safety of food additives is based upon testing various experimental animals. Short-term and long-term tests are carried out using a variety of animals. An additive is considered safe if a level 100 times higher than that used in foods does not harm the test animals. The FDA periodically checks foods for pesticide content. FDA researchers purchase foods in supermarkets and then check for various pesticides.

The FDA also establishes food standards for common widely used food products. These **food standards** establish that certain foods must contain specific ingredients in specified amounts. The specified ingredients do not have to be listed on the label, but any extra ingredients must be listed. For instance, tomato catsup manufactured according to the food standard does not have a listing of ingredients on the label. A bottle of French dressing lists vegetable gum, algin, and calcium disodium EDTA as ingredients. This listing just gives the extra ingredients and not the standard ingredients. If a food is not covered by a food standard, all the ingredients must be listed on the label. Some food manufacturers list all the ingredients in their products even if they are covered by food standards.

Recently, the FDA has announced new **nutritional regulations** which are to apply to packaged foods shipped interstate beginning in 1975. Not all the regulations are definite at this time, but they include the following:

1. Most packaged foods must have nutritional labeling including caloric, protein, carbohydrate, and fat content, along with vitamin and mineral content expressed as percentages of the Recommended Daily Allowances.
2. Certain health claims for food supplements will not be allowed.
3. Foods can be labeled for cholesterol and fat content. Fat content may be expressed in terms of percentages of polyunsaturated, saturated, and other fatty acids.
4. Fortified foods including milk, bread, juices, and diet foods must list vitamin A, vitamin C, thiamine, riboflavin, niacin, calcium, and iron content as percentages of Recommended Daily Allowances.
5. Certain limitations will be imposed on the ingredients in vitamin–mineral products.

The intent of these new regulations is to give the consumer sufficient nutritional information on labels of packaged food. However, labeling will not be mandatory for all foods. Many food manufacturers agree with the FDA regulations, but some threaten to challenge them in court.

The Food, Drug and Cosmetic Act and Fair Packaging and Labeling Act provide for prosecution, imprisonment, fines, seizure of goods, and injunctions against manufacturers violating food processing, labeling, and packaging laws. The FDA attempts to be aware of such violations, and very often will require the removal or recall of food or drug items from retail stores. The FDA has inspectors who visit food plants and warehouses. However, it employs about 200 inspectors for about 60,000 food plants. Consequently, food plants are inspected on the average of once every 6 years. Of course, many states have agencies which also inspect food plants. The Department of Agriculture employs over 7,000 meat and poultry inspectors.

The U.S. Food and Drug Administration is highly respected by similar organizations throughout the world. However, FDA critics in this country feel that it is a weak agency, subject to political pressure by Congress, the executive branch, and powerful industries. Others claim that the FDA does a good job considering the budgeting and other restrictions imposed by the government.

BOOKS

BIBLIOGRAPHY

Agricultural Research Service. *Toward the New: A Report on Better Foods and Nutrition from Agricultural Research*. Washington, D.C.: U.S. Government Printing Office, 1970.

Barclay, G. W. *Techniques of Population Analysis*. New York: Wiley, 1958. 311 pp.

Benarde, M. A. *Our Precarious Habitat*. New York: W. W. Norton, 1970.

Bloom, S. C., and S. E. Degler. *Pesticides and Pollution*. Washington, D.C.: Bureau of National Affairs, Environmental Management Series, 1960. 99 pp.

Brown, L. R. *Seeds of Change: The Green Revolution and Development in the 1970's.* New York: Encyclopaedia Brittanica, Praeger, 1970. 205 pp.

Carson, R. *Silent Spring.* Boston: Houghton Mifflin, 1962.

Ehrlich, P. R. *The Population Bomb.* New York: Ballantine, 1968.

Ehrlich, P., and A. Ehrlich. *Population, Resources, Environment.* 2nd ed. San Francisco: W. H. Freeman, 1973.

Freeman, O. L. *Worlds Without Hunger.* New York: Frederick A. Praeger, 1968.

Graham, F., Jr. *Since Silent Spring.* Boston: Houghton Mifflin, 1970. 333 pp.

Hardin, G., ed. *Population, Evolution, and Birth Control.* 2nd ed. San Francisco: W. H. Freeman, 1969. 386 pp.

Kotz, N. *Let Them Eat Promises: The Politics of Hunger in America.* Englewood Cliffs, N.J.: Prentice-Hall, 1970.

Leisner, R. S., and E. J. Kormondy, eds. *Population and Food.* Dubuque, Iowa: Wm. C. Brown, 1971.

Meyer, L. *Food Chemistry.* New York: Reinhold, 1966.

Neal, H. A. *The Protectors: The Story of the Food and Drug Administration.* New York: Julian Messner, 1969.

Pyke, M. *Man and Food.* New York: McGraw-Hill, 1970.

Slack, A. V. *Defense Against Famine: The Role of the Fertilizer Industry.* New York: Doubleday, 1970.

Winter, R. *A Consumer's Dictionary of Food Additives.* New York: Crown, 1972.

Winter, R. *Poisons in Your Food.* New York: Crown, 1969.

ARTICLES AND PAMPHLETS

Aaronson, T. "World Priorities." *Environment* **14** (6), (1972).

Baldwin, I. L. "Chemicals and Pests." *Science* **137,** 1042–1043 (1962).

Boerma, A. H. "A World Agricultural Plan." *Scientific American* **223** (2), 54–69 (1970).

Brown, L. R. "Human Food Production as a Process in the Biosphere." *Scientific American* **223,** 161 (Sept. 1970).

"Chinese Restaurant Syndrome." *Chemistry* **42** (8), 4 (1969).

Darby, W. J. "A Scientist Looks at *Silent Spring.*" *Chemical and Engineering News* (Oct. 1, 1962).

Frost, J. "Earth, Air, Water." *Environment* **11** (6), 14–33 (1969).

Garcia, R. "The Control of Malaria." *Environment* **14** (5) (1972).

Holcomb, R. W. "Insect Control: Alternatives to the Use of Conventional Pesticides." *Science* **168,** 456–458 (1970).

Jacobson, M., and M. Beroza. "Insect Attractants." *Scientific American* 20–27 (Aug. 1964).

Keller, E. "The DDT Story." *Chemistry* 8–12 (Feb. 1970).

Lykken, L. "Chemical Control of Pests." *Chemistry* **44,** 18–21 (1971).

Manufacturing Chemists Association. *Everyday Facts about Food Additives.* Washington, D.C.: Director of Public Relations, Manufacturing Chemists Association.

Nicholson, H. P. "Pesticide Pollution Control." *Science* **158** (3803), 871–876 (1967).

Pirie, N. W. "Orthodox and Unorthodox Methods of Meeting World Food Needs." *Scientific American* **216** (2), 27 (1970).

Pratt, C. J. "Chemical Fertilizers." *Scientific American* **212,** 62–72 (June 1965).

Scrimshaw, N. S. "Food." *Scientific American* **209,** 72–80 (Sept. 1963).

Shea, K. "Captan and Folpet." *Environment* **14** (1) (1972).

Ugent, D. "The Potato." *Science* **170** (3963), 1161–1166 (1970).

U.S. Department of Agriculture. *Yearbook for 1959: Food.* Washington, D.C.: U.S. Government Printing Office, 1959.

U.S. Department of Agriculture. *Yearbook for 1966: Protecting Our Food*. Washington, D.C.: U.S. Government Printing Office, 1966.

U.S. Department of Agriculture. *Yearbook for 1969: Food for Us All*. Washington, D.C.: U.S. Government Printing Office, 1969.

van den Bosh, R. "The Cost of Poisons." *Environment* **14** (7) (1972).

Wilkes, G. H., and S. Wilkes. "The Green Revolution," *Environment* **14** (8) (1972).

Williams, C. M. "Third-generation Pesticides." *Scientific American* **2**, 1713–1717 (1967).

Woodwell, G. M. "Toxic Substances and Ecological Cycles." *Scientific American* **216** (3), 24–31 (1967).

QUESTIONS AND PROBLEMS

1. Are population predictions of any concern to you? Why?

2. It is obvious that at some time in the future the increase in the rate of the population growth will have to decrease. How do you suppose this decrease will come about?

3. Describe the function of oral contraceptives.

4. What are prostaglandins?

5. What five technological factors are involved in modern agriculture?

6. What food crops account for over 50 percent of the world's food energy?

7. Rank order the following choices according to your feelings (1 = strongly agree, and so on) concerning the use of meat in your diet:

    _____ I want meat in my diet every day, with few exceptions.
    _____ I would be willing not to eat meat a few days out of each month if it would conserve protein.
    _____ I would be willing to give up meat if a balanced supply of non-meat protein products was available.

8. Describe the Green Revolution.

9. What is the major problem associated with the hybrid seeds used in the Green Revolution?

10. What are the three major soil nutrients needed by plants?

11. Are chemical fertilizers necessary? Why?

12. Give the sources of the following fertilizers:

    (a) nitrogen-containing fertilizers
    (b) phosphorus-containing fertilizers
    (c) potassium-containing fertilizers

13. What are chemical pesticides?

14. Is pesticide use necessary? Why?

15. What are chlorinated hydrocarbon pesticides and what environmental problems are caused by their use?

16. What problems are associated with the use of DDT to control rather than eradicate malaria?

17. What is biological control? Give three examples of biological control methods.

18. What are food additives?

19. Are food additives necessary? Why?

20. What is the FDA GRAS List?

21. What is the Delaney Amendment?

22. How are food additives tested?

23. Do you agree with the new FDA nutritive labeling regulations? Why?

24. Rank order the following choices according to your feelings on the use of food additives used in prepared foods (1 = strongly agree, and so on):

    _____ Food additives should not be used.
    _____ Food additives should be used when needed as long as they have been determined to be safe.
    _____ Food additive use should be restricted to cases in which food safety is involved.

INDEX